Oxford Applied Mathematics and Computing Science Series

OXFORD APPLIED MATHEMATICS AND COMPUTING SCIENCE SERIES

I. Anderson: *A First Course in Combinatorial Mathematics* (*Second Edition*)

D. W. Jordan and P. Smith: *Nonlinear Ordinary Differential Equations* (*Second Edition*)

D. S. Jones: *Elementary Information Theory*

B. Carré: *Graphs and Networks*

A. J. Davies: *The Finite Element Method*

W. E. Williams: *Partial Differential Equations*

R. G. Garside: *The Architecture of Digital Computers*

J. C. Newby: *Mathematics for the Biological Sciences*

G. D. Smith: *Numerical Solution of Partial Differential Equations* (*Third Edition*)

J. R. Ullmann: *A Pascal Database Book*

S. Barnett and R. G. Cameron: *Introduction to Mathematical Control Theory* (*Second Edition*)

A. B. Tayler: *Mathematical Models in Applied Mechanics*

R. Hill: *A First Course in Coding Theory*

P. Baxandall and H. Liebeck: *Vector Calculus*

D. C. Ince: *An Introduction to Discrete Mathematics and Formal System Specification*

P. Thomas, H. Robinson, and J. Emms: *Abstract Data Types: Their Specification, Representation, and Use*

D. R. Bland: *Wave Theory and Applications*

R. P. Whittington: *Database Systems Engineering*

J. J. Modi: *Parallel Algorithms and Matrix Computation*

P. Gibbins: *Logic with Prolog*

W. L. Wood: *Practical Time-stepping Schemes*

W. L. Wood

Retired Senior Lecturer
University of Reading

Practical Time-stepping Schemes

CLARENDON PRESS · OXFORD
1990

Oxford University Press, Walton Street, Oxford OX2 6DP
Oxford New York Toronto
Delhi Bombay Calcutta Madras Karachi
Petaling Jaya Singapore Hong Kong Tokyo
Nairobi Dar es Salaam Cape Town
Melbourne Auckland
and associated companies in
Berlin Ibadan

Oxford is a trade mark of Oxford University Press

Published in the United States
by Oxford University Press, New York

British Library Cataloguing in Publication Data
Wood, W. L.
Practical time-stepping schemes.
1. Ordinary differential equations. Numerical solution
I. Title
515.3'52
ISBN 0-19-853208-3
ISBN 0-19-859677-4 (pbk.)

Library of Congress Cataloging in Publication Data
Wood, W. L.
Practical time-stepping schemes / W.L. Wood.
(Oxford applied mathematics and computing science series)
Includes bibliographical references.
1. Differential equations—Numerical solutions. I. Title.
II. Series.
QA372.W75 1990 515'.35—dc20 89-37596
ISBN 0-19-853208-3
ISBN 0-19-859677-4 (pbk.)

Photoset in Northern Ireland by The Universities Press (Belfast) Ltd.
Printed in Great Britain by
Courier International Ltd
Tiptree, Essex

Preface

The aim of this book is to give the mathematical background necessary for the understanding of the algorithms which engineers (civil, mechanical, electrical) and scientists have proposed for the solution of their time-dependent problems. The book includes a study of the equations for modelling aquifers, heat conduction and diffusion, the dynamic vibration of structures, and associated non-linear and coupled problems. I assume the space discretization is done by the finite element method; the results are first and second order differential equations of a particular structure because of their origin. This is the product of my research over the last fifteen years as mathematical consultant to the Consulting Engineers, Sir William Halcrow and Partners, the department of Civil Engineering at Swansea, and the Institute of Hydrology. My emphasis throughout the book is on the methods of direct relevance for engineers, hydrologists, and other applied scientists. A large proportion of the mathematics literature is inaccessible to engineers who tend to develop and use their own algorithms regardless of what mathematicians prove. Engineering intuition, while sometimes brilliant, is not infallible and there is need for an objective mathematical survey of these algorithms and the statements made about them.

The book is meant for postgraduate applied mathematicians and theoretically inclined engineers working on the numerical solution of engineering problems. Some of the material has been given as part of the M.Sc. course on the Numerical Solution of Differential Equations at the University of Reading. I assume the reader has an acquaintance with elementary numerical analysis as for instance in *Numerical Analysis* by Johnson and Riess (1977, Addison Wesley) or *Introduction to Numerical Analysis* (second edition) by Fröberg (1970, Addison Wesley), and some basic knowledge of finite element theory as given, for instance, in *The Finite Element Method: A First Approach* by Davies (1986, Oxford University Press).

This book has evolved from my collaboration with many people and I would particularly like to express appreciation of the help of O. C. Zienkiewicz, R. W. Lewis, B. M. Irons, R. L. Taylor, and A. Calver. Thanks are also due to Alan Muir Wood for general assistance.

Department of Mathematics W. L. W.
University of Reading
December 1988

Acknowledgements

Figures and tables from papers by the author in the *International Journal of Numerical Methods in Engineering* and the *International Journal of Communications in Applied Numerical Methods* are reprinted by permission of John Wiley and Sons Ltd. These are identified in the text. Figures 1 and 3 and Table 1 from Calver and Wood, *Journal of Hydrology*, 1989 are reprinted by permission of Elsevier Science Publishers B.V. Acknowledgement is also due to J. Carroll for permission to publish results from his M.Sc. Dissertation, University of Reading, 1988.

Contents

1 The differential equations

1.1. Introduction

In this chapter we introduce the partial differential equations whose numerical solution is the subject of this book. These partial differential equations are second order in space coordinates, and either first or second order in time. It is assumed that all the differential equations discussed in this book are well-posed. The reader is referred to Richtmyer and Morton (1967) for a more detailed discussion of the expression 'well posed'. Here we shall just express our ideas in words as follows: we are assuming

(1) that the problems we are discussing are formulated so that small changes in the data will make only small changes in the solution; this is essential in engineering problems where the data can frequently be measured only very roughly;

(2) the initial conditions are formulated so that they represent the physics well enough for the solution of the mathematical equation to represent the later stages of the physical situation.

The partial differential equations are first space discretized by using standard finite element methods. The result is then a system of first or second order ordinary differential equations in time with a particular structure because of their origin. In the rest of this chapter we illustrate the kinds of physical problem from which these equations can come.

1.2. Equations which are first order in time

We use as a standard first order in time equation the parabolic equation of diffusion (Crank 1975). With a constant diffusion coefficient and in its non-dimensionalized form this equation is

$$\frac{\partial u}{\partial t} = \nabla^2 u + f \qquad (1.2.1)$$

in some region Ω, where u is the concentration of the diffusing substance and f represents a source or sink term. Taking $u \approx \sum_{j=1}^{N} u_j(t)\Phi_j$, where the $u_j(t)$ are the nodal coordinates and the Φ_j are the basis functions, multiplying eqn (1.2.1) by the test functions Φ_j in turn and integrating by parts in space, we obtain the finite element 'Galerkin' form

$$\boldsymbol{M}\dot{\boldsymbol{u}} + \boldsymbol{K}\boldsymbol{u} = \boldsymbol{f}, \tag{1.2.2}$$

where $\boldsymbol{M}$ is the 'mass matrix' with entries

$$\left\{\int_{\Omega} \Phi_i \Phi_j \, \mathrm{d}\Omega\right\}, \tag{1.2.3}$$

$\boldsymbol{K}$ is the 'stiffness matrix' with entries

$$\left\{\int_{\Omega} \nabla\Phi_i \cdot \nabla\Phi_j \, \mathrm{d}\Omega\right\}, \tag{1.2.4}$$

$\boldsymbol{u}$ is the vector of nodal coordinates $u_j(t)$, and the vector $\boldsymbol{f}$ now contains the boundary conditions. The solution of the system of N equations (1.2.2) consists of the particular integral plus the complementary function solution of the homogeneous equation

$$\boldsymbol{M}\dot{\boldsymbol{u}} + \boldsymbol{K}\boldsymbol{u} = \boldsymbol{0}. \tag{1.2.5}$$

If we substitute a trial function

$$\boldsymbol{u} = \mathrm{e}^{-\lambda t}\boldsymbol{v} \tag{1.2.6}$$

into eqn (1.2.5) we obtain the system of simultaneous equations

$$\lambda \boldsymbol{M}\boldsymbol{v} = \boldsymbol{K}\boldsymbol{v}, \tag{1.2.7}$$

where the matrices $\boldsymbol{M}$ and $\boldsymbol{K}$ are symmetric and positive definite (supposing that Dirichlet and/or non-homogeneous flow boundary conditions are included). Suppose eqn (1.2.7) is satisfied by $\lambda = \lambda_j$, $\boldsymbol{v} = \boldsymbol{v}_j$. Since the matrix $\boldsymbol{M}$ is positive definite we can make a Cholesky decomposition (Wilkinson 1965), $\boldsymbol{M} = \boldsymbol{L}\boldsymbol{L}^{\mathrm{T}}$, where $\boldsymbol{L}$ is lower triangular and non-singular. Substituting into eqn (1.2.7) and multiplying both sides by $\boldsymbol{L}^{-1}$ gives

$$\lambda_j \boldsymbol{L}^{\mathrm{T}}\boldsymbol{v}_j = \boldsymbol{L}^{-1}\boldsymbol{K}(\boldsymbol{L}^{-1})^{\mathrm{T}}\boldsymbol{L}^{\mathrm{T}}\boldsymbol{v}_j.$$

Hence $\boldsymbol{L}^{\mathrm{T}}\boldsymbol{v}_j$ is an eigenvector and λ_j an eigenvalue of the symmetric matrix $\boldsymbol{P}=\boldsymbol{L}^{-1}\boldsymbol{K}(\boldsymbol{L}^{-1})^{\mathrm{T}}$. The λ_j are therefore real and the matrix $\boldsymbol{P}$ possesses a complete set of linearly independent eigenvectors such that, if $\boldsymbol{V}$ is the matrix with the $\boldsymbol{v}_j$ as its columns, $\boldsymbol{L}^{\mathrm{T}}\boldsymbol{V}$ is orthogonal, that is,

$$\boldsymbol{V}^{\mathrm{T}}\boldsymbol{L}\boldsymbol{L}^{\mathrm{T}}\boldsymbol{V}=\boldsymbol{V}^{\mathrm{T}}\boldsymbol{M}\boldsymbol{V}=\boldsymbol{I}. \tag{1.2.8a}$$

Also, from eqn (1.2.7),

$$\lambda_j=\lambda_j\boldsymbol{v}_j^{\mathrm{T}}\boldsymbol{M}\boldsymbol{v}_j=\boldsymbol{v}_j^{\mathrm{T}}\boldsymbol{K}\boldsymbol{v}, \tag{1.2.8b}$$

and hence, if $\boldsymbol{K}$ is positive definite, λ_j is positive. This means that the solution of eqn (1.2.5) is decreasing with time. Substituting $\boldsymbol{u}=\boldsymbol{V}\boldsymbol{x}$ into eqn (1.2.5) and pre-multiplying both sides by $\boldsymbol{V}^{\mathrm{T}}$ produces the result

$$\boldsymbol{V}^{\mathrm{T}}\boldsymbol{M}\boldsymbol{V}\dot{\boldsymbol{x}}+\boldsymbol{V}^{\mathrm{T}}\boldsymbol{K}\boldsymbol{V}\boldsymbol{x}=\boldsymbol{0},$$

that is,

$$\boldsymbol{I}\dot{\boldsymbol{x}}+\boldsymbol{\Lambda}\boldsymbol{x}=\boldsymbol{0},$$

where $\Lambda=\operatorname{diag}(\lambda_j)$. This is known as modal decomposition or discretization because the system can now be split into the scalar equations

$$\dot{x}_j+\lambda_j x_j=0.$$

This is the justification for frequently demonstrating numerical methods which are meant to be applied to the system (1.2.5) on the simple scalar equation

$$\dot{x}+\lambda x=0, \qquad \text{or} \qquad \dot{x}+\lambda x=f, \tag{1.2.9}$$

with λ stated to be real and positive.

Another problem which is useful for illustrating methods for solving first order time dependent equations is that of ground-water flow. Bear (1979) gives all the background details. With no sources or sinks the basic equation for the potential Φ is of the same type as the diffusion equation:

$$S_0\frac{\partial\Phi}{\partial t}=K\nabla^2\Phi, \tag{1.2.10}$$

in three space dimensions (x, y, z), where S_0 is the storativity and we have assumed an isotropic medium with a constant hydraulic conductivity K. For a phreatic (that is, unconfined) aquifer, we can integrate through the vertical, z, dimension to obtain

$$S\frac{\partial h}{\partial t}=K\frac{\partial}{\partial x}\left(h\frac{\partial h}{\partial x}\right)+K\frac{\partial}{\partial y}\left(h\frac{\partial h}{\partial y}\right), \tag{1.2.11}$$

where S is again a storativity term and h is the height of the water table above the bottom of the aquifer which is supposed level. Equation (1.2.11) is frequently linearized by introducing the transmissivity $T=Kh$, which can be estimated from site measurements. The result is the equation:

$$S\frac{\partial h}{\partial t}=T\left(\frac{\partial^2 h}{\partial x^2}+\frac{\partial^2 h}{\partial y^2}\right). \tag{1.2.12}$$

A more complicated example comes from the Richards equation for saturated–unsaturated subsurface flow (Richards 1931), which continues up to the surface instead of being confined to the saturated part of the ground. This can be written in the form:

$$C(\Psi)\frac{\partial \Phi}{\partial t}=\frac{\partial}{\partial x}\left\{K_x(\Psi)\frac{\partial \Phi}{\partial x}\right\}+\frac{\partial}{\partial y}\left\{K_y(\Psi)\frac{\partial \Phi}{\partial y}\right\}+\frac{\partial}{\partial z}\left\{K_z(\Psi)\frac{\partial \Phi}{\partial z}\right\}, \tag{1.2.13}$$

where $0x$, $0y$ are horizontal axes, $0z$ is vertical, Ψ is the pressure potential, $\Phi=\Psi+z$ is the total hydraulic potential, K_x, K_y, and K_z are the components of hydraulic conductivity in the $0x$, $0y$, and $0z$ directions respectively, and $C(\Psi)$ is the specific moisture capacity of the ground. The motivation for producing these examples is that eqn (1.2.10) or eqn (1.2.12), when space-discretized by finite elements, will again give a system of linear equations of the form (1.2.2). Equation (1.2.11) gives a non-linear equation of a particular form:

$$\boldsymbol{M}\dot{\boldsymbol{h}}+\boldsymbol{K}(\boldsymbol{h})\boldsymbol{h}=\boldsymbol{f}, \tag{1.2.14}$$

where $\boldsymbol{h}$ is the vector of nodal values of the finite element solution. Equation (1.2.13), when space discretized, gives a highly non-linear form:

$$\boldsymbol{M}(\boldsymbol{\Phi})\dot{\boldsymbol{\Phi}}+\boldsymbol{K}(\boldsymbol{\Phi})\boldsymbol{\Phi}=\boldsymbol{f}, \tag{1.2.15}$$

where $\boldsymbol{\Phi}$ is the vector of nodal values of the finite element solution for the total hydraulic potential. Equations (1.2.14) and (1.2.15) are used for illustration of methods for solving non-linear equations in Chapter 7. References are given in Chapter 8 for the background theory to further equations which are needed to illustrate methods for dealing with coupled systems.

1.3. The dynamic vibration equation

A large section of this book is concerned with the numerical solution of the equation for the dynamic vibration of structures which starts by being second order in time and is most frequently integrated directly in this form. Many mathematicians do not meet the theory of elasticity in their undergraduate courses. We give here a brief description of the basic ideas in order to introduce the dynamic vibration equation, with the hope that this will enable the applied mathematician to share the interest of the engineers in this subject.

The simplest example is that of the transverse vibration of a stretched string (see, for example, Stakgold 1967). If the transverse deflection of the string is u, its initial mass per unit length and tension are ρ and T respectively, it is subjected to a force $f(t)$, and a damping proportional to the velocity, its equation of motion is of the form

$$\rho \frac{\partial^2 u}{\partial t^2} + \mu \frac{\partial u}{\partial t} - T \frac{\partial^2 u}{\partial x^2} = f(t), \tag{1.3.1}$$

where x is the distance along the string. A more complicated formulation is used for the displacements given by the vector $\boldsymbol{u}$ at a general point in a structure. The strains $\boldsymbol{\epsilon}$ are related to the displacements by the equation $\boldsymbol{\epsilon} = \boldsymbol{L}\boldsymbol{u}$, where what matters about $\boldsymbol{L}$ from our point of view is that it is a matrix whose entries are first order differential operators in the space coordinates. For example, in plane strain where we only need the in-plane displacements u and v, the operator $\boldsymbol{L}$ is given by

$$\boldsymbol{L} = \begin{bmatrix} \dfrac{\partial}{\partial x} & 0 \\ 0 & \dfrac{\partial}{\partial y} \\ \dfrac{\partial}{\partial y} & \dfrac{\partial}{\partial x} \end{bmatrix}.$$

Linear elasticity theory (Hooke's Law) states that there is a linear relation between the vector of stresses, $\boldsymbol{\sigma}$, and the strains given by

$$\boldsymbol{\sigma} = \boldsymbol{D}\boldsymbol{\epsilon}, \tag{1.3.2}$$

where $\boldsymbol{D}$ is a matrix containing the elastic constants of the material. Then the internal force per unit volume due to the stressed state of the material is

$$\boldsymbol{L}^{\mathrm{T}}\boldsymbol{\sigma} = \boldsymbol{L}^{\mathrm{T}}\boldsymbol{D}\boldsymbol{\epsilon} = \boldsymbol{L}^{\mathrm{T}}\boldsymbol{D}\boldsymbol{L}\boldsymbol{u}. \tag{1.3.3}$$

Hence, assuming there is a damping effect linearly proportional to the velocity at a point of the structure, we can use d'Alembert's Principle to write down the equation of motion as

$$\rho \frac{\partial^2 \boldsymbol{u}}{\partial t^2} + \mu \frac{\partial \boldsymbol{u}}{\partial t} - \boldsymbol{L}^{\mathrm{T}}\boldsymbol{D}\boldsymbol{L}\boldsymbol{u} = \boldsymbol{F} \tag{1.3.4}$$

over the whole domain Ω of the structure, where ρ is the density, μ is the damping parameter, and $\boldsymbol{F}$ the vector of external forces acting at a point. We now make the finite element discretization in space:

$$\boldsymbol{u} \approx \sum_j \boldsymbol{u}_j(t)\boldsymbol{\Phi}_j,$$

where $\boldsymbol{\Phi}_j$ are the basis functions, multiply eqn (1.3.4) by the $\boldsymbol{\Phi}_j$ in turn and integrate over Ω. The standard integration by parts operation on the internal force term then gives the system of second order ordinary differential equations

$$\boldsymbol{M}\ddot{\boldsymbol{u}} + \boldsymbol{C}\dot{\boldsymbol{u}} + \boldsymbol{K}\boldsymbol{u} = \boldsymbol{f} \tag{1.3.5}$$

where $\boldsymbol{u}$, $\dot{\boldsymbol{u}}$, and $\ddot{\boldsymbol{u}}$, are the nodal displacements, velocities, and accelerations respectively, $\boldsymbol{M}$ is the mass matrix as given in eqn (1.2.3), and $\boldsymbol{K}$ is a stiffness matrix type with entries

$$\int_\Omega (\boldsymbol{L}\boldsymbol{\Phi}_j)^{\mathrm{T}}\boldsymbol{D}(\boldsymbol{L}\boldsymbol{\Phi}_k)\,\mathrm{d}\Omega, \tag{1.3.6}$$

where $\boldsymbol{\Phi}_j$ is the vector of basis functions attached to the displacement at a point, such that $\boldsymbol{L}\boldsymbol{\Phi}_j$ is a vector of the same order as the matrix $\boldsymbol{D}$. The vector $\boldsymbol{f}$ represents the external forcing function including the boundary conditions. The similar finite element space discretization of eqn (1.3.1) corresponds to the case where $\boldsymbol{D} \equiv T$ and $\boldsymbol{L} \equiv \nabla$, the gradient operator. The damping matrix, $\boldsymbol{C}$, may be difficult to determine in practice and it is frequently assumed to be equal to a linear combination of the matrices $\boldsymbol{M}$ and $\boldsymbol{K}$,

$$\boldsymbol{C} = a\boldsymbol{M} + b\boldsymbol{K}, \tag{1.3.7}$$

known as classical or Rayleigh damping. Estimates of the values of the parameters a and b may be made from experiments.

There are many more details in, for example, Planck (1932), Timoshenko (1934), and, for the finite element approach, Zienkiewicz (1977).

The solution of eqn (1.3.5) consists of a linear combination of the solution of the homogeneous equation

$$\boldsymbol{M}\ddot{\boldsymbol{u}} + \boldsymbol{C}\dot{\boldsymbol{u}} + \boldsymbol{K}\boldsymbol{u} = \boldsymbol{0} \tag{1.3.8}$$

together with the response to the external forcing function $\boldsymbol{f}$. Suppose we assume a trial solution $\boldsymbol{u} = \mathrm{e}^{\lambda t}\boldsymbol{v}$ in eqn (1.3.8). This gives the system of equations

$$(\lambda^2\boldsymbol{M} + \lambda\boldsymbol{C} + \boldsymbol{K})\boldsymbol{v} = \boldsymbol{0}, \tag{1.3.9}$$

where the matrix is a form of Lambda matrix (a matrix whose entries are polynomials in a parameter λ; see, for instance, Frazer *et al.* 1963) and $\boldsymbol{v}$ is the eigenvector corresponding to the eigenvalue λ. Now pre-multiply eqn (1.3.9) by $\boldsymbol{v}^{\mathrm{T}}$ to give the scalar quadratic equation in λ:

$$\lambda^2\boldsymbol{v}^{\mathrm{T}}\boldsymbol{M}\boldsymbol{v} + \lambda\boldsymbol{v}^{\mathrm{T}}\boldsymbol{C}\boldsymbol{v} + \boldsymbol{v}^{\mathrm{T}}\boldsymbol{K}\boldsymbol{v} = \boldsymbol{0} \tag{1.3.10}$$

We know that the mass matrix $\boldsymbol{M}$ is symmetric and positive definite, and the stiffness matrix $\boldsymbol{K}$ is also symmetric and non-negative definite, so provided the damping matrix $\boldsymbol{C}$ is symmetric and also non-negative definite we can quote the Routh–Hurwitz conditions (Appendix 1) to show that the values of λ given by eqn (1.3.10) have non-positive real parts. This means that the solution of eqn (1.3.8) remains bounded and hence the numerical solution should also remain bounded. If the damping matrix $\boldsymbol{C}$ is zero, λ^2 is negative, the solution is oscillatory and it does not decay.

With the assumption of Rayleigh damping, eqn (1.3.7), we can make a modal decomposition like that in § 1.2 with the first order equation. Using again the substitution $\boldsymbol{u} = \boldsymbol{V}\boldsymbol{x}$, where $\boldsymbol{V}$ is the modal matrix of the eigenvectors $\boldsymbol{v}_j$, in eqn (1.3.5), and pre-multiplying both sides of the equation by $\boldsymbol{V}^{\mathrm{T}}$, we have the result:

$$\boldsymbol{V}^{\mathrm{T}}\boldsymbol{M}\boldsymbol{V}\ddot{\boldsymbol{x}} + \boldsymbol{V}^{\mathrm{T}}\boldsymbol{C}\boldsymbol{V}\dot{\boldsymbol{x}} + \boldsymbol{V}^{\mathrm{T}}\boldsymbol{K}\boldsymbol{V}\boldsymbol{x} = \boldsymbol{V}^{\mathrm{T}}\boldsymbol{f}. \tag{1.3.11}$$

Thus if the damping matrix $\boldsymbol{C}$ is simultaneously diagonalized with the matrices $\boldsymbol{M}$ and $\boldsymbol{K}$ (as it certainly is with the assumption of Rayleigh damping (1.3.7)), we have the modal decomposition into the scalar equations usually written in the form

$$m_j\ddot{x}_j + c_j\dot{x}_j + k_jx_j = f_j, \tag{1.3.12}$$

$j = 1, 2, \ldots, N$, the number of unknown nodal coordinates. This is then the justification for demonstrating numerical methods on the scalar equation

$$m\ddot{x} + c\dot{x} + kx = f. \tag{1.3.13}$$

The solution of the homogeneous eqn (1.3.13) with $f = 0$ is given by $x = \mathrm{e}^{\lambda t}$ where

$$m\lambda^2 + c\lambda + k = 0,$$

that is,

$$\lambda = \frac{1}{2m}\{-c \pm i\surd(4km - c^2)\}, \qquad i = \surd{-1}. \tag{1.3.14}$$

The solution must be real, hence with $c^2 < 4km$, if we put $\bar{\omega}^2 = \surd(4km - c^2)/2m$, we can write it as

$$\exp\left(-\frac{c}{2m}\right)(A\cos\bar{\omega}t + B\sin\bar{\omega}t),$$

that is, a damped oscillation of period $2\pi/\bar{\omega}$. Critical damping is given by $c^2 = 4km$, when the roots cease to be complex. Structures usually have very light damping; for example, dynamic tests on a structure done by Galambos and Mayes (1979) show that the damping can be as low as 4% of critical. Hence the emphasis is entirely on the damped oscillatory form of the solution of eqn (1.3.8), and frequently methods are demonstrated simply on the undamped equation.

$$\boldsymbol{M}\ddot{\boldsymbol{u}} + \boldsymbol{K}\boldsymbol{u} = \boldsymbol{f}. \tag{1.3.15}$$

There often seems to be a tacit assumption that any algorithm which is stable with zero natural damping must be stable with it. This is true for the simplest algorithms but not necessarily for the higher order algorithms as shown in § 4.7.

Engineers often calculate the eigenvalues λ_j, and obtain the corresponding eigenvectors for the lowest half dozen or so, solve the scalar equations (1.3.12), and add the results to get an approximate solution. The finite element space discretization gives a good approximation to the modes corresponding to the lowest eigenvalues, but the approximation gets worse as one progresses up the spectrum (Strang and Fix 1973), and the system is 'stiff'. However, the business of finding the eigenvalues and vectors of a very large system is a formidable task, especially where the eigenvalues are close together, and many engineers prefer the direct methods described in this book. These methods are usually meant to be applied to the second order system of eqns (1.3.5), but there are some advocates of first reducing this to the equivalent first order system. Jensen (1974) gives several ways of doing this; the simplest is to put

$$\dot{\boldsymbol{u}} = \boldsymbol{v} \quad \text{and} \quad \ddot{\boldsymbol{u}} = \dot{\boldsymbol{v}}, \tag{1.3.16}$$

into eqn (1.3.5) to give the equivalent system of first order equations:

$$\dot{\boldsymbol{u}} = \boldsymbol{v} \quad \text{and} \quad \boldsymbol{M}\dot{\boldsymbol{v}} = -\boldsymbol{C}\boldsymbol{v} - \boldsymbol{K}\boldsymbol{u} + \boldsymbol{f}, \tag{1.3.17}$$

which can be written as

$$\dot{\boldsymbol{z}} = \boldsymbol{B}\boldsymbol{z} + \boldsymbol{F}, \tag{1.3.18}$$

where

$$\begin{aligned} \boldsymbol{z}^{\mathrm{T}} &= [\boldsymbol{u}, \boldsymbol{v}], \\ \boldsymbol{F}^{\mathrm{T}} &= [\boldsymbol{0}, \boldsymbol{M}^{-1}\boldsymbol{f}], \end{aligned}$$

and

$$\boldsymbol{B} = \begin{bmatrix} \boldsymbol{0} & \boldsymbol{I} \\ -\boldsymbol{M}^{-1}\boldsymbol{K} & -\boldsymbol{M}^{-1}\boldsymbol{C} \end{bmatrix} \tag{1.3.19}$$

Numerical methods for first order equations can be used on the system (1.3.18). The matrix $\boldsymbol{B}$ has $2N$ complex eigenvalues representing the damped oscillations of the original system (1.3.5). But the matrix to be inverted in the numerical solution is still of order N because the first half of the equivalent first order system as shown in eqns (1.3.17) simply consists of scalar equations.

When the deformations of the structure are large, the Hooke's Law expression of linear elasticity (1.3.2) no longer holds (because elasticity theory is based on the assumption of small deformations). Equation (1.3.2) does not hold either when the material of the structure has a non-linear behaviour. The dynamic vibration equation is then usually written in the form

$$\boldsymbol{M}\ddot{\boldsymbol{u}} + \boldsymbol{C}\dot{\boldsymbol{u}} + \boldsymbol{P}(\boldsymbol{u}) = \boldsymbol{f}, \tag{1.3.20}$$

where $\boldsymbol{P}(\boldsymbol{u})$ depends on the displacement $\boldsymbol{u}$ is a non-linear way.

1.4. Eigenvalue bounds

A development of a theorem of Irons (1970) on finite element eigenvalue bounds is useful in this book. (This theorem was only published by Irons in a department report and it has been rediscovered several times.) With the first order problems discussed in §1.2, from eqn (1.2.7) we have the eigenvalue–eigenvector equation:

$$\lambda \boldsymbol{M}\boldsymbol{v} = \boldsymbol{K}\boldsymbol{v}, \tag{1.4.1}$$

and with the second order problems discussed in §1.3, from eqn (1.3.9) with zero damping we have the eigenvalue–eigenvector equation:

$$\lambda^2 \boldsymbol{M}\boldsymbol{v} = -\boldsymbol{K}\boldsymbol{v}. \tag{1.4.2}$$

Hence in either case we are interested in the eigenvalue bounds given by the Rayleigh Quotient (Garabedian 1964),

$$\xi = \frac{\boldsymbol{v}^{\mathsf{T}}\boldsymbol{K}\boldsymbol{v}}{\boldsymbol{v}^{\mathsf{T}}\boldsymbol{M}\boldsymbol{v}}. \tag{1.4.3}$$

Because of the way the stiffness and mass matrices are assembled, we have

$$\xi = \sum_j (\boldsymbol{v}_j^{\mathsf{T}}\boldsymbol{K}_j^{(e)}\boldsymbol{v}_j) / \sum_j (\boldsymbol{v}_j^{\mathsf{T}}\boldsymbol{M}_j^{(e)}\boldsymbol{v}_j), \tag{1.4.4}$$

where $\boldsymbol{K}_j^{(e)}$, $\boldsymbol{M}_j^{(e)}$ are the stiffness and mass matrices respectively of the jth element, and $\boldsymbol{v}_j$ is the corresponding component of the vector $\boldsymbol{v}$. Then if $\xi_j^{(e)}$ is the Rayleigh Quotient for the jth element, we have

$$\xi_j^{(e)} = \boldsymbol{v}_j^{\mathrm{T}} \boldsymbol{K}_j^{(e)} \boldsymbol{v}_j / (\boldsymbol{v}_j^{\mathrm{T}} \boldsymbol{M}_j^{(e)} \boldsymbol{v}_j). \tag{1.4.5}$$

Substituting from (1.4.5) into eqn (1.4.4) we see that

$$\xi = \sum_j \alpha_j \xi_j^{(e)} \Big/ \sum_j \alpha_j, \tag{1.4.6}$$

where $\alpha_j = \boldsymbol{v}_j^{\mathrm{T}} \boldsymbol{M}_j^{(e)} \boldsymbol{v}_j > 0$, because the mass matrix is positive definite. In eqn (1.4.6), ξ is expressed as a weighted mean of the $\xi_j^{(e)}$ with positive weights, and hence

$$\xi_{\min}^{(e)} \leq \xi \leq \xi_{\max}^{(e)}. \tag{1.4.7}$$

Thus we have bounds on the eigenvalues of the global problem from the Rayleigh Quotients of the elements. In particular, we can see what difference it makes to these bounds if the mass matrix is lumped, that is, replaced by a diagonal matrix. An estimate can be obtained by using

$$\xi_{\max}^{(e)} \leq \max\,\{\boldsymbol{v}_j^{\mathrm{T}} \boldsymbol{K}_j^{(e)} \boldsymbol{v}_j\} / \min\{\boldsymbol{v}_j^{\mathrm{T}} \boldsymbol{M}_j^{(e)} \boldsymbol{v}_j\} \tag{1.4.7a}$$

With linear elements in one dimension, the distributed mass matrix is

$$\boldsymbol{M}_{\mathrm{D}}^{(e)} = \frac{h_j}{6}\begin{bmatrix} 2 & 1 \\ 1 & 2 \end{bmatrix}, \tag{1.4.8}$$

where h_j is the ratio of the actual length of the element to the length of the region over which the equation is being solved (Strang and Fix 1973), which has eigenvalues $\frac{h_j}{2}, \frac{h_j}{6}$. The lumped mass matrix is

$$M_{\mathrm{L}}^{(e)} = \frac{h_j}{2}\begin{bmatrix} 1 & 0 \\ 0 & 1 \end{bmatrix}, \tag{1.4.9}$$

which has eigenvalues $\frac{h_j}{2}, \frac{h_j}{2}$. This means that with the first order problems, the upper bound (1.4.7a) to the eigenvalue λ in eqn (1.4.1) is changed in the ratio 3 : 1 in going from a distributed to a lumped mass matrix, and in second order problems where $\lambda^2 = -\omega^2$ in eqn (1.4.2), the bound on the eigenvalue ω is changed in the ratio $\sqrt{3} : 1$ (see Exercise 1.1).

The corresponding result with linear triangular elements is obtained by finding the eigenvalues of the distributed mass matrix

$$\boldsymbol{M}_{\mathrm{D}}^{(e)} = \frac{\Delta^{(e)}}{12}\begin{bmatrix} 2 & 1 & 1 \\ 1 & 2 & 1 \\ 1 & 1 & 2 \end{bmatrix}, \tag{1.4.10}$$

where $\Delta^{(e)}$ is the area of the element. The distributed mass matrix can be calculated by using the area coordinates of the triangle L_1, L_2, L_3 (Strang and Fix 1973) as the basis functions in eqn (1.2.3) together with the useful integration formula

$$\iint_{\Delta^{(e)}} L_1^a L_2^b L_3^c \,\mathrm{d}x\,\mathrm{d}y = \frac{2a!\,b!\,c!}{(a+b+c+2)!}\Delta^{(e)}.$$

The eigenvalues of the matrix $\boldsymbol{M}_{\mathrm{D}}^{(e)}$ are

$$\frac{\Delta^{(e)}}{12}, \frac{\Delta^{(e)}}{12}, \frac{\Delta^{(e)}}{3}, \tag{1.4.11}$$

whereas the lumped mass matrix $\boldsymbol{M}_{\mathrm{L}}^{(e)}$ has three equal eigenvalues $\dfrac{\Delta^{(e)}}{3}$. Hence the bound (1.4.7a) on the value of λ in eqn (1.4.1) is changed in the ratio 4:1, and the bound on the value of ω associated with eqn (1.4.2) is changed in the ratio 2:1 in changing from distributed to lumped mass matrix (see Exercise 1.2). These results give an idea of the magnitude of the effect produced by using a lumped mass matrix and are useful in connection with the conditionally stable algorithms discussed in later chapters.

Summing up, we have that the original partial differential equations (1.2.1), (1.3.1) or (1.3.4) have classical solutions composed of an infinite number of modes (eigenfunctions) and corresponding natural frequencies (Garabedian 1964). The finite element space discretization gives a system of N simultaneous ordinary differential equations, (1.2.2) or (1.3.5). These each have N modes (given by the eigenvectors) and corresponding eigenvalues. The eigenvalues λ_j of the first order form (1.2.5), and the eigenfrequencies $\omega_j = \sqrt{(k/m)}$ of the undamped second order form (1.3.15) have a particular relationship to those of the corresponding partial differential equations (Garabedian 1964;

Strang and Fix 1973). With the distributed mass matrix, the jth eigenvalue/frequency of the discretized form is always greater than the jth eigenvalue/frequency of the partial differential equation. The lumped mass matrix solution gives a lower estimate of this eigenvalue/frequency but we do not know whether it is too large or too small. We know that the distributed mass matrix gives a more accurate solution but we shall see in Chapter 2 that, because the lumped mass matrix gives a lower maximum eigenvalue/frequency, it can be useful to use it in a conditionally stable numerical scheme.

The Rayleigh Quotient in eqn (1.4.3) also demonstrates the 'stiffness' effect introduced by the finite element space discretization. With one dimensional problems, the stiffness matrix from eqn (1.2.4) has entries proportional to h^{-1}, where h is the element length, and the mass matrix from eqn (1.2.3) has entries proportional to h. Hence the highest eigenvalues are proportional to h^{-2}, so that the more the mesh is refined, the higher they become. In two and three dimensions the effect is the same: the highest value of the Rayleigh Quotient is proportional to h^{-2}, where h is a dimension of the elements.

Exercises

1.1. Show that for a one-dimensional problem with linear elements, stiffness matrix

$$\boldsymbol{K}^{(e)} = \frac{1}{h_j}\begin{bmatrix} 1 & -1 \\ -1 & 1 \end{bmatrix}$$

and distributed mass matrix as in eqn (1.4.8), we have equality in (1.4.7a).

1.2. The stiffness matrix for a linear triangle finite element ABC can be expressed as

$$\boldsymbol{K}^{(e)} = \frac{1}{2}\begin{bmatrix} \cot B + \cot C & -\cot C & -\cot B \\ -\cot C & \cot C + \cot A & -\cot A \\ -\cot B & -\cot A & \cot A + \cot B \end{bmatrix},$$

where A, B, C are the angles of the triangle. With the distributed mass matrix given in eqn (1.4.10), show that the eigenvalues of $\boldsymbol{M}_{\mathrm{D}}^{(e)^{-1}}\boldsymbol{K}^{(e)}$ are $\dfrac{12}{\Delta}$ times the eigenvalues of $\boldsymbol{K}^{(e)}$ where Δ is the area of the triangle.

Hence, using the Gerschgorin circle theorem (see, for example, Conte and de Boor 1980), show that the value of $\xi_{\max}^{(e)}$ in §1.4 can be estimated as $24/h^2$ where h is the minimum element height.

1.3. Find the distributed and lumped mass matrices for a bilinear square element and use eqn (1.4.7a) to obtain an estimate of the difference in the bounds on λ in eqn (1.4.1) and ω in eqn (1.4.2).

1.4. Consider the non-dimensional problem of a stretched string, length unity, ends fixed and equation for the transverse deflection u given by

$$\frac{\partial^2 u}{\partial t^2} = \frac{\partial^2 u}{\partial x^2},$$

with $u(x, 0) = f(x)$, $\dfrac{\partial u}{\partial t}(x, 0) = 0$. Show that the classical solution is $u = \sum_{k=1}^{\infty} A_k \cos k\pi t \sin k\pi x$, where $\sum_{k=1}^{\infty} A_k \sin k\pi x = f(x)$, and the natural frequencies of the string are $k/2$, $k = 1, 2, 3, \ldots$. Divide the string into three equal linear finite elements and solve the problem approximately with (a) a distributed mass matrix, and (b) a lumped mass matrix. Show that the approximations to the two lowest frequencies are too high with (a) and too low with (b).

References

Bear, J. (1970). *Hydraulics of groundwater.* McGraw-Hill.

Conte, S. D. and de Boor, C. (1980). *Elementary numerical analysis.* McGraw-Hill.

Crank, J. (1975). *The mathematics of diffusion.* Clarendon Press. Oxford.

Frazer, R. A., Duncan, W. J., and Collar, A. R. (1963). *Elementary matrices.* Cambridge University Press.

Galambos, T. V. and Mayes, R. C. (1979). Lessons from dynamic tests of an eleven storey building. *Engineering Structures,* **1,** 264–73.

Garabedian, P. R. (1964). *Partial differential equations.* Wiley.

Irons, B. M. (1970). Applications of a theorem on eigenvalues to finite element problems. *CR/132/70.* University of Wales, Department of Civil Engineering, Swansea.

Jensen, P. S. (1974). Transient analysis of structures by stiffly stable methods. *Computers and Structures,* **4,** 615–26.

Planck, Max (1932). *The mechanics of deformable bodies* (trans. H. L. Brose). Macmillan.

Richards, L. A. (1931). Capillary conduction of liquids through porous mediums. *Physics,* **1,** 318–33.

Richtmyer, R. D. and Morton, K. W. (1967). *Difference methods for initial value problems*. Wiley.

Stakgold, I. (1967). *Boundary value problems of mathematical physics*, Vol. 1. Macmillan.

Strang, G. and Fix, G. J. (1973). *An analysis of the finite element method*. Prentice-Hall.

Timoshenko, S. (1934). *Theory of elasticity*. McGraw-Hill.

Wilkinson, J. H. (1965). *The algebraic eigenvalue problem*. Oxford University Press.

Zienkiewicz, O. C. (1977). *The finite element method*. McGraw-Hill.

2 *p*-Step and *p*-stage schemes

2.1. Introduction

In this chapter we introduce two particular types of algorithm for the numerical solution of the dynamic vibration equation

$$\boldsymbol{M}\ddot{\boldsymbol{x}} + \boldsymbol{C}\dot{\boldsymbol{x}} + \boldsymbol{K}\boldsymbol{x} = \boldsymbol{f} \tag{2.1.1}$$

which are very popular with engineers. In eqn (2.1.1) $\boldsymbol{x}$, $\dot{\boldsymbol{x}}$, $\ddot{\boldsymbol{x}}$ are the vectors of nodal displacements, velocities and accelerations respectively and we have given starting conditions $\boldsymbol{x}(0)$ and $\dot{\boldsymbol{x}}(0)$. The matrices $\boldsymbol{M}$, $\boldsymbol{C}$, $\boldsymbol{K}$ are the mass, damping, and stiffness matrices respectively and $\boldsymbol{f}$ is the forcing function.

It is useful to link the discussion of the numerical solution of the dynamic vibration equation (2.1.1) with that of the parabolic system of equations

$$\boldsymbol{M}\dot{\boldsymbol{z}} + \boldsymbol{K}\boldsymbol{z} = \boldsymbol{f}, \tag{2.1.2}$$

where $\boldsymbol{z}$ is some physical quantity and $\dot{\boldsymbol{z}}$ its time derivative. Here $\boldsymbol{M}$ and $\boldsymbol{K}$ are the mass and stiffness matrices respectively, $\boldsymbol{f}$ is the forcing function and we have the given starting condition $\boldsymbol{z}(0)$.

We want to be in a position to discuss the p-step and one-step p-stage algorithms proposed by engineers for the numerical solution of the eqn (2.1.1). It is simplest first to make a general approach to these algorithms and to use some particular examples to illustrate key ideas associated with the theory. In Chapter 3 we make a detailed assessment of two simple algorithms which have alternative forms of two-step and two-stage type. Then in Chapter 4 we present the theory for general values of the parameter p.

We start by assuming that we have conditions such that we can make a modal discretization of the system of eqns (2.1.1) as in Chapter 1, eqn (1.3.12), and we discuss the algorithms as applied to the homogeneous form of eqn (1.3.13) with $f \equiv 0$. We consider later ways of dealing with the forcing function.

1. We can use p-step algorithms of the form

$$\sum_{j=0}^{p} (\alpha_j m + \gamma_j \Delta t c + \beta_j \Delta t^2 k) x_{n+j} = 0. \qquad (2.1.3)$$

   Here Δt is the time step, $x_{n+j} \approx x((n+j)\Delta t)$ and α_j, β_j, γ_j are parameters with $\alpha_p = 1$ to normalize. Given values of $x_0, x_1, \ldots, x_{p-1}$ we can use eqn (2.1.3) repeatedly to generate x_p, x_{p+1}, etc. We have $x_0 = x(0)$ and $x_1, x_2, \ldots, x_{p-1}$ can be generated to the required accuracy, consistent with the accuracy of the algorithm being used, from $x(0)$, $\dot{x}(0)$ and the differential equation assumed to hold at time $t = 0$.

2. Alternatively we can consider one-step p-stage algorithms of the form

$$\boldsymbol{X}_{n+1} = \boldsymbol{A}\boldsymbol{X}_n \qquad (2.1.4)$$

   where

$$\boldsymbol{X}_n^{\mathrm{T}} = \left[x_n, \dot{x}_n, \ddot{x}_n, \ldots, \overset{p-1}{x_n} \right], \qquad (2.1.5)$$

   i.e. the displacement and $(p-1)$ derivatives are carried at each time level. The matrix $\boldsymbol{A}$ is the $p \times p$ amplification matrix with terms which are linear combinations of the parameters m, c, and k. Here we have $\boldsymbol{X}_0$ given by $x_0 = x(0)$, $\dot{x}_0 = \dot{x}(0)$ and $\ddot{x}_0, \ldots, \overset{p-1}{x_0}$ may be derived by using the differential equation and the results of differentiating it (assuming this is possible). The question of starting conditions is discussed further later.

The advantage of one-step algorithms is that it is very easy to change the size of time step to control the error.

From eqn (2.1.4), we have

$$\boldsymbol{X}_{n+2} = \boldsymbol{A}^2 \boldsymbol{X}_n, \qquad \boldsymbol{X}_{n+3} = \boldsymbol{A}^3 \boldsymbol{X}_n, \ldots \qquad (2.16)$$

We now form the linear combination

$$\sum_{j=0}^{p} q_j \boldsymbol{X}_{n+j} = (q_p \boldsymbol{A}^p + q_{p-1} \boldsymbol{A}^{p-1} + \cdots + q_1 \boldsymbol{A} + q_0 \boldsymbol{I}) \boldsymbol{X}_n. \quad (2.1.7)$$

Then if $q_0, q_1, \ldots, q_p$ are the coefficients in the characteristic polynomial of the matrix $\boldsymbol{A}$, we have

$$\sum_{j=0}^{p} q_j \boldsymbol{X}_{n+j} = \boldsymbol{0} \tag{2.1.8}$$

by the Cayley–Hamilton theorem (the matrix satisfies its own characteristic polynomial, see, for example, Fröberg 1970).

Hence from eqn (2.1.8) all the components of the vector $\boldsymbol{X}_n$ satisfy the same recurrence relation and thus, with sufficiently accurate starting values, are approximated to the same order of accuracy. (See Lambert (1973) for the solution of recurrence relations of the form of (2.1.3), otherwise known as difference equations.) By simply taking the first equation of the set of p equations in eqn (2.1.8) we have an equivalent p-step algorithm in the displacements x_n alone. The coefficients of the terms in this recurrence relation in the displacements are $q_0, q_1, \ldots, q_p$. Thus the characteristic polynomial of the p-step algorithm is also the characteristic polynomial of the amplification matrix of the corresponding one-step p-stage algorithm. We can also identify the coefficients in the p-step algorithm (2.1.3) as

$$q_j = \alpha_j m + \gamma_j \Delta t c + \beta_j \Delta t^2 k \tag{2.1.9}$$

(remembering that $\alpha_p = 1$). From this we see that a one-step p-stage scheme is always equivalent to some p-step scheme. This means that we can study the properties of a p-step scheme and immediately draw conclusions about the one-step p-stage scheme to which it corresponds. Further, by recombining the modes, as the whole process is linear, we can apply the same reasoning to these p-step and corresponding one-step p-stage schemes applied to the system of eqns (2.1.1).

We can apply the same line of argument as above to the first order form of eqn (2.1.1) with $\boldsymbol{f} = \boldsymbol{0}$

$$\dot{\boldsymbol{z}} + \boldsymbol{B}\boldsymbol{z} = \boldsymbol{0}, \tag{2.1.10}$$

where

$$\boldsymbol{z}^{\mathrm{T}} = [\boldsymbol{x}, \boldsymbol{v}] \quad \text{and} \quad \boldsymbol{B} = \begin{bmatrix} \boldsymbol{0} & -\boldsymbol{I} \\ \boldsymbol{M}^{-1}\boldsymbol{K} & \boldsymbol{M}^{-1}\boldsymbol{C} \end{bmatrix}. \tag{2.1.11}$$

The corresponding p-step algorithm can be written

$$\sum_{j=0}^{p} (\gamma_j \boldsymbol{I} + \beta_j \Delta t \boldsymbol{B}) \boldsymbol{x}_{n+j} = \boldsymbol{0}. \tag{2.1.12}$$

This p-step algorithm can also be applied to the numerical solution of the parabolic eqn (2.1.2) with the matrix $\boldsymbol{B}=\boldsymbol{M}^{-1}\boldsymbol{K}$.

Equation (2.1.12) has the familiar form of the p-step algorithms discussed in mathematics textbooks such as Lambert (1973). We can take a one-step p-stage algorithm of the form given in eqn (2.1.4)

$$\boldsymbol{Z}_{n+1}=\boldsymbol{A}\boldsymbol{Z}_n, \tag{2.1.13}$$

with $\boldsymbol{Z}_n^{\mathrm{T}}=\left[z_n, \dot{z}_n, \ldots, \overset{p-1}{z_n}\right]$. Then following through the same kind of argument as above we see that a one-step p-stage scheme of the form in eqn (2.1.13) is equivalent to a p-step scheme of the form given in eqn (2.1.12). The characteristic polynomial of the p-step algorithm is also the characteristic polynomial of the amplification matrix of the one-step p-stage algorithm and we can identify the coefficients of the polynomial now as

$$\boldsymbol{q}_j=\gamma_j\boldsymbol{I}+\beta_j\Delta t\boldsymbol{B} \tag{2.1.14}$$

(with $\gamma_p=1$ to normalize). We can also see that in particular a one-step one-stage algorithm used on the first order form (2.1.10) must be equivalent to some one-step two-stage algorithm used on the second order form (2.1.1). The θ-method is the standard one-step one-stage algorithm for the first order equation and we shall examine in detail the links between this and the Newmark method (Newmark 1959) in Chapter 3.

When eqn (2.1.10) represents the first order form of the dynamic vibration equation with damping less than critical, the eigenvalues of the matrix B will be in complex conjugate pairs with positive real parts. Equations (2.1.12) to (2.1.14) could equally well apply to the case when eqn (2.1.10) comes from a problem of parabolic type such as in eqn (2.1.2). The eigenvalues of the matrix $\boldsymbol{B}$ are then real and positive.

Programs for the use of the algorithms (2.1.3) or (2.1.4) are sometimes proposed so that by putting the matrix $\boldsymbol{M}\equiv\boldsymbol{0}$ they can also be used for the numerical solution of eqn (2.1.2) with the matrix $\boldsymbol{C}$ replaced by the mass matrix (Zienkiewicz *et al.* 1984).

In order to discuss the choice of parameters in an algorithm such as that in eqn (2.1.3), we introduce in the next sections the ideas of consistency, stability and convergence.

2.2. Consistency

We first note in connection with the p-step algorithm (2.1.3) that this is reminiscent of the p-step algorithms of the form

$$\sum_{j=0}^{p} (\alpha_j y_{n+j} - \Delta t \beta_j f_{n+j}) = 0 \tag{2.2.1}$$

for the numerical integration of $\dot{y} = f(t, y)$, which are discussed, for example, in Lambert (1973) and in many other places in the mathematics literature. By analogy with eqn (2.2.1), it seems natural to treat the forcing function $f(t)$ in eqn (2.1.1) in the same way as the $\boldsymbol{Kx}$ term, i.e. to write the p-step algorithm of the form given in eqn (2.1.3) as applied to the numerical integration of eqn (2.1.1) as

$$\sum_{j=0}^{p} \{(\alpha_j \boldsymbol{M} + \gamma_j \Delta t \boldsymbol{C} + \beta_j \Delta t^2 \boldsymbol{K})\boldsymbol{x}_{n+j} - \beta_j \Delta t^2 \boldsymbol{f}_{n+j}\} = \boldsymbol{0}. \tag{2.2.2}$$

We shall see in Chapter 3 that the most popular algorithm for dealing with the dynamic vibration equation, that based on Newmark's method (Newmark 1959), does logically lead to a p-step algorithm of the form given in eqn (2.2.2). But, possibly because of the way that the elementary theory of differential equations makes the familiar distinction between 'Particular Integral' and 'Complementary Function', in the engineering literature we often find the forcing function treated separately as, for example, in Zienkiewicz *et al.* (1984), or Thomas and Gladwell (1985). The accuracy of the solution must depend on the accuracy of the approximation of the forcing function as well as of the rest of the equation, and Penry and Wood (1985) demonstrate this. If the forcing function is approximated to a lower order of accuracy then the accuracy of the solution is reduced to this order.

We follow here the approach to the concept of consistency given in Richtmyer and Morton (1967). We write the algorithm (2.1.3) so that it is a direct replacement of the differential equation (2.1.1):

$$\boldsymbol{L}_{\Delta t}(x_n) = \sum_{j=0}^{p} (\alpha_j \Delta t^{-2} \boldsymbol{M} + \gamma_j \Delta t^{-1} \boldsymbol{C} + \beta_j \boldsymbol{K})\boldsymbol{x}_{n+j} - \beta_j \boldsymbol{f}_{n+j} = \boldsymbol{0}, \tag{2.2.3}$$

where $f_{n+j}=f((n+j)\Delta t)$. Then we measure how closely the classical solution of the differential equation (2.1.1) satisfies the difference scheme (2.2.3). We define the local truncation error as

$$L_{\Delta t}(x(n\Delta t))=\sum_{j=0}^{p}\{(\alpha_j\Delta t^{-2}M+\gamma_j\Delta t^{-1}C+\beta_j K)$$
$$\times x((n+j)\Delta t)-\beta_j f_{n+j}\}=\mathbf{O}(\Delta t^{\rho}) \quad (2.2.4)$$

as $\Delta t\to 0$, for all sufficiently smooth classical solutions of the differential equation (2.1.1). Then if the local truncation error tends to zero as the time step Δt tends to zero, the algorithm is called consistent. Thus here consistency requires that $\rho\geq 1$. We shall see that engineers' algorithms have an in-built consistency because they are based on rational assumptions. They start for example by assuming that the displacement varies as a polynomial of degree p in each time step (Zienkiewicz *et al.* 1984) or from assuming that the acceleration varies linearly during the time interval $\theta\Delta t$, $\theta\geq 1$ (Bathe and Wilson 1973), or (the most popular method of all) from assuming that the displacement and velocity have independent polynomial variation in the time step (Newmark 1959).

For example, the central difference algorithm is a particular form of Newmark given by

$$\Delta t^{-2}M(x_{n+1}-2x_n+x_{n-1})+\Delta t^{-1}C(x_{n+1}-x_{n-1})/2+Kx_n-f_n=\mathbf{0}.$$

Substituting the parameters α_j, γ_j, β_j, in eqn (2.2.4) and expanding by Taylor series about $n\Delta t$ gives

$$L_{\Delta t}(x(n\Delta t))=M\left\{\ddot{x}(n\Delta t)+\frac{\Delta t^2}{12}\ddddot{x}(n\Delta t)+\cdots\right\}$$
$$+C\left\{\dot{x}(n\Delta t)+\frac{\Delta t^2}{3}\dddot{x}(n\Delta t)+\cdots\right\}+Kx(n\Delta t)-f(n\Delta t).$$

Then, using the differential equation (2.1.1) gives

$$L_{\Delta t}(x(n\Delta t))=\mathbf{O}(\Delta t^2).$$

Thus the central difference algorithm is consistent of order two.

The corresponding definition of the local truncation error for the algorithm (2.1.12) is

$$\sum_{j=0}^{p}(\gamma_j\Delta t^{-1}I+\beta_j B)z((n+j)\Delta t)-\beta_j f_{n+j}=\mathbf{O}(\Delta t^{\rho}). \quad (2.2.5)$$

2.3. Stability

Of the various problems inherent in the choice of the parameters in the p-step or p-stage algorithms probably the easiest for the practical user to understand is that of stability. For physical reasons, the solutions of equations (2.1.1 or 2.1.2) are expected to remain bounded as shown in Chapter 1; if the numerical solution becomes unbounded it is unstable. It is then obvious that there is something wrong. We shall discuss general conditions for stability of the p-step (and hence for the corresponding one-step p-stage algorithms) in Chapter 4. Here we begin by illustrating the concept of stability in the present context by applying some two-step algorithms to the dynamic vibration equation

$$\boldsymbol{M}\ddot{\boldsymbol{x}} + \boldsymbol{C}\dot{\boldsymbol{x}} + \boldsymbol{K}\boldsymbol{x} = \boldsymbol{0}, \tag{2.3.1}$$

where this represents an under-damped system as discussed in Chapter 1. We first consider a simple explicit two-step algorithm

$$\boldsymbol{M}(\boldsymbol{x}_{n+1} - 2\boldsymbol{x}_n + \boldsymbol{x}_{n-1}) + \frac{\Delta t}{2}\boldsymbol{C}(\boldsymbol{x}_{n+1} - \boldsymbol{x}_{n-1}) + \Delta t^2 \boldsymbol{K}\boldsymbol{x}_n = \boldsymbol{0}. \tag{2.3.2}$$

This is called the central difference algorithm because it is centred about the time level $n\Delta t$. The first and second time derivatives are replaced by the familiar centred finite difference expressions (Fröberg 1970). This algorithm is considered to be 'explicit' because the matrix to be inverted to obtain the new value $\boldsymbol{x}_{n+1}$, i.e. $(\boldsymbol{M} + 0.5\Delta t\boldsymbol{C})$ does not involve the stiffness matrix $\boldsymbol{K}$. If the mass matrix $\boldsymbol{M}$ and the damping matrix $\boldsymbol{C}$ are 'lumped' (Chapter 1), the algorithm (2.3.2) is easy to implement and this makes it popular. Of course it is particularly easy to implement if the damping matrix is absent. There are many references to this algorithm in the literature. For example, Krieg and Key (1973) in a paper on transient shell response by numerical time integration take the limiting case of a shell regarded as flat, isotropic, and uniformly thick. The vibration of the shell is then represented by the equation

$$c_m^2 \frac{\partial^2 u}{\partial x^2} - \frac{\partial^2 u}{\partial t} = f(x, t), \tag{2.3.3}$$

where u is the in-plane displacement and the sound speed

$c_m = \sqrt{(E/\rho)}$ (E and ρ are the modulus of elasticity and the density respectively). The authors compare several different methods for the numerical integration of eqn (2.3.3) and conclude that the one with the lumped mass matrix and the explicit central difference algorithm is good in terms of accuracy and economy for this problem. Another example of the use of the central difference algorithm is in a paper by Fu (1972) considering the two-dimensional wave propagation in a semi-infinite plate subjected to a suddenly applied moment along its edge. We mention this paper again later.

In order to analyse the algorithm (2.3.2), it is sufficient to consider its use on the scalar equation

$$m\ddot{x} + c\dot{x} + kx = 0. \tag{2.3.4}$$

Equation (2.3.4) can represent a single degree of freedom problem (Fig. 2.1), or it can represent one mode of a modally discretized system. We begin by noting that the exact solution of eqn (2.3.4) has terms of the form

$$\exp[\{-c \pm \sqrt{(c^2 - 4km)}\}/2m]. \tag{2.3.5}$$

We see that this solution is oscillatory if $c^2 < 4km$. The critical damping c_{cr} is given by $c_{cr}^2 = 4km$. We simplify the eqn (2.3.4) by making the substitutions

$$\omega^2 = k/m \quad \text{and} \quad c = 2\nu\omega, \tag{2.3.6}$$

i.e. $2\pi/\omega$ is the natural period of the undamped system and ν is the fraction of critical damping. We suppose that eqn (2.3.4)

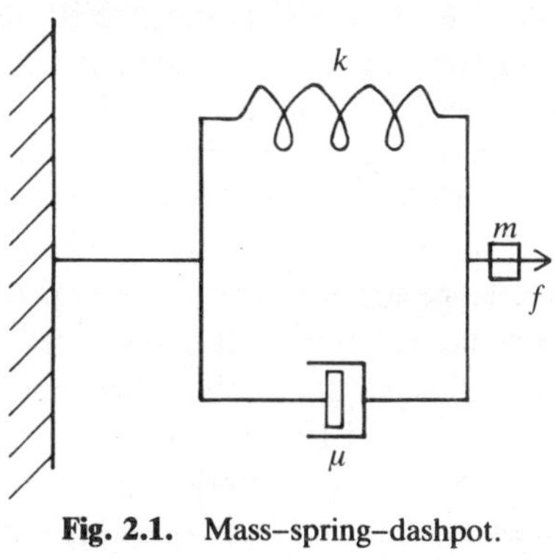

Fig. 2.1. Mass–spring–dashpot.

represents a mode of an under-damped system, i.e. $0 \le \nu < 1$. Typically for a structure the damping is low, 10% or less (Galambos and Mayes 1979). With the substitutions (2.3.6), eqn (2.3.4) becomes

$$\ddot{x} + 2\nu\omega\dot{x} + \omega^2 x = 0 \tag{2.3.7}$$

and the exact solution of (2.3.7) has terms $\exp(-\nu\omega t \pm i\bar{\omega}t)$, where $\bar{\omega} = \omega\sqrt{(1-\nu^2)}$.

The central difference algorithm (2.3.2) applied to eqn (2.3.7) gives the recurrence relation

$$(1+\nu\omega\Delta t)x_{n+1} + (-2+\omega^2\Delta t^2)x_n + (1-\nu\omega\Delta t)x_{n-1} = 0. \tag{2.3.8}$$

By substituting $x_{n+1} = rx_n = r^2 x_{n-1}$, the characteristic polynomial of this recurrence relation is

$$(1+\nu\omega\Delta t)r^2 + (-2+\omega^2\Delta t^2)r + 1 - \nu\omega\Delta t = 0. \tag{2.3.9}$$

The solution of the recurrence relation (2.3.8) is $x_n = Ar_1^n + Br_2^n$ where r_1, r_2, are the roots of eqn (2.3.9) and A and B are chosen to suit the initial conditions, (Lambert 1973; Fröberg 1970). The roots of the quadratic in eqn (2.3.9) are complex conjugates, $\rho\exp(\pm i\Phi)$ if $\rho^2 = (1-\nu\omega\Delta t)/(1+\nu\omega\Delta t)$ and $2\rho\cos\Phi = (2-\omega^2\Delta t^2)/(1+\nu\omega\Delta t)$. The condition for complex roots requires

$$(\omega^2\Delta t^2 - 2)^2 < 4(1 - \nu^2\omega^2\Delta t^2),$$

i.e.

$$\omega^2\Delta t^2 < 4(1-\nu^2). \tag{2.3.10}$$

The exact solution of eqn (2.3.7) is multiplied at each time step by $\exp(-\nu\omega\Delta t \pm i\bar{\omega}\Delta t)$ and the approximate solution given by the recurrence relation (2.3.8) is multiplied at each time step by $\rho\exp(\pm i\Phi)$. Thus ρ is approximating $\exp(-\nu\omega\Delta t)$ and Φ is approximating $\bar{\omega}\Delta t$, so long as the roots remain complex. We know that the solution of eqn (2.3.7) should remain bounded, as shown in Chapter 1, hence we want the approximate solution given by the recurrence relation (2.3.8) to remain bounded, that is, we want the algorithm to be stable. This requires the roots of the characteristic polynomial (2.3.9) to be less than or equal to unity in modulus. To find the condition for this, we use the Routh–Hurwitz transformation (Appendix 1), $r = (1+z)/(1-z)$, and then eqn (2.3.9) becomes:

$$(1+\nu\omega\Delta t)(1+z)^2+(-2+\omega^2\Delta t^2)(1-z^2)$$
$$+(1-\nu\omega\Delta t)(1-z)^2=0,$$

i.e.

$$a_0z^2+a_1z+a_2=0 \qquad (2.3.11)$$

where $a_0=(4-\omega^2\Delta t^2)$, $a_1=4\nu\omega\Delta t$, $a_2=\omega^2\Delta t^2$. The unit circle $|r|=1$ in the r-plane is thus mapped into the imaginary axis $\mathbb{R}z=0$ in the z-plane and the interior of the unit circle $|r|=1$ becomes the half plane $\mathbb{R}z<0$. For stability, we require $|r|\leq 1$ and this corresponds to the condition for the roots of eqn (2.3.11) to have non-positive real parts. The Routh–Hurwitz condition for this is

$$a_0>0,\ a_1\geq 0,\ a_2\geq 0,$$

i.e.

$$4-\omega^2\Delta t^2>0, \qquad \nu\omega\Delta t\geq 0, \qquad \omega^2\Delta t^2\geq 0. \qquad (2.3.12)$$

Hence we have

$$\omega\Delta t<2 \qquad (2.3.13)$$

as the general condition for the stability of the central difference algorithm (2.3.8) irrespective of whether the roots are real or complex. Comparing (2.3.10) with (2.3.13), we see that if $\nu\neq 0$ there is an interval $2\sqrt{(1-\nu^2)}\leq\omega\Delta t<2$ when the algorithm is stable but the roots are real and do not model the true solution. The algorithm (2.3.2) is thus conditionally stable, the condition being given by inequality (2.3.10) if we want to ensure that the roots remain complex. Figure 2.2 sums up these results.

2.4. Effect of mass lumping

We now use the central difference algorithm further to illustrate another point of practical interest. Suppose we take a system of two equations approximating a one-dimensional continuum, a piano wire, for example, ends fixed, divided into three equal linear elements. The system of equations is

$$\boldsymbol{M}\ddot{\boldsymbol{x}}+\boldsymbol{K}\boldsymbol{x}=\boldsymbol{0} \qquad (2.4.1)$$

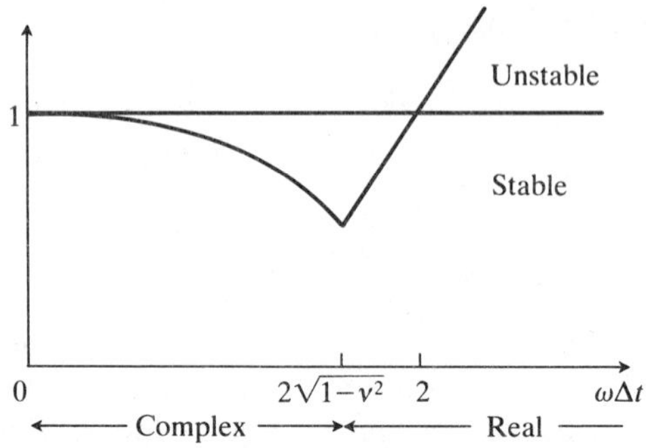

Fig. 2.2. Sketch of the maximum modulus root of the stability polynomial of the central difference algorithm against $\omega\Delta t$.

supposing the damping is zero. Taking the eqn (2.4.1) as non-dimensionalized with h = element length, we have from basic finite element theory (Strang and Fix 1973),

$$\boldsymbol{M} = \frac{h}{6}\begin{bmatrix} 4 & 1 \\ 1 & 4 \end{bmatrix}, \qquad \boldsymbol{K} = \frac{1}{h}\begin{bmatrix} 2 & -1 \\ -1 & 2 \end{bmatrix}. \tag{2.4.2}$$

Then $2\pi/\omega_j$ is the natural period of a mode of this system if ω_j^2 is an eigenvalue of the matrix

$$\begin{aligned} \boldsymbol{M}^{-1}\boldsymbol{K} &= \frac{6}{h^2}\begin{bmatrix} 4 & 1 \\ 1 & 4 \end{bmatrix}^{-1}\begin{bmatrix} 2 & -1 \\ -1 & 2 \end{bmatrix} \\ &= \frac{2}{5h^2}\begin{bmatrix} 4 & -1 \\ -1 & 4 \end{bmatrix}\begin{bmatrix} 2 & -1 \\ -1 & 2 \end{bmatrix} \\ &= \frac{2}{5h^2}\begin{bmatrix} 9 & -6 \\ -6 & 9 \end{bmatrix}. \end{aligned} \tag{2.4.3}$$

The eigenvalues of the matrix $\boldsymbol{M}^{-1}\boldsymbol{K}$ are given by

$$(5h^2\omega^2 - 18)^2 - 144 = 0 \tag{2.4.4}$$

and are thus $6/h^2$ and $6/(5h^2)$. The stability is controlled by the larger value so that from inequality (2.3.13) we have $\Delta t < 2h/\sqrt{6}$. Alternatively if we take the lumped form (when we get the corresponding finite difference expression), the values of ω_j^2 are the eigenvalues of the matrix

$$\frac{1}{h^2}\begin{bmatrix} 2 & -1 \\ -1 & 2 \end{bmatrix}, \tag{2.4.4a}$$

which are $3/h^2$ and $1/h^2$.

The stability condition from (2.3.13) is now $\Delta t < 2h/\sqrt{3}$ so that with a lumped mass matrix we can have stability with a slightly larger time step. In this simple problem, we have considered the effect of mass lumping on the global matrix and found that the time step limit goes up from $2h/\sqrt{6}$ to $2h/\sqrt{3}$. The rough assessment given by using the Rayleigh quotient on a single element in §1.4 and Exercise 1.1 gives the corresponding time step limits as $h/\sqrt{12}$ and $h/2$, respeectively for the distributed and lumped mass matrices which are lower and thus give conservative estimates.

A further example of the effect of mass lumping is given by Fu (1972). Fu considers the problem of a plate in a state of plane strain (Zienkiewicz 1977) with two-dimensional wave propagation represented by a system of equations of the form

$$\boldsymbol{M}\ddot{\boldsymbol{u}} + \boldsymbol{K}\boldsymbol{u} = \boldsymbol{f} \tag{2.4.5}$$

(ignoring damping). Fu treats the finite element approximation as if it were a finite difference approximation along the same lines as in the analysis in Richtmyer and Morton. He first considers a rectangular plate divided into two linear triangular elements. Linear combinations $\boldsymbol{U}$ and $\boldsymbol{V}$ of the nodal displacement are formed and the time-stepping scheme taken as

$$\boldsymbol{U}(t+\Delta t) - \boldsymbol{U}(t) = \mathrm{i}\omega\Delta t\boldsymbol{V}(t) \tag{2.4.6}$$

$$\boldsymbol{V}(t+\Delta t) - \boldsymbol{V}(t) = \mathrm{i}\omega\Delta t\boldsymbol{U}(t+\Delta t) \tag{2.4.7}$$

where $\omega = \sqrt{[4E/\{\rho(1+\nu)\Delta x\Delta y\}]}$ and E, ρ, and ν are the modulus of elasticity, density and Poisson's ratio respectively; Δx and Δy are the lengths of the sides of the rectangle. The eqns (2.4.6) and (2.4.7) can be combined to give

$$\begin{aligned}\{\boldsymbol{U}(t+2\Delta t) - \boldsymbol{U}(t+\Delta t)\} &- \{\boldsymbol{U}(t+\Delta t) - \boldsymbol{U}(t)\}\\ &= \mathrm{i}\omega\Delta t\{\boldsymbol{V}(t+\Delta t) - \boldsymbol{V}(t)\}\\ &= -\omega^2\Delta t^2\boldsymbol{U}(t+\Delta t),\end{aligned}$$

i.e.

$$\boldsymbol{U}(t+2\Delta t) - 2\boldsymbol{U}(t+\Delta t) + \boldsymbol{U}(t) + \omega^2\Delta t^2\boldsymbol{U}(t+\Delta t) = 0 \tag{2.4.8}$$

and a similar equation for $\boldsymbol{V}(t)$, so that the algorithm here is equivalent to the central difference algorithm (2.3.8). Fu then analyses the stability of the algorithm and shows that for the case of the two linear triangles, the critical time step for the lumped mass matrix is larger than that for the distributed mass matrix by a ratio of $3:\sqrt{2}$. For the rectangular plate considered as one bilinear element, the corresponding ratio is $3:1$.

Using the Rayleigh Quotient theory as in Chapter 1, we can bound the effect of a lumped versus a distributed mass matrix in terms of the corresponding smallest eigenvalues of the mass matrices of the individual elements. With the mass matrices for a one-dimensional problem given in Chapter 1, this gives $\sqrt{3}:1$ for the ratio of the largest eigenvalue of $\boldsymbol{M}^{-1}\boldsymbol{K}$ to that of $\boldsymbol{M}_L^{-1}\boldsymbol{K}$. Thus the estimate of the critical time step in the central difference algorithm is increased by the ratio $\sqrt{3}:1$ by using the lumped mass matrix. For the two-dimensional problem using linear triangles, the correspondng mass matrices in Chapter 1 show the estimate of the critical time step increased by the ratio $2:1$. Exercise 1.3 explains the Fu result with the bilinear element.

2.5. Effect of artificial damping

Another idea that we can illustrate in connection with the central difference scheme is that of changing the stability limit on the time step by introducing extra artificial damping. Park (1977) refers to the use of 'artificial viscosity' by von Neumann and Richtmyer (1950) in order to reduce the oscillations around hydrodynamic shocks. (There is an updated discussion of this complicated business in Richtmyer and Morton (1967).) Park quotes Melosh (1975) as proposing to stabilize the numerical solution of the structural dynamic equation by his explicit scheme by adding in some artificial damping. Actually Melosh says 'because manipulative errors may imply negative damping, a small change in the damping percentage (on the order of 10^{-16} for 16 digit precision) may be introduced to insure stability'. Artificial damping is sometimes made proportional to $\omega\Delta t$ so that its effect is greater at the upper end of the $\omega\Delta t$ spectrum. Returning to the central difference algorithm, we can see from eqn (2.3.13) that the presence of the parameter v or its replacement by an artificial term $v_0 + v_1\omega\Delta t$ makes no difference to the stability limit in eqn (2.3.13). If the $\dot{x}$ term in eqn (2.3.7) is replaced by $(x_n - x_{n-1})/\Delta t$ then the characteristic polynomial becomes

$$r^2 + r(-2 + 2\nu\omega\Delta t + \omega^2\Delta t^2) + 1 - 2\nu\omega\Delta t = 0. \quad (2.5.1)$$

The first of the stability conditions in eqn (2.3.13) is then replaced by

$$4 - 4\nu\omega\Delta t - \omega^2\Delta t^2 > 0. \quad (2.5.2)$$

With $\nu = \nu_0 + \nu_1\omega\Delta t$, where ν_0, ν_1 are ≥ 0 inequality (2.5.2) gives

$$\omega^2\Delta t^2(1 + 4\nu_1) + 4\nu_0\omega\Delta t - 4 < 0.$$

This requires

$$\omega\Delta t < \{-2\nu_0 + 2\sqrt{(1 + 4\nu_1 + \nu_0^2)}\}/(1 + 4\nu_1). \quad (2.5.3)$$

The expression on the right of (2.5.3) is equal to 2 if $\nu_0 = \nu_1 = 0$ and is less than 2 if $\nu_0^2 + 1 + 4\nu_1 < (1 + \nu_0 + 4\nu_1)^2$, i.e. $0 < 16\nu_1^2 + 4\nu_1 + 2\nu_0 + 8\nu_1\nu_0$, which is true because these quantities are positive. Hence the stability limit is decreased by adding in damping here. Alternatively, if the first derivative is replaced by $(x_{n+1} - x_n)/\Delta t$, the characteristic polynomial becomes

$$r^2(1 + 2\nu\omega\Delta t) + r(-2 - 2\nu\omega\Delta t + \omega^2\Delta t^2) + 1 = 0 \quad (2.5.4)$$

and the first inequality in (2.3.12) now gives

$$4 + 4\nu\omega\Delta t - \omega^2\Delta t^2 > 0.$$

Substituting $\nu = \nu_0 + \nu_1\omega\Delta t$ gives

$$\omega^2\Delta t^2(1 - 4\nu_1) - 4\nu_0\omega\Delta t - 4 < 0. \quad (2.5.5)$$

This means that if $\nu_1 \geq \frac{1}{4}$ we have unconditional stability. Otherwise, if $\nu_1 < \frac{1}{4}$, the stability limit is given by

$$\omega\Delta t < 2\{\nu_0 + \sqrt{(\nu_0^2 + 1 - 4\nu_1)}\}/(1 - 4\nu_1). \quad (2.5.6)$$

The limit given by (2.5.6) is equal to 2 if $\nu_0 = \nu_1 = 0$ and is greater than 2 if

$$\nu_0^2 + 1 + 4\nu_1 > (1 - 4\nu_1 - \nu_0)^2,$$

i.e. if $\nu_0(1 + 4\nu_1) > 2\nu_1(4\nu_1 - 1)$, which is negative by the above assumption. Hence the algorithm corresponding to the stability polynomial (2.5.4) can have its stability improved by adding in artificial damping.

2.6. Stability of the θ-method

We can also illustrate conditional stability with the θ-method applied to the first order equation

$$\dot{z} + \boldsymbol{B}z = \boldsymbol{0}. \tag{2.6.1}$$

We note that if eqn (2.6.1) is the first order form of the scalar dynamic vibration equation

$$\ddot{x} + 2\nu\omega\dot{x} + \omega^2 x = 0$$

then the matrix

$$\boldsymbol{B} = \begin{bmatrix} 0 & -1 \\ \omega^2 & 2\nu\omega \end{bmatrix}. \tag{2.6.2}$$

The eigenvalues of the matrix $\boldsymbol{B}$ are then given by

$$\xi(\xi - 2\nu\omega) + \omega^2 = 0, \tag{2.6.3}$$

i.e.

$$\xi = \omega(\cos\alpha \pm \mathrm{i}\sin\alpha) \tag{2.6.4}$$

putting $\nu = \cos\alpha$.

But if eqn (2.6.1) represents the parabolic problem (2.1.2), then the matrix $\boldsymbol{B} = \boldsymbol{M}^{-1}\boldsymbol{K}$ and the eigenvalues ξ are real and positive (Chapter 1). The θ-method applied to the scalar equation

$$\dot{z} + \xi z = 0, \tag{2.6.5}$$

which we suppose comes from a modal decomposition of the system in eqn (2.6.1) gives

$$z_{n+1} - z_n + \xi\Delta t\{(1-\theta)z_n + \theta z_{n+1}\} = 0, \tag{2.6.6}$$

where $0 \leq \theta \leq 1$. We can rewrite this equation as

$$(1 + \theta\xi\Delta t)z_{n+1} = \{1 - (1-\theta)\xi\Delta t\}z_n,$$

i.e. $z_{n+1} = rz_n$ where

$$r = \{1 - (1-\theta)\xi\Delta t\}/(1 + \theta\xi\Delta t). \tag{2.6.7}$$

Now if ξ comes from the dynamic vibration equation, we have to substitute from eqn (2.6.4) and then the stability condition $|r| \leq 1$

requires

$$|1-(1-\theta)\omega\Delta t\cos\alpha - \mathrm{i}(1-\theta)\omega\Delta t\sin\alpha| \le |1+\theta\omega\Delta t\cos\alpha + \mathrm{i}\theta\omega\Delta t\sin\alpha|, \quad (2.6.8)$$

i.e.

$$\{1-(1-\theta)\omega\Delta t\cos\alpha\}^2 + (1-\theta)^2\omega^2\Delta t^2\sin^2\alpha \le (1+\theta\omega\Delta t\cos\alpha)^2 + \theta^2\omega^2\Delta t^2\sin^2\alpha.$$

Sorting this out and cancelling the factor $\omega\Delta t$ which is certainly positive we have

$$2\cos\alpha + (2\theta-1)\omega\Delta t \ge 0. \quad (2.6.9)$$

Remembering that $\cos\alpha = \nu$ which is ≥ 0, we can see that inequality (2.6.9) is satisfied if $\theta \ge 0.5$, i.e. we then have unconditional stability. If $\theta < 0.5$ we have stability if

$$\omega\Delta t \le 2\cos\alpha/(1-2\theta) \quad (2.6.10)$$

which of course only makes sense if $\cos\alpha \ne 0$, i.e. the damping is not zero.

If the eigenvalue ξ comes from the parabolic equation (2.1.2), then it is real and positive. We have stability if

$$2 + (2\theta-1)\xi\Delta t \ge 0. \quad (2.6.11)$$

We have unconditional stability with $\theta \ge 0.5$ and if $\theta < 0.5$ we have stability if

$$\xi\Delta t \le 2/(1-2\theta). \quad (2.6.12)$$

In eqns (2.6.11) and (2.6.12), the parameter ξ represents an eigenvalue of $\boldsymbol{M}^{-1}\boldsymbol{K}$. Again using the Rayleigh Quotient estimates and the element mass matrices in Chapter 1, we find these give the effect of mass lumping as increasing the critical time step by 3:1 for one-dimensional problems with linear elements and 4:1 for two-dimensional problems using linear triangles.

Another problem that arises in connection with the numerical solution of the parabolic equation is as follows, illustrated on the scalar equation

$$\dot{x} + \xi x = 0. \quad (2.6.13)$$

The solution of eqn (2.6.13) is some multiple of $\exp(-\xi t)$ and the ratio

$$x((n+1)\Delta t)/x(n\Delta t) = \exp(-\xi\Delta t),$$

which is always positive. But from eqn (2.6.7), the amplification factor r which is approximating $\exp(-\xi\Delta t)$ becomes negative if

$$\xi\Delta t > 1/(1-\theta). \tag{2.6.14}$$

Hence for any value of $\theta \neq 1$ the numerical solution can exhibit a spurious ripple effect for values of $\xi\Delta t$ such that inequality (2.6.14) is true. The Crank–Nicolson (1947) method with $\theta = 0.5$ is unconditionally stable but oscillations may occur for $\xi\Delta t > 2$. Taking $\theta = 2/3$ is sometimes recommended (Zienkiewicz 1977), but this can also produce oscillations when $\xi\Delta t > 3$.

Figure 2.3 (Wood and Lewis 1975) shows an example of the Crank–Nicolson effect with the numerical solution of the non-dimensionalized heat conduction equation

$$\frac{\partial\Phi}{\partial t} = \frac{\partial^2\Phi}{\partial x^2}, \tag{2.6.15}$$

representing the variation of the temperature Φ in a bar which is insulated along its length. The length is taken as four units and

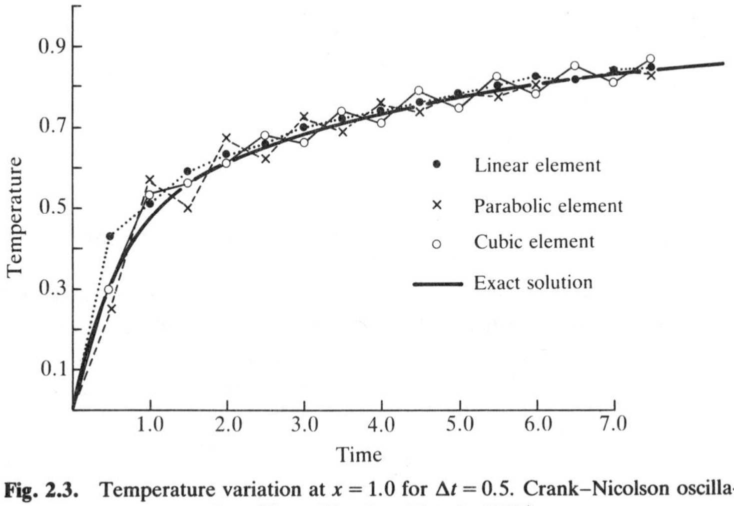

Fig. 2.3. Temperature variation at $x = 1.0$ for $\Delta t = 0.5$. Crank–Nicolson oscillation. (From Wood and Lewis 1975.)

the bar is divided into four equal elements. Initially the temperature is zero everywhere, then the end at which $x = 0$ is suddenly raised to unit temperature. The other end of the bar is insulated, i.e. the boundary condition here is $\partial\Phi/\partial x = 0$. The exact solution can be found in Carslaw and Jaeger (1959). Figure 2.3 shows the ripple effect produced by using the θ-method with $\theta = 0.5$ and 4-node, 8-node, and 12-node elements. The fully implicit algorithm with $\theta = 1$ shows no ripple effect. We recall that ξ is an eigenvalue of the matrix $\boldsymbol{M}^{-1}\boldsymbol{K}$ and by lumping the mass matrix we can obtain lower values for the maximum value of ξ and hence increase the critical time step at which the spurious ripple effect occurs.

Another example of the effect of mass lumping is given by Fujii (1973) in connection with the maximum principle. The maximum principle for the heat conduction equation

$$\frac{\partial u}{\partial t} = \alpha \nabla^2 u \tag{2.6.16}$$

states that, provided there is no heat sink or source,

(1) a negative temperature never occurs when the initial and boundary temperatures are non-negative; and

(2) the maximum temperature must occur at the boundary or at the start.

Fujii shows that with linear triangles for the space discretization and the θ-method for the time-stepping, the discrete maximum principle operates with the lumped mass matrix if the following conditions hold:

1. The triangular elements are of acute type. The acuteness is measured by a parameter σ; $\sigma < 0$ means that the elements have angles all less than $\pi/2$, $\sigma = 0$ means there are angles equal to $\pi/2$, and $\sigma > 0$ means that there are angles greater than $\pi/2$. The triangles are said to be of acute type if $\sigma \le 0$.

2. The time step size Δt is bounded as in the formula

$$\alpha(1-\theta)\Delta t \le \kappa_{\min}^2/(m+1) \tag{2.6.17}$$

where m is the dimension of the space in which the problem is formulated and $\kappa_{\min}$ is the minimum height of an element (the height of an element here is measured as the perpendicular distance from a vertex to the opposite side). When we have uniform right-angled triangles, this corresponds to the finite difference case and κ is the mesh size. When we have the implicit scheme with $\theta = 1$ there is no restriction on the time step from (2.6.17).

With the consistent mass matrix, Fujii shows that the discrete maximum principle holds if we assume

(a) $\sigma < 0$, i.e. right-angled triangles are not allowed;

(b)

$$\alpha(1-\theta)\Delta t \leq 2\kappa_{\min}^2/\{(m+1)(m+2)\}; \qquad (2.6.18)$$

(c)

$$\alpha\theta\Delta t \geq \kappa_{\max}^2/\{|\sigma|(m+1)(m+2)\}. \qquad (2.6.19)$$

Condition (c) thus means that there is a lower bound on the time step size. Fujii does not show that the discrete maximum principle only holds if conditions (a), (b), and (c) are true. This result of Fujii is sometimes quoted as a reason for using the lumped rather than the consistent mass matrix.

Advice in the literature as to whether or not to lump the mass matrix can be very confusing. In linear problems the effect of lumping is evident but most practical problems are non-linear. It is wise to disregard sweeping assertions that lumping always helps convergence and stability. What is optimal for one problem is not necessarily optimal for another. It helps to make tests to get a verdict of what is best from the cost-efficiency point of view. The following is a list of points to be taken into account. It does not pretend to be exhaustive.

1. We expect the solution of a steady-state problem which is equivalent to minimizing a functional to be more accurate with the distributed mass matrix. However, the order of error, e.g. $O(h^2)$ may be the same with the lumped mass matrix; we have to remember that the expression '$O(h^2)$' refers to what happens when h tends to zero and our h may not be small enough for the $O(h^2)$ term to dominate the error.

2. The distributed mass matrix gives eigenvalues of $\boldsymbol{M}_D^{-1}\boldsymbol{K}$ which are too high compared with those of the original problem

which has been discretized (Chapter 1). The eigenvalues of $\boldsymbol{M}_L^{-1}\boldsymbol{K}$ are lower than those of $\boldsymbol{M}_D^{-1}\boldsymbol{K}$ but we do not know whether they are too high or too low. However, this does mean that the critical time step for either instability (2.6.11) or for spurious ripples (2.6.13) is higher with the lumped mass matrix. With the fully implicit $\theta = 1$ method for a non-linear parabolic problem there should not be any trouble from instability or spurious ripples; in the author's experience there is also no significant difference in the number of iterations per time step between the programs with lumped and distributed mass matrices.

3. The distributed mass matrix may be more trouble to program. In a non-linear problem it takes more cpu time per time step if the distributed mass matrix has to be updated and used for operations. In a cost-efficiency study this extra may be balanced by the extra accuracy from the distributed mass matrix (measured for instance from an overall mass-balance error) but this must be very problem-dependent.

2.7. The Newmark algorithm

We next want to consider the Newmark algorithm for the numerical solution of the dynamic vibration equation (Newmark 1959). We first state the Newmark algorithm in its general form. The standard Newmark equations are

$$\boldsymbol{x}_{n+1} = \boldsymbol{x}_n + \Delta t\boldsymbol{v}_n + (1-2\beta)\Delta t^2\boldsymbol{a}_n/2 + \beta\Delta t^2\boldsymbol{a}_{n+1} \quad (2.7.1)$$

$$\boldsymbol{v}_{n+1} = \boldsymbol{v}_n + (1-\gamma)\Delta t\boldsymbol{a}_n + \gamma\Delta t\boldsymbol{a}_{n+1}, \quad (2.7.2)$$

where $\boldsymbol{x}_n$, $\boldsymbol{v}_n$, $\boldsymbol{a}_n$ are the approximations to the displacement, velocity and acceleration respectively at time $t = n\Delta t$ and β and γ are the Newmark parameters. The acceleration $\boldsymbol{a}_{n+1}$ is obtained by substituting for $\boldsymbol{x}_{n+1}$ and $\boldsymbol{v}_{n+1}$ in the equation

$$\boldsymbol{M}\boldsymbol{a}_{n+1} + \boldsymbol{C}\boldsymbol{v}_{n+1} + \boldsymbol{K}\boldsymbol{x}_{n+1} = \boldsymbol{0}. \quad (2.7.3)$$

Then $\boldsymbol{x}_{n+1}$ and $\boldsymbol{v}_{n+1}$ can be obtained from eqns (2.7.1) and (2.7.2).

In Chapter 3, we show that we can eliminate the velocity and acceleration terms and express the Newmark algorithm as a two-step algorithm in the displacements, so that when applied to the numerical solution of the dynamic vibration equation (2.3.1) it gives the recurrence relation:

$$\begin{aligned}\{\boldsymbol{M}+\gamma\Delta t\boldsymbol{C}+\beta\Delta t^2\boldsymbol{K}\}\boldsymbol{x}_{n+1}&+\{-2\boldsymbol{M}+(1-2\gamma)\Delta t\boldsymbol{C}\\&+(0.5+\gamma-2\beta)\Delta t^2\boldsymbol{K}\}\boldsymbol{x}_n\\&+\{\boldsymbol{M}+(\gamma-1)\Delta t\boldsymbol{C}+(0.5-\gamma+\beta)t^2\boldsymbol{K}\}\boldsymbol{x}_{n-1}=0.\end{aligned}\tag{2.7.4}$$

Comparing eqn (2.7.4) with eqn (2.3.2) we see that the central difference algorithm corresponds to Newmark with $\beta=0$ and $\gamma=0.5$. We now take as another example the Newmark algorithm with parameters $\beta=0.25$, $\gamma=0.5$ (the Newmark algorithm is discussed fully in Chapter 3). Applied to eqn (2.3.1), this version of the Newmark algorithm gives

$$\begin{aligned}(\boldsymbol{M}+0.5\Delta t\boldsymbol{C}+0.25\Delta t^2\boldsymbol{K})\boldsymbol{x}_{n+1}&+(-2\boldsymbol{M}+0.5\Delta t^2\boldsymbol{K})\boldsymbol{x}_n\\&+(\boldsymbol{M}-0.5\Delta t\boldsymbol{C}+0.25\Delta t^2\boldsymbol{K})\boldsymbol{x}_{n-1}=0.\end{aligned}\tag{2.7.5}$$

This algorithm has centred differences for the acceleration and velocity terms like the central difference algorithm, but uses a weighted average of the stiffness terms. Hence this algorithm is implicit and the matrix to be inverted at each time step is $(\boldsymbol{M}+0.5\Delta t\boldsymbol{C}+0.25\Delta t^2\boldsymbol{K})$. Applying this Newmark algorithm to the scalar equation (2.3.7) we have

$$\begin{aligned}(1+v\omega\Delta t+0.25\omega^2\Delta t^2)x_{n+1}&+(-2+0.5\omega^2\Delta t^2)x_n\\&+(1-v\omega\Delta t+0.25\omega^2\Delta t^2)x_{n-1}=0.\end{aligned}\tag{2.7.6}$$

The characteristic polynomial of this recurrence relation is

$$\begin{aligned}(1+v\omega\Delta t+0.25\omega^2\Delta t^2)r^2&+(-2+0.5\omega^2\Delta t^2)r+1\\&-v\omega\Delta t+0.25\omega^2\Delta t^2=0.\end{aligned}\tag{2.7.7}$$

The roots of the quadratic (2.7.7) are complex if

$$(-2+0.5\omega^2\Delta t^2)^2<4\{(1+0.25\omega^2\Delta t^2)^2-v^2\omega^2\Delta t^2\}.\tag{2.7.8}$$

This condition is satisfied if

$$4-2\omega^2\Delta t^2+0.25\omega^4\Delta t^4<4+2\omega^2\Delta t^2+0.25\omega^4\Delta t^4-4v^2\omega^2\Delta t^2,$$

i.e. $v^2<1$.

Thus the roots of eqn (2.7.7) are complex for all values of $\omega\Delta t$ for the case when the damping is less than critical. Also the product of the roots is less than or equal to unity, hence the

Newmark algorithm given by eqn (2.7.6) is unconditionally stable, i.e. stable for any value of $\omega\Delta t$. These two examples illustrate how the stability of a p-step algorithm depends on the roots of its characteristic polynomial. They also illustrate how what we expect of the behaviour of the roots of the characteristic polynomial depends on the eigenvalues of the Lambda matrix, eqn (1.3.9), which determine the form of the solution of the original differential equation.

2.8. Stability—general considerations

In the simple algorithms considered so far, i.e. Newmark and the θ-method, all the roots of the characteristic polynomial of the difference scheme can be approximating the true solution of the differential equation. In the approximation to the dynamic vibration equation, there are two roots and in the approximation to the parabolic equation there is only one root. When we have an algorithm for the numerical solution of the dynamic vibration equation with $p>2$ or for the parabolic equation with $p>1$, we have a characteristic polynomial some of whose roots are spurious. We suppose that the physical problems we are discussing have bounded solutions and we require that any numerical solution should have the same property, then we say that it is stable. The solution of the recurrence relation of the type (2.1.3) or (2.1.12) remains bounded if all the roots of the characteristic polynomial lie in or on the unit circle in the Argand diagram with any repeated roots strictly inside (Lambert 1973). This is the general condition for a stable solution. But we also have to consider how to choose the parameters in the algorithm so that the root or roots which dominate the numerical solution, i.e. those of maximum modulus, approximate the true solution and the spurious roots are small enough to die away quickly and give no trouble. There are other considerations in the choice of the parameters which we shall discuss in later chapters.

There is a bewildering array of labels of kinds of stability in the mathematics literature. We only want six labels here. The first three depend on the eigenvalues of the differential equation given by $\det P(\lambda)=0$ where $P(\lambda)$ is the Lambda matrix of the differential equation (Chapter 1).

1. Algorithms which give stable solutions for differential equations with real negative eigenvalues are called A_0-stable (Cryer 1973). For any values of $\lambda\Delta t$ on the negative real axis in Fig. 2.4, these algorithms give stable solutions. This applies to the parabolic equation (2.1.2) where $\lambda = -\xi$ and ξ is an eigenvalue of $M^{-1}K$. From the theory above the θ-method for the parabolic equation is A_0-stable with $\theta \geq 0.5$.

2. Dahlquist (1963) introduces the concept of A-stability. The original definition specifies that a p-step method

$$\sum_{j=0}^{p} \alpha_j x_{n+j} = \Delta t \sum_{j=0}^{p} \beta_j f_{n+j} \tag{2.8.1}$$

where $f_n = f(n\Delta t, x_n)$, for the numerical solution of the equation

$$\frac{\mathrm{d}x}{\mathrm{d}t} = f(t, x) \tag{2.8.2}$$

is called A-stable if all these numerical solutions tend to zero as n tends to infinity for fixed positive values of the time step Δt when the method is applied to any differential equation of the form

$$\frac{\mathrm{d}x}{\mathrm{d}t} = \lambda x \tag{2.8.3}$$

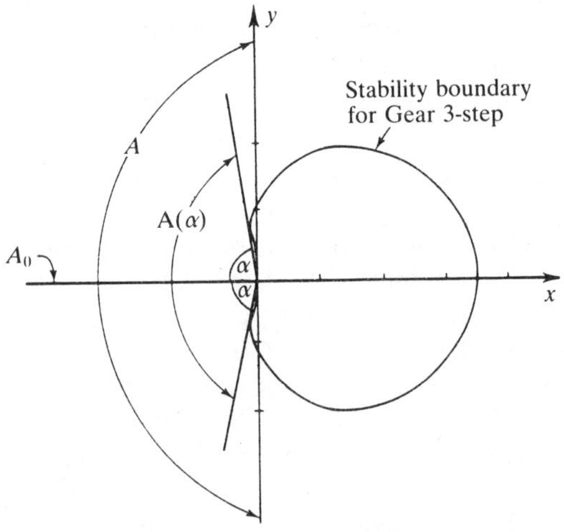

Fig. 2.4. Argand diagram showing stability regions. (From Penry and Wood 1985.)

where λ is a complex constant with negative real part. A note added in proof at the end of the Dahlquist (1963) paper adds that if an A-stable method is applied to eqn (2.8.3) where λ is pure imaginary then the numerical solution is bounded. To explain this further, we see that the method (2.8.1) applied to eqn (2.8.3) gives a recurrence relation

$$\sum_{j=0}^{p} (\alpha_j - \Delta t \lambda \beta_j) x_{n+j} = 0. \tag{2.8.4}$$

The solution of eqn (2.8.4) (Lambert 1973) is of the form

$$x_n = \sum_{l=1}^{p} a_l r_l^n \tag{2.8.5}$$

where the r_l are the p roots of the polynomial

$$\sum_{j=0}^{p} (\alpha_j - \Delta t \lambda \beta_j) r^j = 0. \tag{2.8.6}$$

Then by the definition above A-stability requires $|r_l| < 1$ for $\mathrm{Re}(\lambda) < 0$, i.e. the numerical solution behaves like the exact solution $\exp(\lambda t)$ of eqn (2.8.3) and decays as the time t increases. If the method is A-stable then applied to the solution of eqn (2.8.3) with $\mathrm{Re}(\lambda) = 0$ the solution is bounded. The exact solution of eqn (2.8.3) with $\mathrm{Re}(\lambda) = 0$ is of constant amplitude. The numerical solution by an A-stable method may be of constant amplitude or it may decay because of algorithmic damping but anyway it is bounded. Thus A-stable algorithms give stable solutions for values of $\lambda \Delta t$ anywhere in the left-hand half plane in Fig. 2.4 including the imaginary axis. For the dynamic vibration equation we have $\lambda = -\xi = -\omega(\cos\alpha \pm \mathrm{i} \sin\alpha)$ from eqn (2.3.24). Hence from the above theory (eqn (2.6.9)), the θ-method with $\theta \geq 0.5$ and the Newmark algorithm (eqn (2.7.7)) with $\beta = 0.25$, $\gamma = 0.5$ are each A-stable.

3. Algorithms which give stable solutions for values of $\lambda \Delta t$ anywhere in the wedge of angle 2α as shown in Fig. 2.4 are called $A(\alpha)$-stable (Widlund 1967). We shall have examples of $A(\alpha)$-stable algorithms in Chapter 4.

Further we have

4. Algorithms which are stable for some restricted range of values of $\lambda \Delta t$ are called conditionally stable. We have already

introduced the conditions for stability of the central difference method in (2.3.12) and for the θ-method in inequalities (2.6.9) and (2.6.11).

5. *L*-stability (left-stability) Ehle (1969). A one-step method is said to be *L*-stable if it is *A*-stable and when applied to the numerical solution of eqn (2.8.3) with $\text{Re}(\lambda) < 0$ it gives $x_{n+1} = r(\lambda\Delta t)x_n$ where $|r(\lambda\Delta t)|$ tends to zero as $\text{Re}(\lambda\Delta t)$ tends to infinity. This is relevant in connection with the stiff systems of equations coming from finite element discretization where we are liable to have high values of $\lambda\Delta t$.

We also introduce

6. *C*-stability. Algorithms for the solution of the dynamic vibration equation which are *A*-stable and have a pair of roots of maximum modulus which are complex conjugates for all $\lambda\Delta t$ we shall call *C*-stable. This distinguishes algorithms which are *A*-stable with real roots of modulus less than or equal to unity from those which have a pair of roots actually modelling the true solution. We shall need this distinction in Chapter 4. We showed from eqn (2.7.7) that Newmark with $2\beta = \gamma = 0.5$ gives a *C*-stable algorithm with the damping factor $\nu < 1$. Lambert (1976) uses the term *P*-stability for the case which would correspond here to the pair of complex conjugate roots having a modulus equal to unity. In the context of the engineering problems considered here, we need the wider definition to include complex conjugate roots with modulus less than or equal to unity.

2.9. Lax Equivalence Theorem

In the mathematical development of the subject the Lax Equivalence Theorem comes in naturally here. We state this theorem as:

Given a well posed initial value problem and a consistent difference approximation to it, stability is necessary and sufficient for convergence.

The implications of well-posedness are mentioned in §1.1. As indicated there, the differential equations discussed in this book are assumed to represent well-posed initial value problems. We have defined consistency as meaning that the local truncation error in the approximation of the differential equation must tend to zero with the time step size where the local truncation error is defined in eqn (2.2.4) for the second order problem and in eqn (2.2.5) for the first order problem. In the problems considered here, stability means that the solution of the difference equation remains bounded.

We now demonstrate how to produce the result we particularly want here, i.e. that consistency plus stability implies convergence for the difference schemes we have introduced. It is sufficient to demonstrate on scalar equations.

We start with the first order equation

$$\dot{z} + \xi z = f \tag{2.9.1}$$

and a corresponding difference scheme given as

$$L_{\Delta t}(z_n) \equiv \sum_{j=0}^{p} \{(\gamma_j \Delta t^{-1} + \beta_j \xi) z_{n+j} - \beta_j f_{n+j}\} = 0. \tag{2.9.2}$$

Then as before the local truncation error is obtained by applying the difference operator $L_{\Delta t}$ to the true solution of the differential equation

$$L_{\Delta t}(z(n\Delta t)) = \tau_{n+p}. \tag{2.9.3}$$

If the error $z(n\Delta t) - z_n = \epsilon_n$, we have

$$L_{\Delta t}(z(n\Delta t)) - L_{\Delta t}(z_n) = \sum_{j=0}^{p} a_j \epsilon_{n+j} = \tau_{n+p}, \tag{2.9.4}$$

where $a_j = \gamma_j \Delta t^{-1} + \beta_j \xi$, $j = 0, 1, \ldots, p$.

We now write eqn (2.9.4) in the equivalent form

$$\begin{bmatrix} \epsilon_{n+1} \\ \epsilon_{n+2} \\ \vdots \\ \epsilon_{n+p} \end{bmatrix} = \begin{bmatrix} 0 & 1 & 0 & \cdots & 0 \\ 0 & 0 & 1 & \cdots & 0 \\ \vdots & & & & \\ -b_0 & -b_1 & \cdots & & -b_{p-1} \end{bmatrix} \begin{bmatrix} \epsilon_n \\ \vdots \\ \\ \epsilon_{n+p-1} \end{bmatrix} + \frac{\Delta t}{c_p} \begin{bmatrix} 0 \\ 0 \\ \vdots \\ \tau_{n+p} \end{bmatrix}, \tag{2.9.5}$$

where $b_j = a_j / a_p$, $j = 0, 1, \ldots, p-1$, $c_p = \Delta t a_p = \gamma_p + \beta_p \Delta t \xi$. We rewrite eqn (2.9.5) as

$$\boldsymbol{e}_{n+1} = \boldsymbol{P} \boldsymbol{e}_n + \Delta t \hat{\boldsymbol{\tau}}_{n+1} \tag{2.9.6}$$

where $\hat{\boldsymbol{\tau}}$ is of the same order as τ_{n+p}.

The solution of this recurrence relation is obtained from the operations:

$$\begin{aligned} \boldsymbol{e}_n &= \boldsymbol{P}\boldsymbol{e}_{n-1} + \Delta t \hat{\boldsymbol{\tau}}_n \\ &= \boldsymbol{P}(\boldsymbol{P}\boldsymbol{e}_{n-2} + \Delta t \hat{\boldsymbol{\tau}}_{n-1}) + \Delta t \hat{\boldsymbol{\tau}}_n \\ &= \cdots \\ &= \boldsymbol{P}^n \boldsymbol{e}_0 + \Delta t \left(\sum_{j=1}^{n} \boldsymbol{P}^{n-j} \hat{\boldsymbol{\tau}}_j \right). \end{aligned} \tag{2.9.7}$$

Then we want to be able to say

$$\|\boldsymbol{e}_n\| \leq \|\boldsymbol{P}\|^n \, \|\boldsymbol{e}_0\| + \Delta t \sum_{j=1}^{n} \|\boldsymbol{P}\|^{n-j} \|\hat{\boldsymbol{\tau}}_j\| \tag{2.9.8}$$

with some convenient norm (see, for example, Conte and de Boor 1980) and deduce convergence from this. This works very simply with the θ-method because P is a scalar here. From eqn (2.6.7)

$$P = \{1 - (1 - \theta)\xi \Delta t\}/(1 + \theta \xi \Delta t) \tag{2.9.9}$$

and as shown previously $|P| \leq 1$ when $\theta \geq 0.5$.

Thus we can deduce from eqn (2.9.8)

$$|e_n| \leq |P|^n \, |e_0| + n \Delta t \tilde{\tau} \tag{2.9.10}$$

(where $\hat{\tau}$ is the maximum value of $\hat{\tau}_j$) and for $n\Delta t = T$, a fixed time, we have

$$|e_n| \leq |P|^n \, |e_0| + T\hat{\tau}. \tag{2.9.11}$$

We have to be sure that $|e_0|$ is sufficiently small not to spoil the convergence; it might not be possible to make it precisely zero but this shows the importance of starting condition accuracy. Provided the starting conditions are suitably accurate then, we have convergence, i.e. the error as given by $|e_n| = |\epsilon_n|$ here tends to zero as the time step size Δt tends to zero and this error is of the same order as the local truncation error τ_n.

We now see what happens when we try a similar operation for the second order differential equation

$$\ddot{x} + 2\nu\omega\dot{x} + \omega^2 x = f. \tag{2.9.12}$$

We define the difference operator

$$L_{\Delta t}(x_n) = \sum_{j=0}^{p} \{(\alpha_j \Delta t^{-2} + 2\gamma_j \Delta t^{-1} \nu\omega + \beta_j \omega^2) x_{n+j} - \beta_j f_{n+j}\}. \tag{2.9.13}$$

The local truncation error is again given by

$$L_{\Delta t}(x(n\Delta t)) = \tau_{n+p}. \tag{2.9.14}$$

We put $x(n\Delta t) - x_n = \epsilon_n$, then we have

$$\sum_{j=0}^{p} a_j \epsilon_{n+j} = \tau_{n+p}, \tag{2.9.15}$$

where $a_j = \alpha_j \Delta t^{-2} + 2\gamma_j \Delta t^{-1} \nu\omega + \beta_j \omega^2$, $j = 0, 1, \ldots, p$. We again have eqn (2.9.5) with the b_j now in terms of the new a_j and Δt replaced by Δt^2. The equivalent of inequality (2.9.8) is now

$$\|\boldsymbol{e}_n\| \leq \|\boldsymbol{P}\|^n \, \|\boldsymbol{e}_0\| + \Delta t^2 \sum_{j=1}^{n} \|\boldsymbol{P}\|^{n-j} \, \|\boldsymbol{\tau}_j\|. \tag{2.9.16}$$

We can see from eqn (2.9.5) that the characteristic polynomial of the matrix $\boldsymbol{P}$ is the characteristic polynomial of the difference scheme. Hence if the difference scheme is stable, the eigenvalues of $\boldsymbol{P}$ have modulus less than or equal to unity. Frequently in the literature, the impression is given that all we have to do now is to conclude that the spectral norm $\|\boldsymbol{P}\|_2 \leq 1$ and hence

$$\Delta t^2 \sum_{j=1}^{n} \|\boldsymbol{P}\|_2^{n-j} \, \|\tau\|_2 \leq n\Delta t^2 \, \|\boldsymbol{\tau}\|_2 = \Delta t T \, \|\boldsymbol{\tau}\|_2$$

at the fixed time $T = n\Delta t$.

But this leads to a result which is wrong; as we see from a completely different approach in Chapter 3 the error from this term is actually the same order as the local truncation error τ and not one order better. The matrix $\boldsymbol{P}$ here is not a normal matrix (Wilkinson 1965), so we cannot put $\|\boldsymbol{P}\|_2 = \rho(\boldsymbol{P})$, the spectral radius of P. We have to rely on being able to use

$$\boldsymbol{P} = \boldsymbol{S\Lambda S}^{-1}, \tag{2.9.17}$$

where $\boldsymbol{S}$ is the modal matrix of P and $\boldsymbol{\Lambda}$ is the diagonal matrix of the eigenvalues (see Richtmyer and Morton 1967 for further discussion of this). With the particular form of the matrix $\boldsymbol{P}$ in eqn (2.9.5), we can see that if the eigenvalues of $\boldsymbol{P}$ are distinct and are $r_1, r_2, \ldots, r_p$, then the modal matrix of $\boldsymbol{P}$ is

$$\boldsymbol{S} = \begin{bmatrix} 1 & 1 & \cdots & 1 \\ r_1 & r_2 & & r_p \\ \vdots & \vdots & & \vdots \\ r_1^{p-1} & r_2^{p-1} & \cdots & r_p^{p-1} \end{bmatrix}. \tag{2.9.18}$$

Hence the determinant of $\boldsymbol{S}$ is Vandermonde's determinant and is given by

$$\det(\boldsymbol{S}) = \pm \prod_{i \neq j} (r_i - r_j), \tag{2.9.19}$$

with the sign adjusted by observing that of the principal diagonal. This means that $\boldsymbol{S}^{-1}$ always has a factor $(r_1 - r_2)^{-1}$, where r_1, r_2 are the principal roots of the characteristic polynomial, i.e. those which approximate the solution of the differential equation. If we call these roots $\rho \exp(\pm \mathrm{i}\Phi)$ as in §2.3, then ρ is approximating $\exp(-\nu\omega\Delta t)$ and Φ is approximating $\omega\sqrt{(1-\nu^2)}\Delta t$. The difference $r_1 - r_2 = 2\mathrm{i}\rho \sin \Phi$ which must be $\mathrm{O}(\Delta t)$. Thus with the consistency conditions this factor produces the Δt^{-1} which causes the error norm $\|\boldsymbol{e}_n\|$ to be of the same order as the local truncation error τ. As Δt tends to zero, these two principal roots coalesce to become two roots equal to unity. This demonstrates the correspondence between the consistency conditions used here and the concept of zero-stability used, for instance, in Lambert (1973). We now know $\boldsymbol{S}^{-1}$ exists if the eigenvalues are distinct and then we have

$$\|\boldsymbol{P}^{n-j}\| = \|\boldsymbol{S}\boldsymbol{\Lambda}^{n-j}\boldsymbol{S}^{-1}\| \leq \|\boldsymbol{S}\|\ \|\boldsymbol{\Lambda}\|^{n-j}\ \|\boldsymbol{S}^{-1}\|. \tag{2.9.20}$$

The stability condition ensures that $\|\boldsymbol{\Lambda}\| \leq 1$. Hence from (2.9.16) we have

$$\|\boldsymbol{e}_n\| \leq \|\boldsymbol{S}\|\ \|\boldsymbol{S}^{-1}\|\ \|\boldsymbol{e}_0\| + \|\boldsymbol{S}\|\ \|\Delta t\boldsymbol{S}^{-1}\|\ T\ \|\boldsymbol{\tau}\| \tag{2.9.21}$$

at the fixed time $T = n\Delta t$. Thus, provided the starting accuracy is adequate, the cumulative error is of the same order as the local truncation error. The starting error e_0 must be of accuracy one order higher than the local truncation error.

Exercise 2.4 uses the Newmark algorithm with $\beta = 0.25$ and $\gamma = 0.5$ given in eqn (2.7.6) to illustrate the importance of checking the effect of $\|S^{-1}\|$.

We have now demonstrated how consistency and stability combine to give convergence for the algorithm examples we have used in this chapter. As we stated before, engineers' algorithms are formulated in such a way that they have automatic consistency and hence stability tends to be the primary consideration. The danger with unconditionally stable algorithms is that because they never blow up they may be used with unsuitably large time steps, quite regardless of the accuracy of the results obtained.

Exercises

2.1. Find the characteristic polynomial of the matrix $\boldsymbol{A}$ given by

$$\boldsymbol{A} = \begin{bmatrix} 1 & 1 & 0.5 \\ A_{21} & A_{22} & 0.5A_{22} \\ 2A_{21} & A_{32} & 0.5A_{32} \end{bmatrix},$$

where

$$A_{21} = -\frac{\Delta t^2 k}{2D}, \qquad A_{22} = \frac{1}{2D}(2m - \Delta t^2 k), \qquad A_{32} = -\frac{\Delta t}{D}(\mu + \Delta t k),$$

and

$$D = m + \tfrac{1}{2}\Delta t \mu.$$

Hence show that the time-stepping scheme

$$\boldsymbol{X}_{n+1} = \boldsymbol{A}\boldsymbol{X}_n,$$

where $\boldsymbol{X}_n^{\mathrm{T}} = [x_n, \Delta t v_n, \Delta t^2 a_n]$, applied to the numerical solution of the equation $m\ddot{x} + \mu\dot{x} + kx = 0$, is the one-step three-stage equivalent of the central difference scheme in eqn 2.3.2.

2.2. Show that the fully explicity ($\theta = 0$) scheme used on the first order form of the equation $m\ddot{x} + \mu\dot{x} + kx = 0$ is given by

$$\begin{bmatrix} x_{n+1} \\ v_{n+1} \end{bmatrix} = \begin{bmatrix} 1 & \Delta t \\ -\dfrac{\Delta t k}{m} & 1 - \dfrac{\Delta t \mu}{n} \end{bmatrix} \begin{bmatrix} x_n \\ v_n \end{bmatrix}.$$

Hence show that the corresponding two-step scheme in the displacements is given by

$$m(x_{n+1} - 2x_n + x_{n-1}) + \mu\Delta t(x_n - x_{n-1}) + k\Delta t^2 x_{n-1} = 0.$$

Use the Routh–Hurwitz conditions (Appendix 1) to show that, if the damping is less than critical, the scheme is stable if $\Delta t \leq \mu/k$ and that this is equivalent to the result in (2.6.10).

2.3. Find the stiffness matrix for a bilinear square element and its eigenvalues. Use the eigenvalue bound (1.4.7) with lumped mass matrices in each case and the matrix given in Exercise 1.2 to show why taking the squares divided by diagonals into triangular elements appears to give a stiffer system than with the bilinear squares.

2.4. Show that for Newmark with $2\beta = \gamma = 0.5$, the matrix $\boldsymbol{P}$ in eqn (2.9.6) is given by

$$\boldsymbol{P} = \begin{bmatrix} 0 & 1 \\ -b_0 & -b_1 \end{bmatrix}.$$

If the eigenvalues of $\boldsymbol{P}$ are $p \pm iq$, show that the modal matrix of $\boldsymbol{P}$ is given by

$$\boldsymbol{S} = \begin{bmatrix} 1 & 1 \\ p - iq & p + iq \end{bmatrix}.$$

Show how $\|\boldsymbol{S}^{-1}\|$ introduces a factor Δt^{-1} into the inequality

$$\|\boldsymbol{e}_n\| \leq \|\boldsymbol{S}\|\ \|\boldsymbol{S}^{-1}\|\ \|\boldsymbol{e}_0\| + \Delta t^2 n\ \|\boldsymbol{S}\|\ \|\boldsymbol{S}^{-1}\|\ \hat{\tau}$$

so that (provided the starting conditions are sufficiently accurate) the error at fixed time $T = n\Delta t$ is of the same order as the local truncation error.

References

Bathe, K. J. and Wilson, E. L. (1973). Stability and accuracy analysis of direct integration methods. *Earthquake Engineering and Structural Dynamics,* **1,** 283–291.

Carslaw, H. S. and Jaeger, J. C. (1949). *Heat conduction in solids* (2nd edn). Oxford University Press.

Conte, S. D. and de Boor, C. (1980). *Elementary numerical analysis.* McGraw-Hill.

Crank, J. and Nicolson, P. (1949). A practical method for the numerical integration of solutions of partial differential equations of heat conduction type. *Proceedings of the Cambridge Philosophical Society,* **43,** 50–67.

Cryer, C. W. (1973). A new class of highly stable methods: A_0 stable methods. *BIT,* **13,** 153–159.

Dahlquist, G. G. (1963). A special stability problem for linear multistep methods. *BIT,* **3,** 27–43.

Ehle, B. L. (1969). On Padé approximations to the exponential function and *A*-stable methods for the numerical solution of initial value problems. *Research Report no. CSRR 2010.* University of Waterloo Department of Applied Analysis and Computer Science.

Fröberg, C.-E. (1970). *Introduction to numerical analysis* (2nd edn). Addison-Wesley.

Fu, C. C. (1972). On the stability of explicit methods for the numerical integration of the equations of motion in finite element methods. *International Journal of Numerical Methods in Engineering,* **4,** 95–107.

Fujii, H. (1973). Some remarks on finite element analysis of time-dependent field problems. In *Theory and practice in finite element structural analysis.* University of Tokyo Press.

Galambos, T. V. and Mayes, R. C. (1979). Lessons from dynamic tests of an eleven storey building. *Engineering Structures,* **1,** 264–273.

Krieg, R. D. and Key, S. W. (1973). Transient shell response by numerical time integration. *International Journal of Numerical Methods in Engineering,* **7,** 273–286.

Lambert, J. D. (1973). *Computational methods in ordinary differential equations.* Wiley.

Lambert, J. D. (1976). Symmetric multistep methods for periodic initial value problems. *Journal of the Institute of Mathematics and its Applications* **18,** 189-202.

Melosh, R. J. (1975). Integration of linear equations of motion. *Journal of Structural Division, Proc. ASCE,* **101** (ST7), 1551–1558.

Newmark, N. M. (1959). A method of computation for structural dynamics. *Proc. ASCE, Journal of Engineering Mechanics* (EM3) **85,** 67–94.

Park, K. C. (1977). Practical aspects of numerical time integration. *Journal of Computers and Structures,* **7,** 343–353.

Penry, S. N. and Wood, W. L. (1985). Comparison of some single-step methods for the numerical solution of the structural dynamics equation. *International Journal of Numerical Methods in Engineering,* **21,** 1941–1955.

Richtmyer, R. D. and Morton, K. W. (1967). *Difference methods for initial value problems* (2nd edn). Wiley.

Strang, G. and Fix, G. J. (1973). *Analysis of the finite element method.* Prentice-Hall.

Thomas, R. M. and Gladwell, I. (1985). Variable-order variable-step algorithms for second order systems. *Numerical Analysis Report no. 109.* University of Manchester/UMIST.

Von Neumann, J. and Richtmyer, R. D. (1950). A method for the numerical calculation of hydrodynamic shocks. *Journal of Applied Physics,* **21,** 232–257.

Widlund, O. B. (1967). A note on unconditionally stable linear multistep methods. *BIT,* **7,** 65–70.

Wilkinson, J. H. (1965). *The algebraic eigenvalue problem.* Oxford University Press.

Wood, W. L. and Lewis, R. W. (1975). A comparison of time marching schemes for the transient heat conduction equation. *International Journal of Numerical Methods in Engineering,* **9,** 679–689.

Zienkiewicz, O. C. (1977). *The finite element method.* McGraw-Hill.

Zienkiewicz, O. C., Wood, W. L., Hine, N. W., and Taylor, R. L. (1984). A unified set of single-step algorithms. Part 1 General formulation and applications. *International Journal of Numerical Methods in Engineering,* **20,** 1529–1552.

3 Newmark and the θ-method

3.1. Introduction

This chapter concentrates on the numerical solution of the dynamic vibration equation represented by the second order system of N equations

$$\boldsymbol{M}\ddot{\boldsymbol{x}} + \boldsymbol{C}\dot{\boldsymbol{x}} + \boldsymbol{K}\boldsymbol{x} = \boldsymbol{f}, \qquad t \geq 0. \tag{3.1.1}$$

In eqn (3.1.1), the vectors $\boldsymbol{x}, \dot{\boldsymbol{x}}, \ddot{\boldsymbol{x}}$ are the vectors of the nodal displacement, velocity and acceleration respectively. The matrices $\boldsymbol{M}$, $\boldsymbol{C}$ and $\boldsymbol{K}$ are the (Nth order) mass, damping and stiffness matrices respectively and $\boldsymbol{f} = \boldsymbol{f}(t)$ is the exterior forcing function. We expect to have starting conditions given, i.e. $\boldsymbol{x}(0)$ and $\dot{\boldsymbol{x}}(0)$. We are assuming here that we have a linear system of equations such that the matrices $\boldsymbol{M}$, $\boldsymbol{C}$, and $\boldsymbol{K}$ remain constant. As explained in Chapter 1, the mass matrix, $\boldsymbol{M}$ and the stiffness matrix, $\boldsymbol{K}$, from the finite element space discretization are symmetric; the matrix $\boldsymbol{M}$ is positive definite, the matrix $\boldsymbol{K}$ is non-negative definite and we assume that this is true of the damping matrix also.

Engineers often have strong preferences either for applying a numerical algorithm to solve the dynamic vibration equation in its second order form (3.1.1) (see, for example, Zienkiewicz 1977) or for first reducing it to an equivalent first order form (see, for example, Park 1975). One way of writing the first order form is

$$\begin{aligned} \dot{\boldsymbol{x}} &= \boldsymbol{v} \\ \dot{\boldsymbol{v}} &= -\boldsymbol{M}^{-1}\boldsymbol{K}\boldsymbol{x} - \boldsymbol{M}^{-1}\boldsymbol{C}\boldsymbol{v} + \boldsymbol{M}^{-1}\boldsymbol{f}. \end{aligned} \tag{3.1.2}$$

We show here that the two simplest algorithms for the numerical solution of these two forms of the dynamic vibration equation are basically very similar and in certain circumstances give exactly the same results.

The two methods for the solution of eqn (3.1.1) are the following.

1. The Newmark method (Newmark 1959). This method solves the second order differential equation directly by a time-stepping scheme described in §3.2. As far as we know, this scheme has not been analysed in a mathematics textbook before. Mathematics textbooks generally concentrate on the first order system of the form

$$\dot{z} = f(z, t) \tag{3.1.3}$$

on the assumption that any equation of higher order can be reduced to this. Lambert (1973), for example, is almost entirely concerned with the first order form; the last chapter only mentions the second order equation of the form

$$\ddot{z} = f(z, t) \tag{3.1.4}$$

approached via a multistep method like that in eqn (2.1.3) with the parameters $\gamma_j = 0$. Henrici (1962) also has a chapter entitled 'Multistep method for equations of the second order' which concentrates on the form (3.1.4).

Why is Newmark so popular with engineers? The author of this book, being a mathematician, surmises that it is because:

(a) With problems of the vibration of structures it may be necessary to have the nodal values of the displacement, velocity and acceleration; with Newmark it is easy to understand how these are being approximated.

(b) Newmark is versatile because, as shown in §3.2, it can be used as a single-step algorithm in the displacement, velocity and acceleration, or as a single-step algorithm in the displacement and velocity or as a two-step algorithm in the displacement alone. The Newmark algorithm can also be used in predictor–corrector form.

(c) There are two parameters at choice, but only two, so not so many as to be confusing.

(d) It has been in use since 1959.

2. The θ-method. The other very simple method analysed in this chapter, §3.3, is the θ-method. We are inclined to call this the 'well-known θ-method' because we have been teaching it at Reading University for many years. But many textbooks on

elementary numerical analysis and differential equations only mention the particular cases of the explicit Euler ($\theta = 0$) or Crank–Nicolson ($\theta = 0.5$). The formula for the θ-method is given in Richtmyer and Morton (1967), page 189, in a table of finite difference approximations to the one-dimensional heat flow equation (1.2.1). As shown in Chapters 7 and 8, the θ-method is the basis for many numerical methods for solving non-linear first order problems.

3.2. The Newmark method

For engineers, the Newmark method (Newmark 1959) is the most popular algorithm for the numerical solution of the dynamic vibration equation represented by the second order system (3.1.1). Newmark's method takes

$$\boldsymbol{x}_{n+1} = \boldsymbol{x}_n + \Delta t \boldsymbol{v}_n + \frac{\Delta t^2}{2}(1 - 2\beta)\boldsymbol{a}_n + \Delta t^2 \beta \boldsymbol{a}_{n+1} \tag{3.2.1}$$

and

$$\boldsymbol{v}_{n+1} = \boldsymbol{v}_n + \Delta t(1 - \gamma)\boldsymbol{a}_n + \Delta t \gamma \boldsymbol{a}_{n+1}, \tag{3.2.2}$$

where $\boldsymbol{x}_n$, $\boldsymbol{v}_n$ and $\boldsymbol{a}_n$ are the approximations to the nodal displacement, velocity and acceleration vectors respectively at time $t = n\,\Delta t$, Δt is the time step and β and γ are the Newmark parameters. Newmark described his parameters β and γ as 'introduced to indicate how much of the acceleration at the end of the interval enters into the relations for velocity and displacement at the end of the interval'. The expressions for $\boldsymbol{x}_{n+1}$ and $\boldsymbol{v}_{n+1}$ can be substituted into the equation

$$\boldsymbol{M}\boldsymbol{a}_{n+1} + \boldsymbol{C}\boldsymbol{v}_{n+1} + \boldsymbol{K}\boldsymbol{x}_{n+1} = \boldsymbol{f}_{n+1}, \tag{3.2.3}$$

where $\boldsymbol{f}_{n+1} = \boldsymbol{f}((n+1)\,\Delta t)$, to give an equation for $\boldsymbol{a}_{n+1}$ in terms of the values of $\boldsymbol{x}_n$, $\boldsymbol{v}_n$ and $\boldsymbol{a}_n$ at the beginning of the time step:

$$\begin{aligned}\boldsymbol{D}\boldsymbol{a}_{n+1} = \boldsymbol{f}_{n+1} &- \boldsymbol{C}[\boldsymbol{v}_n + \Delta t(1 - \gamma)\boldsymbol{a}_n] \\ &- \boldsymbol{K}\left[\boldsymbol{x}_n + \Delta t \boldsymbol{v}_n + \frac{\Delta t^2}{2}(1 - 2\beta)\boldsymbol{a}_n\right],\end{aligned} \tag{3.2.4}$$

where

$$\boldsymbol{D} = \boldsymbol{M} + \gamma\,\Delta t \boldsymbol{C} + \beta\,\Delta t^2 \boldsymbol{K} \tag{3.2.5}$$

is the matrix whose inverse is to be used at each time step. The new values of the displacement and velocity vectors are then found by substituting for $\boldsymbol{a}_{n+1}$ in the systems of scalar equations (3.2.1) and (3.2.2). From the initial conditions, we have the starting values of $\boldsymbol{x}_0$ and $\boldsymbol{v}_0$ and it is usual to obtain the value of $\boldsymbol{a}_0$ by assuming that the differential equation system (3.1.1) is satisfied at $t = 0$. Alternative proposals for starting conditions will be discussed later.

It is evident that, stated in this way, Newmark's method can be regarded as a single-step three-stage algorithm calculating $\boldsymbol{x}_n$, $\boldsymbol{v}_n$ and $\boldsymbol{a}_n$ at each time step. It can also be regarded as a single-step two-stage algorithm calculating $\boldsymbol{x}_n$ and $\boldsymbol{v}_n$ at each time step or as a two-step algorithm calculating the displacements $\boldsymbol{x}_n$ only (Wood 1984). However, we must emphasize that in each of these cases, the algorithm is programmed so that the matrix whose inverse is used at each time step is the same, i.e. the Nth order matrix $\boldsymbol{D}$ given in eqn (3.2.5). For convenience, in the following analysis we demonstrate these three forms of Newmark's method on the scalar equation

$$m\ddot{x} + \mu\dot{x} + kx = f(t). \tag{3.2.6}$$

This represents a single degree of freedom problem as illustrated in Fig. 2.1. A mass m vibrates on the end of a spring with spring constant k under the influence of a dashpot producing damping proportional to the velocity and an external forcing function $f = f(t)$. Frequently in the engineering literature the damping term and the forcing function are left out of the following analysis. The reason often given for leaving out the damping term is that the damping for structures is extremely low. It is also tacitly assumed that any algorithm which is stable with no natural damping is bound to be stable with natural damping. We shall see later that this is not necessarily true. The forcing function frequently seems to be put in afterwards in some arbitrary averaged manner; it is more reasonable to leave it in as it naturally comes into the algorithm through the substitution into eqn (3.2.3). We also want to compare its treatment in the Newmark algorithm with the result of applying the θ-method to the first order form of eqn (3.2.6).

We now look at three forms of Newmark's algorithm as applied to eqn (3.2.6).

1. Single-step three-stage. We can write the Newmark algorithm as

$$\boldsymbol{X}_{n+1} = \boldsymbol{A}\boldsymbol{X}_n + \boldsymbol{F}, \tag{3.2.7}$$

where the vector $\boldsymbol{X}_n$ can conveniently be written as

$$\boldsymbol{X}_n = (x_n, \Delta t v_n, \Delta t^2 a_n)^{\mathrm{T}} \tag{3.2.8}$$

(so that each component has the same dimensions).

The matrix $\boldsymbol{A}$ in eqn (3.2.7) is the amplification matrix given by $\boldsymbol{A} = \{A_{ij}\}$, where

$$\begin{aligned} &A_{11} = (m + \gamma\Delta t\mu)/D, \qquad A_{12} = \{m + (\gamma - \beta)\Delta t\mu\}/D,\\ &A_{13} = \{(0.5 - \beta)m + (0.5\gamma - \beta)\Delta t\mu\}/D, \qquad A_{21} = -\gamma\Delta t^2 k/D\\ &A_{22} = \{m + (\beta - \gamma)\Delta t^2 k\}/D,\\ &A_{23} = \{(1 - \gamma)m + (\beta - 0.5\gamma)\Delta t^2 k\}/D\\ &A_{31} = -\Delta t^2 k/D, \qquad A_{32} = -\Delta t(\mu + \Delta t k)/D,\\ &A_{33} = -\{(1 - \gamma)\Delta t\mu + (0.5 - \beta)\Delta t^2 k\}/D \end{aligned} \tag{3.2.9a}$$

and the denominator

$$D = m + \gamma\Delta t\mu + \beta\Delta t^2 k. \tag{3.2.9b}$$

The vector F is given by

$$\boldsymbol{F} = \frac{\Delta t^2}{D}[\beta f_{n+1}, \gamma f_{n+1}, f_{n+1}]^{\mathrm{T}}. \tag{3.2.10a}$$

In the next section we show that the Newmark algorithm can also be expressed as a single-step two-stage algorithm. It may seem strange at first sight that the same algorithm can have two forms, one where the amplification matrix is of order three and one where it is of order two. But we can see from eqns (3.2.9a) which give the entries $\{A_{ij}\}$ in the third order amplification matrix, that we have

$$(0.5 - \gamma + \beta)A_{j1} + (\gamma - 1)A_{j2} + A_{j3} = 0 \tag{3.2.10b}$$

for $j = 1, 2, 3$. Hence the third order amplification matrix has a determinant equal to zero and thus has a zero eigenvalue for any values of the Newmark parameters β and γ. This means that there are actually only two independent equations here. It is useful to have the third order amplification matrix to compare with the results of introducing another parameter into the Newmark algorithm which we shall discuss in Chapter 4.

2. Single-step two-stage. Since the Newmark algorithm uses the approximate form of the differential equation

$$ma_n + \mu v_n + kx_n = f_n \tag{3.2.11}$$

at every time step we can substitute for a_n from eqn (3.2.11) into the first two equations of the system (3.2.7) to give

$$\boldsymbol{Y}_{n+1} = \tilde{\boldsymbol{A}}\boldsymbol{Y}_n + \tilde{\boldsymbol{f}}, \tag{3.2.12}$$

where the vector $\boldsymbol{Y}_n = [x_n, \Delta t v_n]^{\mathrm{T}}$ and the amplification matrix here is $\tilde{\boldsymbol{A}} = \{\tilde{A}_{ij}\}$, which is given by

$$\begin{aligned}
\tilde{A}_{11} &= [m^2 + \gamma\Delta t m\mu - (0.5-\beta)\Delta t^2 mk + (\beta - 0.5\gamma)\Delta t^3\mu k]/mD\\
\tilde{A}_{12} &= [m^2 + (\gamma - 0.5)\Delta t m\mu + (\beta - 0.5\gamma)\Delta t^2\mu^2]/mD,\\
\tilde{A}_{21} &= [-m + (0.5\gamma - \beta)\Delta t^2 k]\Delta t^2 k/mD,\\
\tilde{A}_{22} &= [m^2 + (\beta - \gamma)\Delta t^2 mk\\
&\quad + (\gamma - 1)\Delta t m\mu + (0.5\gamma - \beta)\Delta t^3\mu k]/mD,
\end{aligned} \tag{3.2.13}$$

D being the same as in eqn (3.2.9). The vector $\tilde{\boldsymbol{f}}$ in eqn (3.2.12) is given by the components

$$\begin{aligned}
\tilde{f}_1 &= [(0.5-\beta)m + (0.5\gamma - \beta)\Delta t\mu]\Delta t^2 f_n/mD\\
&\quad + \beta\Delta t^2 f_{n+1}/D\\
\tilde{f}_2 &= [(1-\gamma)m + (\beta - 0.5\gamma)\Delta t^2 k]\Delta t^2 f_n/mD\\
&\quad + \gamma\Delta t^2 f_{n+1}/D.
\end{aligned} \tag{3.2.14}$$

Equations (3.2.7) and (3.2.12) are just two ways of regarding the Newmark algorithm. In practice, the programing of the algorithm is likely to be the same, that is, we calculate the new value of the acceleration $\boldsymbol{a}_{n+1}$ (which with the system of equations (3.1.1) requires the inversion of the matrix $\boldsymbol{D}$ in eqn (3.2.5)), then obtain $\boldsymbol{x}_{n+1}$ and $\boldsymbol{v}_{n+1}$. This is the way that Newmark's algorithm is used when the velocities or the velocities and the accelerations are required at each node as well as the displacements. When only the displacements are required we can reformulate the algorithm as follows.

3. Two-step algorithm in the displacements only. Taking the two equations in (3.2.12) together with the first equation written

for the next time step:

$$x_{n+2} = \tilde{A}_{11}x_{n+1} + \tilde{A}_{12}\Delta t v_{n+1} + \tilde{f}_{n+1,1}, \tag{3.2.15}$$

we can eliminate v_n and v_{n+1} from these three equations to give a two-step recurrence in the displacements x_n, x_{n+1} and x_{n+2}. Rather than perform the algebra necessary for this, however, it is simpler to proceed as follows.

The basic system of Newmark equations from which the two-step recurrence in the displacements can be obtained consists of

$$ma_j + \mu v_j = \hat{f}_j, \qquad j = n-1, n, n+1 \tag{3.2.16}$$

(where $\hat{f}_j = f_j - kx_j$), together with eqns (3.2.1) and (3.2.2) and the corresponding equations for x_n and v_n, i.e. seven equations altogether. Eliminating the six quantities v_{n+1}, v_n, v_{n-1} and a_{n+1}, a_n, a_{n-1} will give the required two-step recurrence. We are not proposing to do this complicated task but merely to note that in the result the $\hat{f}_j$ terms appear in a certain combination and this means that the f_j terms are combined in precisely the same way as the kx_j terms. Thus if we get the equivalent two-step recurrence from the homogeneous form of eqn (3.2.6)

$$m\ddot{x} + \mu\dot{x} + kx = 0, \tag{3.2.17}$$

we can immediately deduce the correct combination for the f_n, f_{n+1} and f_{n+2} terms for the corresponding form for eqn (3.2.6) used for example in Wood (1981).

The easiest way to get the two-step recurrence from the eqns (3.2.12) with $\tilde{f} \equiv 0$ is to put $x_{n+1} = rx_n$ and $v_{n+1} = rv_n$ to give

$$\begin{aligned} rx_n &= \tilde{A}_{11}x_n + \tilde{A}_{12}\Delta t v_n \\ r\Delta t v_n &= \tilde{A}_{21}x_n + \tilde{A}_{22}\Delta t v_n. \end{aligned} \tag{3.2.18}$$

We then eliminate x_n and suppose non-trivial v_n, or vice versa. These must give the same result, that is, the characteristic polynomial of the amplification matrix $\tilde{A}$ is given by

$$\begin{aligned} &\{m + \gamma\Delta t\mu + \beta\Delta t^2 k\}r^2 \\ &\quad + \{-2m + (1-2\gamma)\Delta t\mu + (0.5 + \gamma - 2\beta)\Delta t^2 k\}r \\ &\quad + \{m + (\gamma - 1)\Delta t\mu + (0.5 - \gamma + \beta)\Delta t^2 k\} = 0. \end{aligned} \tag{3.2.19}$$

We now use the general result in §2.1 which enables us to identify the coefficients in the characteristic polynomial of the amplification matrix with the coefficienct in the two-step recurrence in the displacements. Thus we can now write down the Newmark two-step recurrence for the numerical solution of the displacements from eqn (3.2.6):

$$\begin{aligned}\{m + \gamma\Delta t\mu + \beta\Delta t^2 k\}x_{n+1}& \\ + \{-2m + (1-2\gamma)\Delta t\mu + (0.5+\gamma-2\beta)\Delta t^2 k\}x_n& \\ + \{m + (\gamma-1)\Delta t\mu + (0.5-\gamma+\beta)\Delta t^2 k\}x_{n-1}& \\ = \Delta t^2\{\beta f_{n+1} + (0.5+\gamma-2\beta)f_n + (0.5-\gamma+\beta)f_{n-1}\}.& \end{aligned} \quad (3.2.20)$$

We see that the denominator $D = m + \gamma\Delta t\mu + \beta\Delta t^2 k$ which appears in the equations for the single-step two-stage and three-stage forms of the Newmark algorithm also multiplies the highest subscript term in the two-step single-stage form given by eqn (3.2.20). This means that when the Newmark algorithm is applied to the system of eqns (3.1.1), whether it is used in the form (1), (2) or (3) the matrix whose inverse is used is the same:

$$\boldsymbol{M} + \gamma\Delta t\boldsymbol{C} + \beta\Delta t^2\boldsymbol{K}. \quad (3.2.21)$$

We note that in order to start using the recurrence relation (3.2.20), we need the values of x_0 and x_1. We expect to have starting values $x(0)$ and $\dot{x}(0)$ given with the differential equation (3.2.6). We take $x_0 = x(0)$ and x_1 must be calculated so that it does not impair the accuracy of the algorithm. We come back to this point in §3.4.

3.3. The θ-method applied to the dynamic vibration equation

The θ-method is the most widely used algorithm for the numerical solution of first order systems of equations because it is so simple. We now look at the effect of applying the θ-method to a first order form of the dynamic vibration equation with a single degree of freedom:

$$m\ddot{x} + \mu\dot{x} + kx = f(t). \quad (3.3.1)$$

1. Single-step two-stage. We replace eqn (3.3.1) with the system

$$\begin{aligned}\dot{x} &= v \\ \dot{v} &= -(\mu v + kx - f)/m,\end{aligned} \quad (3.3.2)$$

i.e.

$$\dot{z} = -Bz + F, \tag{3.3.3}$$

where $z^{\mathrm{T}} = [x, v]$, $F = [0, f(t)/m]$ and

$$B = \begin{bmatrix} 0 & -1 \\ k/m & \mu/m \end{bmatrix}. \tag{3.3.4}$$

The θ-method for approximating eqn (3.3.3) then gives

$$\frac{1}{\Delta t}(z_{n+1} - z_n) = -B[\theta z_{n+1} + (1-\theta)z_n] + \theta F_{n+1} + (1-\theta)F_n, \tag{3.3.5}$$

where z_n, F_n are the approximations for $z(n\Delta t)$, $F(n\Delta t)$ respectively.

We can reformulate eqn (3.3.5) as

$$[I + \theta\Delta t B]z_{n+1} = [I - (1-\theta)\Delta t B]z_n + \Delta t[\theta F_{n+1} + (1-\theta)F_n]. \tag{3.3.6}$$

In order to compare the result with the Newmark single-step two-stage, we again use $Y_n = [x_n, \Delta t v_n]^{\mathrm{T}}$. We then have

$$Y_{n+1} = A_\theta Y_n + F_\theta, \tag{3.3.7}$$

where the θ-method amplification matrix A_θ is obtained from

$$A_\theta = \begin{bmatrix} 1 & -\theta \\ \theta\Delta t^2 k & m + \theta\Delta t\mu \end{bmatrix}^{-1} \begin{bmatrix} 1 & 1-\theta \\ -(1-\theta)\Delta t^2 k & m - (1-\theta)\Delta t\mu \end{bmatrix}.$$

Substituting for the inverse matrix this is equal to

$$D_\theta^{-1} \begin{bmatrix} m + \theta\Delta t\mu & \theta \\ -\theta\Delta t^2 k & 1 \end{bmatrix} \begin{bmatrix} 1 & 1-\theta \\ -(1-\theta)\Delta t^2 k & m - (1-\theta)\Delta t\mu \end{bmatrix}.$$

Then multiplying out we have

$$A_\theta = D_\theta^{-1} \begin{bmatrix} m + \theta\Delta t\mu - \theta(1-\theta)\Delta t^2 k & m \\ -\Delta t^2 k & m - (1-\theta)\Delta t\mu - \theta(1-\theta)\Delta t^2 k \end{bmatrix}, \tag{3.3.8}$$

where

$$D_\theta = m + \theta\Delta t\mu + \theta^2\Delta t^2 k. \tag{3.3.9}$$

The vector $\boldsymbol{F}_\theta$ in eqn (3.3.7) is given by

$$\boldsymbol{F}_\theta = m\Delta t^2 D_\theta^{-1}\begin{bmatrix} \theta[\theta f_{n+1} + (1-\theta)f_n] \\ [\theta f_{n+1} + (1-\theta)f_n] \end{bmatrix}. \tag{3.3.10}$$

Comparing the eqns (3.2.13) and (3.2.14) for the Newmark single-step two-stage with eqns (3.3.8) and (3.3.10) for the θ-method, we see that they are precisely the same if $\beta = 0.25$, $\gamma = 0.5$ and $\theta = 0.5$. This is when each of the algorithms is equivalent to the trapezium rule. We are then integrating eqn (3.3.3) by taking

$$\begin{aligned} \boldsymbol{z}_{n+1} - \boldsymbol{z}_n &= \int_{n\Delta t}^{(n+1)\Delta t} (-\boldsymbol{B}\boldsymbol{z} + \boldsymbol{F})\,\mathrm{d}t \\ &\approx \frac{\Delta t}{2}\{-\boldsymbol{B}(\boldsymbol{z}_{n+1} + \boldsymbol{z}_n) + \boldsymbol{F}_{n+1} + \boldsymbol{F}_n\}. \end{aligned} \tag{3.3.11}$$

2. Two-step algorithm in the displacements only. We can obtain the corresponding two-step single-stage algorithm in the same way as that used for the Newmark algorithm in §3.2 paragraph 3. We can conveniently write the amplification matrix in the form

$$\boldsymbol{A}_\theta = \begin{bmatrix} 1 - \theta\Delta t^2 k/D_\theta & m/D_\theta \\ -\Delta t^2 k/D_\theta & 1 - (\Delta t\mu + \theta\Delta t^2 k)/D_\theta \end{bmatrix}. \tag{3.3.12}$$

The characteristic polynomial of the matrix $\boldsymbol{A}_\theta$ is thus given by

$$\begin{aligned} &r^2 - r\{2 - (\Delta t\mu + 2\theta\Delta t^2 k)/D_\theta\} \\ &\qquad + 1 - \{\Delta t\mu + (2\theta - 1)\Delta t^2 k\}/D_\theta = 0. \end{aligned} \tag{3.3.13}$$

As before, we can deduce that the corresponding two-step algorithm in the displacements only is given by

$$\begin{aligned} &D_\theta x_{n+1} + \{-2m - (2\theta - 1)\Delta t\mu + 2\theta(1-\theta)\Delta t^2 k\}x_n \\ &\qquad + \{m + (\theta - 1)\Delta t\mu + (1-\theta)^2\Delta t^2 k\}x_{n-1} \\ &\qquad = \Delta t^2\{\theta^2 f_{n+1} + 2\theta(1-\theta)f_n + (1-\theta)^2 f_{n-1}\}. \end{aligned} \tag{3.3.14}$$

Comparing eqn (3.3.14) with eqn (3.2.20), we can see that if $\mu = 0$, i.e. there is no natural damping in the system, then the Newmark and the θ-method algorithms are the same if $\beta = \theta^2$ and $\gamma = 2\theta - 0.5$. Given the same starting conditions, they will produce exactly the same results. If the natural damping is not zero then as with the single-step two-stage algorithms, the Newmark and θ-method will only be exactly the same if $\theta = 0.5$ and $2\beta = \gamma = 0.5$ when each is equivalent to using the trapezium rule as shown in eqn (3.3.11). With low natural damping, the results will be very close with $\beta = \theta^2$ and $\gamma = 2\theta - 0.5$.

3.4. Accuracy and stability

As shown in Chapter 2 for numerical solutions of the dynamic vibration equation, considerations of accuracy and stability are closely connected. We obtain the information about the accuracy and stability of the Newmark and θ-methods from the homogeneous forms of eqns (3.2.19) and (3.3.14) respectively. We recall from Chapter 2 that the accuracy of the velocities and the accelerations (when they are calculated in the algorithm) is exactly the same as that of the displacement.

Newmark

Accuracy The homogeneous form of eqn (3.2.20) is the p-step type of equation discussed in Chapter 2:

$$L_{\Delta t}x_{n+j} = \sum_{j=0}^{j=p} (\alpha_j m + \gamma_j \Delta t \mu + \beta_j \Delta t^2 k) x_{n+j} = 0 \qquad (3.4.1)$$

with $p = 2$.

As in Chapter 2, we get the local truncation error by applying the difference operator $L_{\Delta t}$ to the classical solution of the differential equation $x((n+j)\Delta t)$. We thus have the local truncation error τ_2 given by

$$\tau_2 = \sum_{j=0}^{2} (\alpha_j \Delta t^{-2} m + \gamma_j \Delta t^{-1} \mu + \beta_j k) x((n+j)\Delta t).$$

Expanding $x((n+j)\Delta t)$ by Taylor series we then have

$$\tau_2 = \sum_{j=0}^{2} (\alpha_j \Delta t^{-2} m + \gamma_j \Delta t^{-1} \mu + \beta_j k)$$

$$\times \left\{ x(n\Delta t) + j\Delta t \dot{x}(n\Delta t) + j^2 \frac{\Delta t^2}{2!} \ddot{x}(n\Delta t) + \cdots \right\}. \qquad (3.4.2)$$

Collecting together terms with like powers of Δt we have

$$\begin{aligned}\tau_2 &= \Delta t^{-2} m x(n\Delta t)\sum_{j=0}^{2}\alpha_j\\
&+\Delta t^{-1}\Big\{\mu x(n\Delta t)\sum_{j=0}^{2}\gamma_j + m\dot{x}(n\Delta t)\sum_{j=0}^{2} j\alpha_j\Big\}\\
&+ m\ddot{x}(n\Delta t)\sum_{j=0}^{2} j^2\alpha_j/2! + \mu\dot{x}(n\Delta t)\sum_{j=0}^{2} j\gamma_j + kx(n\Delta t)\sum_{j=0}^{2}\beta_j\\
&+\Delta t\Big\{m\dddot{x}(n\Delta t)\sum_{j=0}^{2} j^3\alpha_j/3! + \mu\ddot{x}(n\Delta t)\sum_{j=0}^{2} j^2\gamma_j/2!\\
&+ k\dot{x}(n\Delta t)\sum_{j=0}^{2} j\beta_j\Big\} + \mathrm{O}(\Delta t^2).\end{aligned}\tag{3.4.3}$$

Comparing the coefficients in eqns (3.2.20) and (3.4.1) we have

$$\begin{aligned}&\alpha_2 = 1, && \alpha_1 = 2, && \alpha_0 = 1\\
&\gamma_2 = \gamma, && \gamma_1 = (1-2\gamma), && \gamma_0 = (\gamma - 1)\\
&\beta_2 = \beta, && \beta_1 = 0.5+\gamma-2\beta, && \beta_0 = 0.5-\gamma+\beta.\end{aligned}\tag{3.4.4}$$

Hence the general Newmark method gives an algorithm with

$$\sum_{j=0}^{2}\alpha_j = \sum_{j=0}^{2} j\alpha_j = \sum_{j=0}^{2}\gamma_j = 0 \tag{3.4.5}$$

and

$$\frac{1}{2!}\sum_{j=0}^{2} j^2\alpha_j = \sum_{j=0}^{2} j\gamma_j = \sum_{j=0}^{2}\beta_j = 1. \tag{3.4.6}$$

Substituting these results into eqn (3.4.3) and using the original differential equation (3.2.17), we see that from the definition of consistency in §2.2, the Newmark algorithm is consistent, i.e. it has accuracy at least $\mathrm{O}(\Delta t)$ for any values of the parameters β and γ. This is because the algorithm is based on the assumption of polynomial variation of the displacement and the velocity in the time interval Δt by means of the eqns (3.2.1) and (3.2.2).

Further we have

$$\begin{aligned}B_1 &= \sum_{j=0}^{2} j\beta_j = B_2 = \frac{1}{2!}\sum_{j=0}^{2} j^2\gamma_j = 0.5+\gamma,\\
B_3 &= \frac{1}{3!}\sum_{j=0}^{2} j^3\alpha_j = 1.\end{aligned}\tag{3.4.7}$$

Substituting these expressions into eqn (3.4.3), we see that when $\gamma = 0.5$, supposing we can differentiate the differential equation to produce

$$m\dddot{x} + \mu\ddot{x} + k\dot{x} = 0, \tag{3.4.8}$$

the coefficient of Δt in the right-hand side of eqn (3.4.3) vanishes and the Newmark algorithm then has accuracy $O(\Delta t^2)$. We recall that the exact solution of the eqn (3.2.17) is oscillatory provided that $\mu^2 < 4mk$, i.e. the damping is less than critical. This oscillatory solution is damped or not damped according to whether or not μ is greater than zero. Since the solution of the recurrence relation (3.2.20) in its homogeneous form is a linear combination of powers of the roots of the quadratic equation (3.2.19), we want these roots to be complex conjugates of modulus less than or equal to unity.

Stability This latter requirement brings us immediately to the question of stability. So we look next at the stability requirements for the Newmark algorithm. As in Chapter 2, in order to use the Routh–Hurwitz stability conditions, we put $r = (1 + z)/(1 - z)$ in the characteristic eqn (3.2.19) to produce an equation of the form

$$a_0 z^2 + a_1 z + a_2 = 0, \tag{3.4.9}$$

where

$$\begin{aligned} a_0 &= 4m + 2(2\gamma - 1)\Delta t\mu + 2(2\beta - \gamma)\Delta t^2 k \\ a_1 &= \Delta t\mu + (\gamma - 0.5)\Delta t^2 k \\ a_2 &= \Delta t^2 k. \end{aligned} \tag{3.4.10}$$

Then the conditions for the characteristic polynomial in eqn (3.2.19) to have roots of modulus not greater than unity are $a_0 > 0$, $a_1 \geq 0$, $a_2 \geq 0$ (see Appendix 1). Applied to the expressions in eqns (3.4.10), assuming the terms m, μ and k are each positive, these conditions are satisfied if

$$2\beta \geq \gamma \geq 0.5. \tag{3.4.11}$$

We note that if $\mu = 0$, that is, no natural damping, then $\gamma = 0.5$ becomes marginal. From eqn (3.2.19), we see that the product of the roots is then unity. Hence the stability conditions for Newmark are the same whether or not there is natural damping. The condition for the roots of the quadratic characteristic polynomial (3.2.19) to be complex is

$$\{-2m+(1-2\gamma)\Delta t\mu+(0.5-\gamma-2\beta)\Delta t^2k\}^2 < 4\{m+\gamma\Delta t\mu+\beta\Delta t^2k\}\{m+(\gamma-1)\Delta t\mu + (0.5-\gamma+\beta)\Delta t^2k\}. \quad (3.4.12)$$

After sorting out the algebra and cancelling a factor Δt^2, this condition reduces to

$$4mk-\mu^2+(2\gamma-1)\Delta t\mu k+\{4\beta-(\gamma+0.5)^2\}\Delta t^2k^2>0. \quad (3.4.13)$$

Thus we see from inequality (3.4.13) that if the solution of the differential equation (3.2.17) is oscillatory, that is, $4mk-\mu^2>0$, and we have $\gamma \geq 0.5$ (one of the stability conditions (3.4.10)) and we also have

$$4\beta-(\gamma+0.5)^2\geq 0, \quad (3.4.14)$$

then the Newmark algorithm certainly gives an oscillatory approximate solution (the combination of two complex conjugates).

Optimum values of the parameters Returning to the analysis of the accuracy of the Newmark algorithm, we can see from the pattern developing in eqn (3.4.3) that the coefficients of $k\ddot{x}(n\Delta t)$, $\mu\dddot{x}(n\Delta t)$, and $m\ddddot{x}(n\Delta t)$ in the $O(\Delta t^2)$ term will be respectively

$$C_1=\frac{1}{2!}\sum_{j=0}^{2} j^2\beta_j=0.25\beta+0.5\gamma+0.25 \quad (3.4.15)$$

$$C_2=\frac{1}{3!}\sum_{j=0}^{2} j^3\gamma_j=\gamma+1/6 \quad (3.4.16)$$

$$C_3=\frac{1}{4!}\sum_{j=0}^{2} j^4\alpha_j=7/12. \quad (3.4.17)$$

For general values of the Newmark parameters, the error in the approximation obtained from the Newmark algorithm has terms in Δt, Δt^2, Δt^3. But this error is only dominated by the Δt term for small values of the time step. Likewise as we saw above, when $\gamma=0.5$ the Δt term vanishes but the error is now only dominated by the Δt^2 term again for small values of the time step. This effect is shown clearly in the numerical results in Penry and Wood (1985). For the larger values of Δt which are frequently used in practice, we look at the optimum value of the parameter β as indicated by the coefficients C_1, C_2, C_3 in eqns (3.4.15) to (3.4.17). These coefficients are equal if $\gamma = 0.5$ and $\beta = 1/12$, but this is not compatible with the requirement for unconditional stability: $2\beta \geq \gamma \geq 0.5$. We can choose to make the coefficients C_1, C_2, C_3 as nearly equal as possible and thus reduce the size of the Δt^2 term by taking the smallest value of β compatible with the condition (3.4.14) for complex roots for the characteristic polynomial, i.e. for C-stability:

$$4\beta = (\gamma + 0.5)^2. \tag{3.4.18}$$

Hilber *et al.* (1977) use a closely related formula in connection with an extension of Newmark's algorithm which is discussed in §4.3.

We note that:

(a) with this value of β in terms of γ we effectively reduce Newmark to a one-parameter algorithm and with this we will be able to make a direct comparison with the effect of the θ-method used on the first order form of eqn (3.3.1);

(b) since $(\gamma + 0.5)^2 \equiv (\gamma - 0.5)^2 + 2\gamma$, we have

$$\beta = (\gamma + 0.5)^2/4 \geq 0.5\gamma, \tag{3.4.19}$$

hence the condition (3.4.18) is still compatible with unconditional stability as we should expect.

We now consider the choice of the value for the parameter γ. We saw above that taking $\gamma = 0.5$ makes the error $O(\Delta t^2)$ but this is not necessarily the best value to use. The mathematics literature tends to place a lot of emphasis on the order of error of a numerical solution but in practice there are other considerations. We saw in §1.4 that a finite element space discretization of a structure tends to create a stiff system of ordinary differential equations and from the theory in Strang and Fix (1973), the modes corresponding to the higher frequencies become more and more inaccurate as we move up the spectrum. From the characteristic polynomial in eqn (3.2.19), we see that with $\mu = 0$, i.e. with no natural damping and $\gamma = 0.5$, the product of the roots is unity. This means that there is no error in the amplitude and the modes of all frequencies including the highly inaccurate ones are preserved intact. Hence it is preferable to use a value of γ only slightly larger than 0.5 in order to damp out the higher frequencies but preserve the more accurate lower ones. The product of the roots of the characteristic polynomial in eqn (3.2.19) with $\mu = 0$ is

$$1-(\gamma-0.5)\omega^2\Delta t^2/(1+\beta\omega^2\Delta t^2), \qquad (3.4.20)$$

where $\omega^2 = k/m$ (the frequency of oscillation of the system is $\omega/2\pi$). Thus with β given by eqn (3.4.18) to maintain complex roots we can choose γ to give a particular value to the spectral radius of the amplification matrix, i.e. the modulus of the roots of the characteristic polynomial for large values of $\omega\Delta t$. For example $\gamma = 0.55$ gives a value for this modulus approximately 0.98 for $\omega\Delta t = \pi/5$ (i.e. $10\Delta t = 2\pi/\omega$ which is ten time steps per period) and approximately 0.9 for very large values of $\omega\Delta t$.

In Chapter 4 we discuss the errors in the p-step algorithms applied to the dynamic vibration equation in more detail distinguishing between the amplitude and the phase errors.

We saw above that we get a higher accuracy from the Newmark algorithm with $\gamma = 0.5$, $\beta = 1/12$ but then we cannot have unconditional stability. The stability conditions from eqns (3.4.10) now give

$$m - \Delta t^2 k/6 > 0,$$

i.e.

$$\omega^2\Delta t^2 \le 6 \qquad (3.4.21)$$

which is an awkward limitation with the high values of ω to be expected from the finite element discretization of a structure. A conditionally stable method is only going to be popular if there is some compensation. The central difference method discussed in Chapter 2 with $\beta = 0$ is popular because with $\mu = 0$ the matrix to be inverted at each time step is only the mass matrix M (eqn (3.2.5)). If this matrix is lumped (§1.4) then the work per time step is very small.

θ-method

Accuracy Treating the two-step form of the θ-method given by the homogeneous form of eqn (3.3.14) in the same way as we have analysed the Newmark algorithm we find that the parameters α_0, α_1, α_2 are the same as in the Newmark equations (3.4.4) and the other parameters are given by

$$\gamma_2 = \theta, \qquad \gamma_1 = 1 - 2\theta, \qquad \gamma_0 = \theta - 1 \qquad (3.4.22)$$

$$\beta_2 = \theta^2, \qquad \beta_1 = 2\theta(1-\theta), \qquad \beta_0 = (1-\theta)^2. \qquad (3.4.23)$$

Hence eqns (3.4.5) and (3.4.6) are satisfied and thus this θ-method is consistent, i.e. it has accuracy $O(\Delta t)$ for any value of θ. Further, by substituting into equations (3.4.7) we have $B_3 = 1$, $B_2 = \theta + 0.5$, $B_1 = 2\theta$, and hence this θ-method has accuracy $O(\Delta t^2)$ with $\theta = 0.5$. This corresponds to the method introduced by Crank and Nicolson (1947).

Stability To use the Routh–Hurwitz stability conditions, we again use the transformation $r = (1+z)/(1-z)$ this time in the characteristic equation of the θ-method as applied to the dynamic vibration equation (3.3.13). The equation corresponding to eqn (3.4.9) has coefficients given by

$$\begin{aligned} a_0 &= 4m + 2(2\theta - 1)\Delta t\mu + (1-2\theta)^2\Delta t^2 k \\ a_1 &= \Delta t\mu + (2\theta - 1)\Delta t^2 k \\ a_2 &= \Delta t^2 k. \end{aligned} \qquad (3.4.24)$$

Thus we have unconditional stability with $\theta \geq 0.5$; if the damping term $\mu = 0$ then $\theta = 0.5$ gives marginal stability. The product of the roots of the quadratic characteristic polynomial in (3.3.13) is now unity. This is the same situation as with the Newmark algorithm with $\gamma = 0.5$, the modes of all frequencies including the inaccurate higher frequencies are preserved. Algorithmic damping is introduced by using $\theta > 0.5$. The condition for the roots of the characteristic polynomial in eqn (3.3.13) to be complex is

$$\begin{aligned} &\{-2m - (2\theta - 1)\Delta t\mu + 2\theta(1-\theta)\Delta t^2 k\}^2 \\ &\quad < 4\{m + \theta\Delta t\mu + \theta^2\Delta t^2 k\}\{m + (\theta - 1)\Delta t\mu \\ &\qquad + (1-\theta)^2\Delta t^2 k\}. \end{aligned} \qquad (3.4.25)$$

This reduces to the simple condition

$$\Delta t^2(4mk - \mu^2) > 0. \qquad (3.4.26)$$

Hence if the differential equation has an oscillatory solution (i.e. damping less than critical), the θ-method will give an oscillatory solution also.

Putting $\theta = 0$ gives an explicit method. The matrix to be inverted is now the mass matrix only but the stability conditions from eqns (3.4.24) now give

$$4m - 2\Delta t\mu + \Delta t^2 k > 0 \tag{3.4.27}$$

and

$$\Delta t\mu - \Delta t^2 k > 0. \tag{3.4.28}$$

Provided that $k > 0$, inequality (3.4.27) can be written as

$$(k\Delta t - \mu)^2 + (4mk - \mu^2) > 0, \tag{3.4.29}$$

which is satisfied if the system is underdamped. But inequality (3.4.28) cannot be satisfied if $\mu = 0$.

Summary

In this chapter, we have compared the effects of using the Newmark algorithm on the second order form of the dynamic vibration equation with the θ-method used on the equivalent first order form in order to show how closely related they are. Newmark starts with two parameters but the condition for an oscillatory solution for all values of $\omega\Delta t$ effectively makes it into a one-parameter algorithm closely resembling the θ-method. For some values of the parameters, they can give exactly the same results.

Exercises

3.1. Find the condition for complex roots of the characteristic polynomial with the Newmark method using $\gamma = 1/2$ and $\beta = 1/12$.

3.2. Is there a θ-method for the first order form of eqn (3.2.6) equivalent to Newmark with $\gamma = 1/2$ and $\beta = 1/12$?

3.3. Show that the algorithm given by eqn (3.3.14) with $\theta = 0$ has a characteristic polynomial with complex conjugate roots if the damping is less than critical. If $\mu \neq 0$ there is conditional stability but if $\mu = 0$ the algorithm is unconditionally unstable. What happens if the algorithm is used for a case where the damping is greater than critical?

References

Crank, J. and Nicolson, P. (1947). Numerical evaluation of solutions of partial differential equations of heat conduction type. *Proceedings of the Cambridge Philosophical Society*, **43**, 50–67.

Henrici, P. (1962). *Discrete variable methods in ordinary differential equations.* Wiley.

Hilber, H. M., Hughes, T. R. J., and Taylor, R. L. (1977). Improved numerical dissipation for time integration algorithms in structural dynamics. *Earthquake Engineering and Structural Dynamics* **5,** 283–292.

Lambert, J. D. (1973). *Computational methods in ordinary differential equations.* Wiley.

Newmark, N. M. (1959). A method of computation for structural dynamics. *Proceedings of ASCE, Journal of Engineering Mechanics* (EM3), **85,** 67–94.

Park, K. C. (1975). An improved stiffly stable method for direct integration of non-linear structural dynamics equations. *Transactions of ASME, Journal of Applied Mechanics,* **42,** 464–470.

Penry, S. N. and Wood, W. L. (1985). Comparison of some single-step methods for the numerical solution of the structural dynamic equation. *International Journal of Numerical Methods in Engineering,* **21,** 1941–1955.

Richtmyer, R. D. and Morton, K. W. (1967). *Difference methods for initial value problems.* Interscience.

Strang, G. and Fix, G. J. (1973). *An analysis of the finite element method.* Prentice-Hall.

Wood, W. L. (1981). Numerical integration of structural dynamics equations including natural damping and periodic forcing terms. *International Journal of Numerical Methods in Engineering,* **17,** 281–289.

Wood, W. L. (1984). A further look at Newmark, Houbolt, etc., time-stepping formulae. *International Journal of Numerical Methods in Engineering,* **20,** 1009–1017.

Zienkiewicz, O. C. (1977) *The finite element method.* McGraw-Hill.

4 General theory and higher order p-step and p-stage methods

4.1. Introduction

The Newmark method is not completely satisfactory for the numerical solution of the dynamic vibration equation because as shown in Chapter 3, second order accuracy can only be obtained taking the parameter $\gamma = 0.5$ which means no algorithmic damping of the inaccurate higher frequencies produced by the finite element space discretization. Ideally what is required is that the lower frequencies should be preserved and the higher frequencies damped in a controllable way. The curve of the spectral radius of the amplification matrix against $\Delta t/T$ (where T is the undamped natural period and Δt is the time step), see Figs. 4.3.1, 4.7.1, for example, should stay close to the unit level as long as possible and decrease to about 0.5 or 0.8 as $\Delta t/T$ tends to infinity. As shown in §4.8, this requires C-stability not merely A-stability. Dahlquist (1963) shows that it is not possible to have A-stability with a higher than second order error. Hence the emphasis on finding a method with second order error and controllable algorithmic damping to cope with the very low natural damping of structures. There is the possibility of obtaining third order error with $A(\alpha)$ stability illustrated in Fig. 2.4 for the Gear (1969) three-step backward difference algorithm. We show results with a single-step equivalent of this in §4.9 but it requires about 7% natural damping which is rather high; it also has the disadvantage of too heavy algorithmic damping. We are also interested in the phase error which determines whether the signal arrives early or late. Hence it is necessary to distinguish between the damping and phase errors and obtain general formulae for them and this is done in §4.2. Section 4.2 uses these results to conclude the discussion of the Newmark and θ-methods begun in Chapter 3.

The search for an improvement on the Newmark method has led to ideas of using the Newmark algorithm with an extra parameter; these are described as Newmark Variations in §4.4.

Section 4.5 describes the generalization of the Newmark method and §4.6 the closely allied SS*pj* family of algorithms. SS*pj* stands for Single-Step algorithm using an approximation of degree p for the unknown function in the time step and is applicable to differential equations of order $j = 1$ or 2. The idea is that a very general package program can be used to generate new algorithms as well as including many old ones. The emphasis is on single-step methods because these give the facility for easy control of the time step according to some error criterion (discussed in §7.8) and adjustment to a changing forcing function. The SS*pj* algorithms can also be regarded as predictor–corrector methods.

Section 4.7 analyses the link between the SS*pj* and generalized Newmark algorithms. They are very closely connected for linear problems but, as shown in Chapter 7, can give different effects when used on non-linear problems. Section 4.8 shows how the SS*p*2 family of algorithms includes other popular methods for solving the dynamic vibration equation such as the Wilson-θ (1968) method for the dynamic stress analysis of underground structures and the Houbolt (1950) method for the solution of the dynamic response of elastic aircraft. Section 4.9 describes the Penry and Wood (1985) experiments with SS*p*1 used on the first order form of the dynamic vibration equation. The results from Penry and Wood also demonstrate the importance of approximating the forcing function with sufficient accuracy. This frequently seems to be regarded as a secondary consideration.

The SS*pj* family is not so general that it includes all other methods of a similar type. In particular it does not include a method introduced by Park (1975). This is a combination of the Gear (1969) two-step and three-step algorithms which results in a three-step method of only second order. Park claims that this has advantages for the solution of non-linear dynamic vibration problems and it is discussed further in Chapter 7. However, we point out (§4.12) that the generalization of SS*p*2 by Thomas and Gladwell (1988), described in §4.10, can be particularized to a generalization of SS*p*1 that does include Park's method. Section 4.11 shows that SS*p*1 includes a number of algorithms that have been separately proposed for the solution of first order equations.

4.2 The dynamic vibration equation—general formulae for amplitude and phase errors

We now make a more detailed study of the errors arising from the use of a p-step algorithm on the test equation

$$\ddot{x} + 2\nu\omega\dot{x} + \omega^2 x = 0. \tag{4.2.1}$$

The test equation (4.2.1) has exact solutions of the form $\exp(vt)$ where

$$v = -\nu\omega \pm i\omega\sqrt{(1-\nu^2)},$$

i.e.

$$\exp(vt) = \exp[\omega(-\cos\alpha \pm i\sin\alpha)t] \tag{4.2.2}$$

where $\nu = \cos\alpha$ (we again suppose the damping to be less than critical so that $\nu < 1$). We want to have an assessment of the error in the numerical solution obtained by using a p-step algorithm of the form

$$\sum_{j=0}^{p} (\alpha_j \Delta t^{-2} + 2\gamma_j \Delta t^{-1}\nu\omega + \beta_j\omega^2)x_{n+j} = 0. \tag{4.2.3}$$

We showed in Chapter 2 that this error will be the same as that from the corresponding one-step p-stage algorithm which has the same characteristic polynomial (given matching starting conditions). Also in Chapter 2 we defined the local truncation error and showed how the error in the numerical solution at a fixed time $T = n\Delta t$ is of the same order as the local truncation error. Here we go straight for the error in the numerical solution by supposing that for small values of the time step Δt, the characteristic polynomial of the p-step algorithm (4.2.3) given by

$$\sum_{j=0}^{p} (\alpha_j \Delta t^{-2} + 2\gamma_j \nu\omega\Delta t^{-1} + \beta_j\omega^2)r^j = 0 \tag{4.2.4}$$

has an approximate principal root

$$\exp(\tilde{v}\Delta t) = \exp[(v+\xi)\Delta t]. \tag{4.2.5}$$

This means that the approximate solution is multiplied at each step by $\exp\{(v+\xi)\Delta t\}$ instead of $\exp(v\Delta t)$. Hence after n time steps where $n\Delta t = T$ the accumulated relative error is

$$1 - \exp(\xi T). \tag{4.2.6}$$

For a fixed value of the time T, $1-\exp(\xi T) = O(\xi)$.

Substituting $r = \exp(\tilde{v}\Delta t)$ into eqn (4.2.4) and expanding the exponential terms we have

$$\sum_{j=0}^{p} (\alpha_j \Delta t^{-2} + 2\gamma_j \nu\omega \Delta t^{-1} + \beta_j \omega^2)$$
$$\times \left(1 + j\Delta t\tilde{v} + j^2 \frac{\Delta t^2}{2!}\tilde{v}^2 + j^3 \frac{\Delta t^3}{3!}\tilde{v}^3 + \cdots\right) = 0. \tag{4.2.7}$$

Collecting together the terms with like powers of Δt in eqn (4.2.7) we then have

$$\Delta t^{-2} \sum_{j=0}^{p} \alpha_j + \Delta t^{-1}\left\{\tilde{v} \sum_{j=0}^{p} j\alpha_j + 2\nu\omega \sum_{j=0}^{p} \gamma_j\right\}$$
$$+ \tilde{v}^2 \sum_{j=0}^{p} j^2\alpha_j/2! + 2\nu\omega\tilde{v} \sum_{j=0}^{p} j\gamma_j + \omega^2 \sum_{j=0}^{p} \beta_j$$
$$+ \tilde{v}\Delta t\left\{\tilde{v}^2 \sum_{j=0}^{p} j^3\alpha_j/3! + 2\nu\omega\tilde{v} \sum_{j=0}^{p} j^2\gamma_j/2!\right.$$
$$\left. + \omega^2 \sum_{j=0}^{p} j\beta_j\right\} + \cdots = 0. \tag{4.2.8}$$

Then if the parameters α_j, β_j, γ_j satisfy the following conditions

$$\sum_{j=0}^{p} \alpha_j = \sum_{j=0}^{p} j\alpha_j = \sum_{j=0}^{p} \gamma_j = 0$$

and

$$\sum_{j=0}^{p} j^2\alpha_j/2! = \sum_{j=0}^{p} j\gamma_j = \sum_{j=0}^{p} \beta_j = A, \text{ say} \tag{4.2.9}$$

(these are the consistency conditions again as in Chapter 3), and we substitute

$$B_1 = \sum_{j=0}^{p} j\beta_j, \quad B_2 = \sum_{j=0}^{p} j^2\gamma_j/2!, \quad B_3 = \sum_{j=0}^{p} j^3\alpha_j/3! \tag{4.2.10}$$

into eqn (4.2.8) we have

$$A(\tilde{v}^2 + 2\nu\omega\tilde{v} + \omega^2) + \Delta t(B_3\tilde{v}^3 + 2\nu\omega B_2\tilde{v}^2 + \omega^2 B_1\tilde{v}) + O(\Delta t^3) = 0. \tag{4.2.11}$$

Now we can substitute $\tilde{v} = v + \xi$ into eqn (4.2.11) and use the fact that

$$v^2 + 2\nu\omega v + \omega^2 = 0 \tag{4.2.12}$$

(because $\exp(vt)$ is an exact solution of the test equation (4.2.1)). Looking separately at the first term on the right-hand side of eqn (4.2.11), we have

$$A(\tilde{v}^2 + 2\nu\omega\tilde{v} + \omega^2) = A\{(v + \xi)^2 + 2\nu\omega(v + \xi) + \omega^2\}.$$

Expanding and using eqn (4.2.12) we find that this expression is equal to

$$A\{2(v + \nu\omega)\xi + \xi^2\}.$$

But from eqn (4.2.12) we have

$$v = \omega(-\nu \pm i \sin\alpha), \tag{4.2.13}$$

i.e.

$$v + \nu\omega = \pm i\omega \sin\alpha.$$

Without loss of generality, we take the plus sign here and then eqn (4.2.11) reduces to

$$2Ai\omega\xi \sin\alpha + A\xi^2 + \Delta t\{B_3(v^3 + 3v^2\xi) + 2\nu\omega B_2(v^2 + 2v\xi) + \omega^2 B_1(v + \xi)\} + O(\Delta t^2) = 0. \tag{4.2.14}$$

It will become apparent later why we keep the $A\xi^2$ and the $\Delta t\xi$ terms in here.

Thus with the conditions (4.2.9), we can conclude that the error ξ in the numerical solution is $O(\Delta t)$, i.e. the method is first order accurate. If we also have $B_1 = B_2 = B_3$, then

$$B_3 v^3 + 2\nu\omega B_2 v^2 + \omega^2 B_1 v = B_1 v(v^2 + 2\nu\omega v + \omega^2) = 0$$

from eqn (4.2.12). Hence ξ is then $O(\Delta t^2)$, i.e. the method is second order accurate. By introducing

$$C_1 = \sum_{j=0}^{p} j^2\beta_j/2!, \quad C_2 = \sum_{j=0}^{p} j^3\gamma_j/3!, \quad c_3 = \sum_{j=0}^{p} j^4\alpha_j/4! \tag{4.2.15}$$

and

$$D_1 = \sum_{j=0}^{p} j^3\beta_j/3!, \quad D_2 = \sum_{j=0}^{p} j^4\gamma_j/4!, \quad D_3 = \sum_{j=0}^{p} j^5\alpha_j/5!, \text{etc.} \tag{4.2.16}$$

we can include further terms in eqn (4.2.13) and see that if $C_1 = C_2 = C_3$, the error ξ is $O(\Delta t^3)$, again using eqn (4.2.12), and so on. But because of the Dahlquist theorem (Dahlquist 1963) which proves that the order of accuracy of a linear multistep A-stable method cannot exceed two, there is little interest in algorithms of third or higher order accuracy for the dynamic vibration equation. As mentioned before, the combintion of very low natural damping and the necessity to have unconditional stability without having to worry about the high frequencies produced by the finite element discretization means that A-stability is desirable.

For the particular examples considered later in this chapter, the stability will always be tackled generally using the Routh–Hurwitz conditions from Appendix 1. In this piece of theory we are looking at effects which apply for small values of $\omega \Delta t$ i.e. we are looking in detail at the lower end of the $\omega \Delta t$ spectrum.

We now want to sort out the amplitude and phase errors by expressing the error as $\xi = -\delta + i\epsilon$. This means that the approximate principal root of the polynomial in (4.2.4) is

$$\exp\{-(\omega v + \delta)\Delta t + i(\omega \sin \alpha + \epsilon)\Delta t\}. \qquad (4.2.17)$$

We are thus supposing that the numerical solution has the effect of

(a) increasing the damping by a small quantity δ;

(b) increasing the frequency of the oscillation by a small quantity $\epsilon/2\pi$, which we can also express as giving a period increase of

$$2\pi/(\omega \sin \alpha + \epsilon) - 2\pi/\omega \sin \alpha \approx -2\pi\epsilon/(\omega^2 \sin^2\alpha) \qquad (4.2.18)$$

or a fractional period increase of $-\epsilon/(\omega \sin \alpha)$.

The stability will be indicated by the sign of δ; a negative value of δ signifies a growing amplitude of the oscillation. It is also frequently of interest to know whether the algorithm will give an increase or decrease in the period of the oscillation.

Substituting $\xi = -\delta + i\epsilon$ into eqn (4.2.14) and using de Moivre's theorem:

$$\begin{aligned}(-\cos \alpha + i \sin \alpha)^k &= (-1)^k(\cos \alpha - i \sin \alpha)^k \\ &= (-1)^k(\cos k\alpha - i \sin k\alpha), \qquad (4.2.19)\end{aligned}$$

so that $v^k = \omega^k(-1)^k(\cos k\alpha - i \sin k\alpha)$, we have the result

$$\begin{aligned}
&-2A\omega \sin \alpha(\epsilon + \mathrm{i}\delta) + A(\epsilon^2 - \delta^2 - 2\mathrm{i}\epsilon\delta) \\
&\quad + \omega^3 \Delta t\{-B_3(\cos 3\alpha - \mathrm{i} \sin 3\alpha) + 2B_2 \cos \alpha \\
&\quad \times (\cos 2\alpha - \mathrm{i} \sin 2\alpha) - B_1(\cos \alpha - \mathrm{i} \sin \alpha)\} \\
&\quad + \omega^2 \Delta t(-\delta + \mathrm{i}\epsilon)\{3B_3(\cos 2\alpha - \mathrm{i} \sin 2\alpha) \\
&\quad - 4B_2 \cos \alpha(\cos \alpha - \mathrm{i} \sin \alpha) + B_1\} \\
&\quad + \omega^4 \Delta t^2\{C_3(\cos 4\alpha - \mathrm{i} \sin 4\alpha) \\
&\quad - 2C_2 \cos \alpha(\cos 3\alpha - \mathrm{i} \sin 3\alpha) \\
&\quad + C_1(\cos 2\alpha - \mathrm{i} \sin 2\alpha)\} \\
&\quad + \omega^5 \Delta t^3\{-D_3(\cos 5\alpha - \mathrm{i} \sin 5\alpha) \\
&\quad + 2D_2 \cos \alpha(\cos 4\alpha - \mathrm{i} \sin 4\alpha) \\
&\quad - D_1(\cos 3\alpha - \mathrm{i} \sin 3\alpha)\} + \mathrm{O}(\Delta t^4) = 0. \qquad (4.2.20)
\end{aligned}$$

Now we separate the real and imaginary parts in eqn (4.2.20) to give

$$\begin{aligned}
&-2A\omega\epsilon \sin \alpha + A(\delta^2 - \epsilon^2) + \omega^3 \Delta t \\
&\quad \times [-B_3 \cos 3\alpha + 2B_2 \cos \alpha \cos 2\alpha - B_1 \cos \alpha) \\
&\quad - \delta\omega^2 \Delta t(3B_3 \cos 2\alpha - 4B_2 \cos^2 \alpha + B_1) \\
&\quad + \epsilon\omega^2 \Delta t(3B^2 \sin 2\alpha - 4B_2 \cos \alpha \sin \alpha) \\
&\quad + \omega^4 \Delta t^2(C_3 \cos 4\alpha - 2C_2 \cos \alpha \cos 3\alpha + C_1 \cos 2\alpha) \\
&\quad + \mathrm{O}(\Delta t^3) = 0. \qquad (4.2.21)
\end{aligned}$$

and

$$\begin{aligned}
&-2A\delta\omega \sin \alpha - 2A\epsilon\delta + \omega^3 \Delta t \\
&\quad \times (B_3 \sin 3\alpha - 2B_2 \cos \alpha \sin 2\alpha + B_1 \sin \alpha) \\
&\quad + \omega^2 \epsilon \Delta t(3B_3 \cos 2\alpha - 4B_2 \cos^2 \alpha + B_1) \\
&\quad - \omega^2 \delta \Delta t(-3B_3 \sin 2\alpha + 4B_2 \sin \alpha \cos \alpha) \\
&\quad - \omega^4 \Delta t^2(C_3 \sin 4\alpha - 2C_2 \cos \alpha \sin 3\alpha \\
&\quad + C_1 \sin 2\alpha) + \mathrm{O}(\Delta t^3) = 0. \qquad (4.2.22)
\end{aligned}$$

Hence as noted in Wood (1981), if $v \neq 0$, i.e. $\alpha \neq \pi/2$, and B_1, B_2, B_3 are not equal then the amplitude error δ and the phase error ϵ are each $\mathrm{O}(\Delta t)$. It is convenient to have the equations which apply when the damping ratio v is zero written out separately as follows.

With $\nu = 0$, that is, $\alpha = \pi/2$, we can extend eqn (4.2.14) to include extra terms:

$$\begin{aligned} -2A\omega(\epsilon + \mathrm{i}\delta) + A(\delta^2 - \epsilon^2 - 2\mathrm{i}\epsilon\delta) + \mathrm{i}\omega^3\Delta t(B_1 - B_3) \\ + \omega^2\Delta t(-\delta + \mathrm{i}\epsilon)(B_1 - 3B_3) + \omega^2\Delta t^2(C_3 - C_1) \\ + \mathrm{i}\omega^5\Delta t^3(D_3 - D_1) + \mathrm{O}(\Delta t^4) = 0. \end{aligned} \tag{4.2.23}$$

The relevant terms from the real and imaginary parts of eqn (4.2.23) then give

$$\begin{aligned} 2A\omega\epsilon = -A\delta^2 - \omega^2\Delta t\delta(B_1 - 3B_3) \\ + \omega^4\Delta t^2(C_3 - C_1) + \mathrm{O}(\Delta t^4) \end{aligned} \tag{4.2.24}$$

$$\begin{aligned} 2A\omega\delta = \omega^3\Delta t(B_1 - B_2) + \omega^2\Delta t\epsilon(B_1 - 3B_3) \\ + \omega^5\Delta t^3(D_3 - D_1) + \mathrm{O}(\Delta t^5). \end{aligned} \tag{4.2.25}$$

Hence the phase error ϵ is $\mathrm{O}(\Delta t^2)$ while the amplitude error δ remains $\mathrm{O}(\Delta t)$ unless $B_1 = B_3$. This shows why it is necessary to allow for the amplitude and phase errors to have different orders and to retain the extra terms in eqns (4.2.20) to (4.2.22) until we can be sure which are required. When ϵ is $\mathrm{O}(\Delta t^2)$ and δ is $\mathrm{O}(\Delta t)$, in order to allow for all terms of $\mathrm{O}(\Delta t^2)$ in eqn (4.2.24) we must include those with δ^2 and $\delta\Delta t$. Thus when $\nu = 0$ and $B_1 \neq B_3$, we have from eqn (4.2.25)

$$\delta = \omega^2\Delta t(B_1 - B_3)/(2A) \text{ to } \mathrm{O}(\Delta t) \tag{4.2.26}$$

and, substituting for δ into eqn (4.2.24) we have

$$\begin{aligned} 2A\omega\epsilon = \omega^4\Delta t^2(B_1 - B_3)^2/(4A) - \omega^4\Delta t^2(-3B_3 + B_1) \\ \times (B_1 - B_3)/(2A) + \omega^4\Delta t^2(C_3 - C_1) + \mathrm{O}(\Delta t^4). \end{aligned} \tag{4.2.27}$$

Rearranging eqn (4.2.27), we have the phase error ϵ given in this case by

$$\epsilon = \frac{\omega^3\Delta t^2}{8A^2}\{(B_3 - B_1)(B_1 - 5B_3) + 4A(C_3 - C_1)\} + \mathrm{O}(\Delta t^4). \tag{4.2.28}$$

If we have $B_1 = B_3$ then eqn (4.2.28) gives

$$\epsilon = \frac{\omega^3\Delta t^2}{2A}(C_3 - C_1) + \mathrm{O}(\Delta t^4) \tag{4.2.29}$$

and eqn (4.2.25) gives

$$\delta = \frac{\omega \Delta t}{2A}\{\epsilon(B_1 - 3B_3) + \omega^3 \Delta t^2 (D_3 - D_1)\} + \cdots. \quad (4.2.30)$$

We then have to substitute for ϵ from eqn (4.2.29) into eqn (4.2.30) to give the result:

$$\delta = \frac{\omega^4 \Delta t^3}{4A^2}\{(B_1 - 3B_3)(C_3 - C_1) + 2A(D_3 - D_1)\} + \mathrm{O}(\Delta t^5). \quad (4.2.31)$$

In the following sections of this chapter we apply these results to various p-step and one-step p-stage algorithms proposed in the engineering literature and discuss the motivation for the formation of these algorithms.

4.3. Newmark and the θ-method (continued)

Newmark

We can now finish off the analysis of the Newmark algorithm (Newmark 1959) begun in Chapter 3, by using the results on the amplitude and phase errors obtained in §4.2. We have a two-step algorithm of the form (4.2.3) with parameters α_j, γ_j, β_j, $j = 0, 1, 2$ given in terms of the Newmark parameters β and γ by eqns (3.4.4). Equations (3.4.5) and (3.4.6) show that the consistency conditions (4.2.9) are satisfied and $A = 1$. Equations (3.4.7) show that $B_1 = B_2 = 0.5 + \gamma$ and $B_3 = 1$. From eqns (3.4.4) and (4.2.15) we also have

$$\begin{aligned} C_1 &= \beta + 0.25 + \gamma/2 \\ C_2 &= \gamma + 1/6 \\ C_3 &= 7/12. \end{aligned} \quad (4.3.1)$$

Substituting into eqn (4.2.21) and leaving out the terms which are not needed here we now have

$$\begin{aligned} 2\omega\epsilon \sin\alpha = {} & \delta^2 + \omega^3 \Delta t(\gamma - 0.5)\cos 3\alpha - \delta\omega^2 \Delta t \\ & \times \{4\cos^2\alpha(1-\gamma) - 2.5 + \gamma\} + \omega^4 \Delta t^2 \\ & \times \{C_3 \cos 4\alpha - 2C_2 \cos\alpha \cos 3\alpha \\ & + C_1 \cos 2\alpha\} + \mathrm{O}(\Delta t^3). \end{aligned} \quad (4.3.2)$$

Similarly the relevant terms in eqn (4.2.22) give

$$2\delta\omega \sin \alpha = \omega^3 \Delta t\{\sin 3\alpha + (0.5+\gamma)(\sin \alpha - 2\cos \alpha \sin 2\alpha)\} \\ - \omega^4 \Delta t^2 (C_3 \sin 4\alpha - 2C_2 \cos \alpha \sin 3\alpha + C_1 \sin 2\alpha) \\ + \mathrm{O}(\Delta t^3). \quad (4.3.3)$$

The first term on the right hand side of eqn (4.3.3) is equal to

$$\omega^3 \Delta t\{\sin \alpha \cos 3\alpha + \cos \alpha \sin 2\alpha + \gamma \sin \alpha (1 - 4\cos^2 \alpha) \\ + 0.5 \sin \alpha - \cos \alpha \sin 2\alpha\} \\ = \omega^3 \Delta t \sin \alpha (\gamma - 0.5)(1 - 4\cos^2 \alpha).$$

Hence eqn (4.3.3) can be simplified into

$$\delta = \frac{\omega^2 \Delta t}{2}(\gamma - 0.5)(1 - 4\cos^2 \alpha) - B\Delta t^2 + \mathrm{O}(\Delta t^3), \quad (4.3.4)$$

where

$$B = \omega^3 \cos \alpha\{2C_3 \cos 2\alpha - C_2(4\cos^2 \alpha - 1) + C_1\} \\ = \omega^3 \cos \alpha\{4\cos^2 \alpha (C_3 - C_2) - 2C_3 + C_2 + C_1\}.$$

Substituting for C_1, C_2, C_3 from eqns (4.3.1) we have

$$B = \nu\{\nu^2(5 - 12\gamma)/3 + (4\beta + 6\gamma - 3)/4\} \quad (4.3.5)$$

in terms of $\nu = \cos \alpha$.

Thus we can see from eqns (4.3.4) and (4.3.5) that for the small values of the damping ratio $\nu = \cos \alpha$ we expect for structures, the formula for δ is dominated by the first term on the right-hand side of eqn (4.3.4) and is positive when the parameter γ is greater than 0.5, i.e. the damping is increased. We recall that the Routh–Hurwitz general stability conditions for the Newmark algorithm are $2\beta \geq \gamma \geq 0.5$.

Now substituting for δ from eqn (4.3.4) into eqn (4.2.21) and keeping all the terms up to and including $\mathrm{O}(\Delta t^2)$ we have the following:

$$2\epsilon\omega \sin \alpha = \omega^3 \Delta t(\gamma - 0.5)\cos 3\alpha \\ + 0.25\omega^4 \Delta t^2 (1 - 4\cos^2 \alpha)^2 (\gamma - 0.5)^2 \\ - 0.5\omega^4 \Delta t^2 (1 - 4\cos^2 \alpha)(\gamma - 0.5) \\ \times \{4\cos^2 \alpha (1 - \gamma) - 2.5 + \gamma\} \\ + \omega^4 \Delta t^2 (C_3 \cos 4\alpha - 2C_2 \cos \alpha \cos 3\alpha + C_1 \cos 2\alpha) \\ + \mathrm{O}(\Delta t^3). \quad (4.3.6)$$

Using eqn (4.2.18) we can now write down a formula for the phase error:

$$-\epsilon = -\omega^2\Delta t(\gamma - 0.5)\frac{\cos 3\alpha}{2\sin\alpha} + O(\Delta t^2). \qquad (4.3.7)$$

Remembering that $\cos 3\alpha < 0$ if $3\alpha > \pi$, i.e. $\alpha > \pi/3$, we see that for $\gamma > 0.5$ and small values of the damping ratio ($\cos\alpha = \nu \le 0.1$ implies $\alpha \ge 84$ degrees approximately), we have a period increase for Δt small enough for the first term to dominate in the right hand side of eqn (4.3.7). Figure 4.3.1 (from Wood 1986) illustrates this for the test problem (4.2.1) with $\omega = 1$ and $\nu = 0.1$, taking starting conditions $x(0) = 1$ and $\dot{x}(0) = 0$ and with $\Delta t = 0.5$. The exact period here is equal to 2π so this gives about 12 time steps per period. Figure 4.3.1 shows the resultant slight damping and period increase.

When the Newmark parameter $\gamma = 0.5$, we have from eqns (3.4.7) and (4.3.1)

$$B_1 = B_2 = B_3, \quad \text{and} \quad C_1 = \beta + 0.5, \qquad C_2 = 2/3, \qquad C_3 = 7/12. \qquad (4.3.8)$$

Equations (4.3.4) and (4.3.5) now give

$$\delta = \omega^3\Delta t^2\nu(\nu^2 - 3\beta)/3. \qquad (4.3.9)$$

Thus there is an increase in damping for the central difference algorithm given by $\beta = 0$ (Chapter 2), and a decrease for small

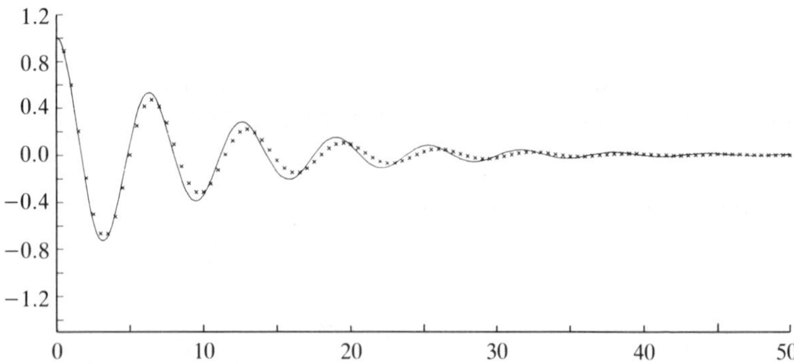

Fig. 4.3.1. Newmark with $\gamma = 0.6$, $2\beta = 0.605$, $\Delta t = 0.5$, $\nu = 0.1$, shows slight damping and period increase.

values of $\cos\alpha$ and $\beta \geq 0.25$ which is required for unconditional stability (Chapter 3).

Substituting these results for $\gamma = 0.5$ into eqn (4.2.21) we now have the phase error given by

$$-\epsilon = \omega^3 \Delta t^2 \{\beta(1-2\nu^2) + (8\nu^4 - 4\nu^2 - 1)/12\}/\{2\sqrt{(1-\nu^2)}\} + \mathrm{O}(\Delta t^3). \quad (4.3.10)$$

With the parameter $\beta = 0$, we have

$$-\epsilon = \omega^3 \Delta t^2 (8\nu^4 - 4\nu^2 - 1)/\{24\sqrt{(1-\nu^2)}\}. \quad (4.3.11)$$

Since we have $8\nu^4 - 4\nu^2 - 1 = 8(\nu^2 - 0.25)^2 - 1.5 < 0$ for $\nu^2 < (1+\sqrt{3})/4 \approx 0.68$, i.e. $\nu \leq 0.83$, approximately, for the conditionally stable central difference method, we expect a period decrease for the values of ν we usually have with structures.

For the unconditionally stable algorithms given by $\beta \geq 0.25$ together with $\gamma = 0.5$, we see from eqn (4.3.10) that we have a period increase if

$$(1-2\nu^2)/4 + (8\nu^4 - 4\nu^2 - 1)/12 > 0.$$

This is equivalent to

$$4\nu^4 - 5\nu^2 + 1 > 0,$$

i.e. $(1-\nu^2)(1-4\nu^2) > 0$, which is satisfied if $\nu < 0.5$. Hence for small values of the damping ratio ν we certainly expect a period increase with these unconditionally stable methods. Figure 4.3.2 (from Wood 1986) illustrates this for the same test problem as in Fig. 4.3.1.

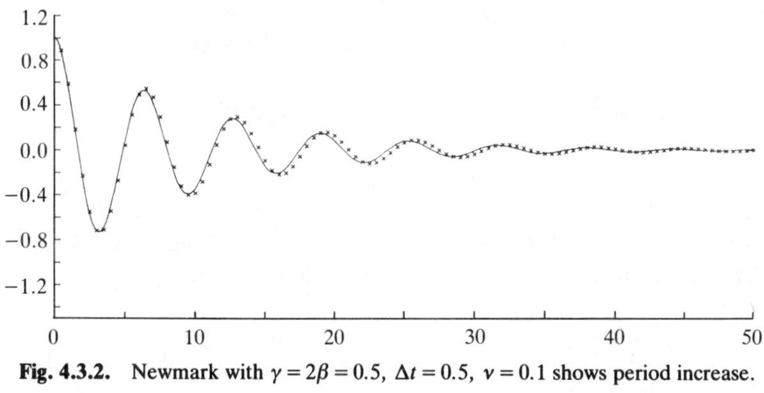

Fig. 4.3.2. Newmark with $\gamma = 2\beta = 0.5$, $\Delta t = 0.5$, $\nu = 0.1$ shows period increase.

When the damping ratio ν is zero and $\gamma > 0.5$, i.e. $B_1 \neq B_3$, we can use eqn (4.2.24) to give

$$\delta = \omega^2 \Delta t(\gamma - 0.5)/2 + O(\Delta t^3) \tag{4.3.12}$$

and from eqn (4.2.28) we have the phase error given by

$$-\epsilon = \frac{\omega^3 \Delta t^2}{8}(\gamma^2 - 3\gamma + 4\beta + 11/12) + O(\Delta t^4). \tag{4.3.13}$$

For unconditional stability, we know from §3.4 that we require $2\beta \geq \gamma \geq 0.5$ and we can see how this links with the sign of δ in eqn (4.3.12) as expected. Algorithmic damping is introduced by making $\gamma > 0.5$; negative damping will cause instability. We also see that the factor on the right-hand side of eqn (4.3.13), for $2\beta \geq \gamma$, is greater than

$$\gamma^2 - \gamma + 11/12 = (\gamma - 0.5)^2 + 2/3$$

and hence unconditionally stable schemes always have a period increase for time step sizes such that the $O(\Delta t^2)$ term dominates in eqn (4.3.13). For this term to dominate we require $\omega^2 \Delta t^2 \ll 1$. With twenty time steps per period, for example, i.e. $20\Delta t = 2\pi/\omega$, we have $\omega^2 \Delta t^2 = (\pi/10)^2 \approx 0.1$ and this is sufficient to make the $O(\Delta t^2)$ term dominant with the usual values of the parameters β and γ. The lowest period increase is with $2\beta = \gamma = 0.5$.

For an example of an explicit scheme we again take the central difference scheme with $\beta = 0$. The phase error is then from eqn (4.3.13) given by

$$-\epsilon = \frac{\omega^2 \Delta t^2}{8}(\gamma^2 - 3\gamma + 11/12) + O(\Delta t^4). \tag{4.3.14}$$

Since the expression $\gamma^2 - 3\gamma + 11/12$ is negative for values of γ between the roots of

$$\gamma^2 - 3\gamma + 11/12 = 0,$$

we have a period decrease for $0.34 < \gamma < 2.65$ approximately. There is a period decrease together with a non-negative value of δ if $0.5 \leq \gamma < 2.65$.

Figure 4.3.3 (from Wood 1986) shows results for the test problem with zero natural damping and (a) $2\beta = \gamma = 0.5$ which

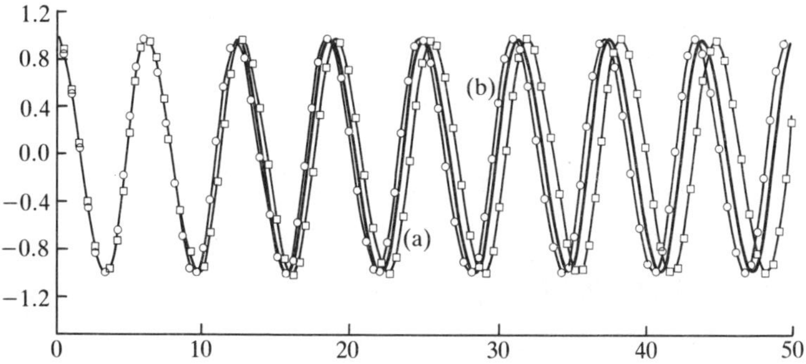

Fig. 4.3.3. Newmark with $\nu = 0$, $\Delta t = 0.5$, $f = 0$ and (a) $\gamma = 2\beta = 0.5$ shows zero algorithmic damping and minimum period increase for unconditional stability, (b) $\gamma = 0.5$, $\beta = 0$, shows zero algorithmic damping with a period decrease. The full curve shows the exact solution.

gives zero algorithmic damping and the minimum period increase for unconditionally stable schemes and (b) $\beta = 0$, $\gamma = 0.5$ which again gives zero algorithmic damping but now with a period decrease. These results both use the time step $\Delta t = 0.5$ for comparison. The stability limit for (b) is actually given by $\omega \Delta t < 2$ from eqn (2.3.12) and $\omega = 1$ for this problem but we are here keeping the value of Δt sufficiently small for the first term on the right of eqn (4.3.14) to dominate.

We also note in connection with the Newmark algorithm that eqns (3.4.7) show that we must have the parameter $\gamma = 0.5$ for a second order method and from eqns (4.3.1) this means that

$$C_1 = \beta + 0.5, \qquad C_2 = 2/3, \qquad C_3 = 7/12.$$

Thus it is impossible to have a third order method with the natural damping ratio $\nu \neq 0$. If $\nu = 0$ we can have third order accuracy with $C_1 = C_3$ which means $\beta = 1/12$. This is incompatible with the conditions for unconditional stability, i.e. $2\beta \geq \gamma \geq 0.5$, thus verifying the Dahlquist theorem in Dahlquist (1963).

θ-method

We can also now finish the analysis begun in Chapter 3 of the θ-method applied to the first order form of the dynamic vibration equation. We have an equivalent two-step algorithm of the form given in eqn (4.2.3). The α_j parameters are the same as in Newmark's equations (3.4.4) and the γ_j and β_j come from eqns (3.4.22) and (3.4.23) to give

$$\alpha_2 = 1, \qquad \alpha_1 = -2, \qquad \alpha_0 = 1$$

$$\gamma_2 = \theta, \qquad \gamma_1 = 1 - 2\theta, \qquad \gamma_0 = \theta - 1$$

$$\beta_2 = \theta^2, \qquad \beta_1 = 2\theta(1-\theta), \qquad \beta_0 = (1-\theta)^2.$$

From eqns (4.2.9) the method is consistent as expected and $A = 1$. From eqns (4.2.10)

$$B_1 = 2\theta, \qquad B_2 = \theta + 0.5, \qquad B_3 = 1.$$

Hence from eqns (4.2.21) and (4.2.22),

if $v \neq 0$ and $\theta \neq 0.5$, then δ and ϵ are each $O(\Delta t)$

if $v \neq 0$ and $\theta = 0.5$, then δ and ϵ are each $O(\Delta t^2)$

if $v = 0$ and $\theta \neq 0.5$, then δ is $O(\Delta t)$ and ϵ is $O(\Delta t^2)$

if $v = 0$ and $\theta = 0.5$, then δ and ϵ are each $O(\Delta t^2)$.

If we look for an explicit method with $\theta = 0$ (corresponding to Newmark with $\beta = 0$), we see from eqn (3.4.29) with μ replaced by $2v\sqrt{(mk)}$, that this method is stable if $\omega \Delta t < 2v$ and without natural damping stability is impossible.

Table 4.3.1 gives a summary of accuracy and stability results for Newmark and the θ-method applied to the dynamic vibration equation.

4.4. Newmark variations

As shown in Strang and Fix (1973), the finite element discretization of a structure gives a very good approximation to the lower frequency modes of vibration but a bad approximation to the higher frequency modes with the errors getting worse as we move up the frequency spectrum. Using the engineer's rule of thumb that there should be at least ten time steps per period, and as with zero natural damping the period is $2\pi/\omega$, this means $\omega \Delta t \leq \pi/5$. The ideal is the best accuracy with $\omega \Delta t \leq \pi/5$ approximately, combined with damping of the inaccurate higher frequency modes as $\omega \Delta t$ tends to infinity. The disadvantage of Newmark's algorithm is that the higher accuracy of the $O(\Delta t^2)$

Table 4.3.1. Accuracy and stability of Newmark and the θ-method.

Method		Cumulative error $\mu=0$ Phase	Cumulative error $\mu=0$ Amplitude	Cumulative error $\mu\neq 0$ Phase	Cumulative error $\mu\neq 0$ Amplitude	Stability
Newmark	$\gamma\neq 0.5$	$O(\Delta t^2)$	$O(\Delta t)$	$O(\Delta t)$	$O(\Delta t)$	$2\beta\geq\gamma\geq 0.5$
	$\gamma=0.5$	$O(\Delta t^2)$	Zero	$O(\Delta t^2)$	$O(\Delta t^2)$	$\beta\geq 0.25$
	$\gamma=0.5$, $\beta=1/12$	$O(\Delta t^4)$	Zero	$O(\Delta t^2)$	$O(\Delta t^2)$	$\Delta t^2<6m/k$
Explicit:	$\gamma=0.5$, $\beta=0$	$O(\Delta t^2)$	Zero	$O(\Delta t^2)$	$O(\Delta t^2)$	$\Delta t^2<4m/k$
θ-method	$\theta\neq 0.5$	$O(\Delta t^2)$	$O(\Delta t)$	$O(\Delta t)$	$O(\Delta t)$	$\theta\geq 0.5$
	$\theta=0.5$	$O(\Delta t^2)$	Zero	$O(\Delta t^2)$	$O(\Delta t^2)$	Stable
Explicit:	$\theta=0$			$O(\Delta t)$	$O(\Delta t)$	$\Delta t<\mu/k$

error is only obtained by taking $\gamma=0.5$ which means that all frequencies are preserved. We can see from the characteristic polynomial in eqn (3.2.19) that with the damping term $\mu=0$ and $\gamma=0.5$ the product of the roots is unity and $\delta=0$. Hence the search for an improvement on Newmark to combine the $O(\Delta t^2)$ error with algorithmic damping as $\omega\Delta t$ tends to infinity has led to several proposals for bringing in an extra parameter. We call these Newmark variations. We now discuss two of these and because they each use the symbol alpha for the extra parameter we distinguish the two alphas by subscripts.

1. Hilber *et al.* (1977) consider the equation (using our notation)

$$\boldsymbol{M}\ddot{\boldsymbol{x}}+\boldsymbol{K}\boldsymbol{x}=\boldsymbol{f} \tag{4.4.1}$$

and apply the standard Newmark equations for $\boldsymbol{x}_{n+1}$ and $\boldsymbol{v}_{n+1}$ as given by eqns (3.2.1) and (3.2.2), but with $\boldsymbol{a}_{n+1}$ obtained from

$$\boldsymbol{M}\boldsymbol{a}_{n+1}+(1+\alpha_{\mathrm{H}})\boldsymbol{K}\boldsymbol{x}_{n+1}-\alpha_{\mathrm{H}}\boldsymbol{K}\boldsymbol{x}_n=\boldsymbol{f}_{n+1} \tag{4.4.2}$$

instead of eqn (3.2.3) as in the original Newmark algorithm. This proposal is then modified in Hughes (1983) to deal with the equation

$$\boldsymbol{M}\ddot{\boldsymbol{x}}+\boldsymbol{C}\dot{\boldsymbol{x}}+\boldsymbol{K}\boldsymbol{x}=\boldsymbol{f} \tag{4.4.3}$$

by obtaining $\boldsymbol{a}_{n+1}$ from the equation

$$\boldsymbol{M}\boldsymbol{a}_{n+1}+(1+\alpha)(\boldsymbol{C}\boldsymbol{v}_{n+1}+\boldsymbol{K}\boldsymbol{x}_{n+1}-\boldsymbol{f}_{n+1}) \\ -\alpha(\boldsymbol{C}\boldsymbol{v}_n+\boldsymbol{K}\boldsymbol{x}_n-\boldsymbol{f}_n)=\boldsymbol{0}, \quad (4.4.4)$$

thus bringing the treatment of the forcing term $\boldsymbol{f}$ into line with that of $\boldsymbol{Kx}$. We note in Chapter 3 that this seems to be the natural development of an algorithm of this type. The algorithm using eqn (4.4.4) is referred to hereafter as the HHT algorithm.

2. Bossak's method is published in Wood *et al.* (1980) and further analysed in Adams and Wood (1983). This method again uses the standard Newmark equations for $\boldsymbol{x}_{n+1}$ and $\boldsymbol{v}_{n+1}$ but now obtains the value of $\boldsymbol{a}_{n+1}$ for the solution of eqn (4.4.3) by substituting into

$$(1-\alpha_{\mathrm{B}})\boldsymbol{M}\boldsymbol{a}_{n+1}+\alpha_{\mathrm{B}}\boldsymbol{M}\boldsymbol{a}_n+\boldsymbol{C}\boldsymbol{v}_{n+1}+\boldsymbol{K}\boldsymbol{x}_{n+1}=\boldsymbol{f}_{n+1}. \quad (4.4.5)$$

For comparison we set out the amplification matrices for these two algorithms applied to the equation

$$\ddot{x}+\omega^2 x=0. \quad (4.4.6)$$

The Bossak algorithm gives the amplification matrix:

$$\boldsymbol{A}_{\mathrm{B}}=\frac{1}{D_{\mathrm{B}}}\begin{bmatrix} 1-\alpha_{\mathrm{B}} & 1-\alpha_{\mathrm{B}} & 0.5-\beta-0.5\alpha_{\mathrm{B}} \\ -\gamma\Omega^2 & 1-\alpha_{\mathrm{B}}+(\beta-\gamma)\Omega^2 & 1-\gamma-\alpha_{\mathrm{B}}+(\beta-0.5\gamma)\Omega^2 \\ -\Omega^2 & -\Omega^2 & -\alpha_{\mathrm{B}}-(0.5-\beta)\Omega^2 \end{bmatrix}, \quad (4.4.7)$$

where $D_{\mathrm{B}}=1-\alpha_{\mathrm{B}}+\beta\Omega^2$, $\Omega=\omega\Delta t$, and the HHT algorithm gives the amplification matrix:

$$\boldsymbol{A}_{\mathrm{H}}=\frac{1}{D_{\mathrm{H}}}\begin{bmatrix} 1+\alpha_{\mathrm{H}}\beta\Omega^2 & 1 & 0.5-\beta \\ -\gamma\Omega^2 & 1-(1+\alpha_{\mathrm{H}})(\gamma-\beta)\Omega^2 & 1-\gamma-(1+\alpha_{\mathrm{H}})(0.5\gamma-\beta)\Omega^2 \\ -\Omega^2 & -(1+\alpha_{\mathrm{H}})\Omega^2 & -(1+\alpha_{\mathrm{H}})(0.5-\beta)\Omega^2 \end{bmatrix}, \quad (4.4.8)$$

where $D_{\mathrm{H}}=1+(1+\alpha_{\mathrm{H}})\beta\Omega^2$.

Comparing eqns (4.4.7) and (4.4.8), we can see that these two algorithms must be closely related for small values of the α parameter and Ω. We can also compare these amplification matrices with that for the standard Newmark expressed as a one-step three-stage algorithm whose entries are given in eqns (3.2.9a). It happens that in each case the useful values of the parameter α are numerically small and Fig. 4.4.1 from Adams and Wood (1983) shows that the results are very close but the HHT algorithm looks slightly better. Also the Bossak algorithm when made second order is only conditionally stable. We concentrate on analysing the HHT algorithm in detail here and drop the subscript on the parameter α in the theory which follows. We want to bring out the effect of the algorithm on preserving the lower frequency modes and damping out the inaccurate higher frequency modes and we also want to observe the effect of natural damping, hence we apply the algorithm to the scalar equation

$$\ddot{x} + 2\nu\omega\dot{x} + \omega^2 x = 0. \tag{4.4.9}$$

For convenience we set out again here the standard Newmark

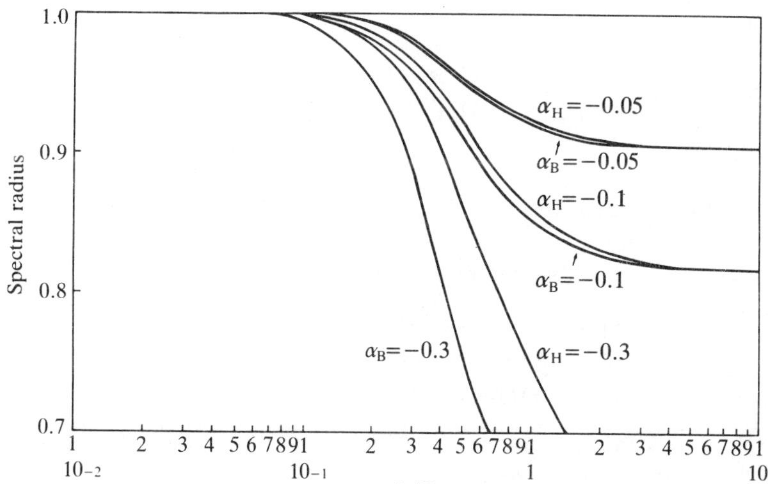

Fig. 4.4.1. Spectral radius of A_B, A_H for $\alpha = -0.05, -0.1, -0.3$ plotted aganst $\Delta t/T$.

equations:

$$x_{n+1} = x_n + \Delta t v_n + \beta \Delta t^2 a_{n+1} + (0.5 - \beta)\Delta t^2 a_n \quad (4.4.10)$$

$$v_{n+1} = v_n + \gamma \Delta t a_{n+1} + (1 - \gamma)\Delta t a_n \quad (4.4.11)$$

and the corresponding HHT algorithm equation with the extra parameter α is given by

$$a_{n+1} + (1+\alpha)(2\nu\omega v_{n+1} + \omega^2 x_{n+1}) - \alpha(2\nu\omega v_n + \omega^2 x_n) = 0. \quad (4.4.12)$$

In order to obtain the characteristic polynomial of the algorithm and hence the corresponding three-step difference scheme in the displacements the simplest method is the one used in Chapter 3: we put $x_{n+1} = rx_n$, $v_{n+1} = rv_n$, and $a_{n+1} = ra_n$ into eqns (4.4.10) to (4.4.12). This gives

$$\begin{aligned} (r-1)x_n - \Delta t v_n + \{-\beta r - (0.5-\beta)\}\Delta t^2 a_n &= 0 \\ (r-1)\Delta t v_n + \{-\gamma r - (1-\gamma)\}\Delta t^2 a_n &= 0 \\ \{(1+\alpha)r - \alpha\}\Omega^2 x_n + 2\nu\Omega\{(1+\alpha)r - \alpha\}\Delta t v_n + r\Delta t^2 a_n &= 0. \end{aligned} \quad (4.4.13)$$

Then the condition that the eqns (4.4.13) should have a non-trivial solution is that the determinant of the system should vanish, i.e.

$$\begin{vmatrix} r-1 & -1 & \{-\beta r - (0.5-\beta)\} \\ 0 & r-1 & \{-\gamma r - (1-\gamma)\} \\ \{r(1+\alpha) - \alpha\}\Omega^2 & 2\nu\Omega\{(1+\alpha)r - \alpha\} & r \end{vmatrix} = 0. \quad (4.4.14)$$

Multiplying out this determinant we have

$$\begin{aligned} &(r-1)[r(r-1) + 2\nu\Omega\{\gamma r + (1-\gamma)\}\{(1+\alpha)r - \alpha\}] \\ &+ \Omega^2\{\gamma r + (1-\gamma)\}\{(1+\alpha)r - \alpha\} \\ &+ \Omega^2(r-1)\{\beta r + (0.5-\beta)\}\{1+\alpha)r - \alpha\} = 0. \end{aligned} \quad (4.4.15)$$

Now we can pick out the coefficients of the powers of r in the characteristic polynomial in eqn (4.4.15):
the coefficient of r^3 is

$$1 + 2\nu\Omega\gamma(1+\alpha) + \Omega^2\beta(1+\alpha) \quad (4.4.16)$$

that of r^2 is

$$-2+2v\Omega(1+\alpha-2\gamma-3\alpha\gamma)$$
$$+\Omega^2\{0.5(1+\alpha)+\gamma-2\beta-\alpha(\gamma-3\beta)\} \quad (4.4.17)$$

that of r is

$$1+2v\Omega\{-1+\gamma+\alpha(-2+3\gamma)\}$$
$$+\Omega^2\{0.5-\gamma+\beta+\alpha(-2\gamma+3\beta)\} \quad (4.4.18)$$

and that of r^0 is

$$2v\Omega\alpha(1-\gamma)+\alpha\Omega^2(-0.5-\beta+\gamma). \quad (4.4.19)$$

(We can check that these coefficients agree with those in eqn (3.2.19) when $\alpha=0$.)

We can now use the fact that the coefficient of r^j in the characteristic polynomial is equal to

$$\alpha_j+2v\Omega\gamma_j+\Omega^2\beta_j$$

to identify the parameters α_j, γ_j, β_j in the HHT algorithm as follows

$$\begin{aligned}
&\alpha_3=1, && \gamma_3=\gamma(1+\alpha), && \beta_3=\beta(1+\alpha)\\
&\alpha_2=-2, && \gamma_2=1-2\gamma+\alpha(1-3\gamma), && \beta_2=0.5+\gamma-2\beta\\
& && && \quad +\alpha(0.5+\gamma-3\beta)\\
&\alpha_1=1, && \gamma_1=\gamma-1+\alpha(3\gamma-2), && \beta_1=0.5-\gamma+\beta\\
& && && \quad +\alpha(3\beta-2\gamma)\\
&\alpha_0=0, && \gamma_0=\alpha(1-\gamma), && \beta_0=\alpha(\gamma-0.5-\beta).
\end{aligned} \quad (4.4.20)$$

Thus we have

$$\sum_{j=0}^{3}\alpha_j=\sum_{j=0}^{3}j\alpha_j=\sum_{j=0}^{3}\gamma_j=0$$

and

$$\tfrac{1}{2}\sum_{j=0}^{3}j^2\alpha_j=\sum_{j=0}^{3}j\gamma_j=\sum_{j=0}^{3}\beta_j=1.$$

Hence the algorithm is consistent for any values of the parameters. In terms of the parameters introduced in §4.2 we have

$A = 1$ and further we have

$$B_1 = \sum_{j=0}^{3} j\beta_j = 1.5 + \alpha + \gamma = B_2 = \sum_{j=0}^{3} j^2\gamma_j/2 \qquad (4.4.21)$$

and

$$B_3 = \tfrac{1}{6}\sum_{j=0}^{3} j^3\alpha_j = 2. \qquad (4.4.22)$$

Since the motivation here is to produce a second order algorithm which will improve upon the standard Newmark result with $2\beta = \gamma = 0.5$, we make $B_1 = B_2 = B_3$ by taking $\gamma = 0.5 - \alpha$. (The corresponding result with the Bossak algorithm is $\gamma = 0.5 - \alpha_B$ for a second order algorithm.) The α_j remain the same and the γ_j and β_j are given by

$$\begin{aligned} \gamma_3 &= 0.5(1-\alpha) - \alpha^2, & \beta_3 &= \beta(1+\alpha) \\ \gamma_2 &= 1.5\alpha + 3\alpha^2, & \beta_2 &= 1 - 2\beta - 3\alpha\beta - \alpha^2 \\ \gamma_1 &= -0.5(1+3\alpha) - 3\alpha^2, & \beta_1 &= \beta + 2\alpha^2 + 3\alpha\beta \\ \gamma_0 &= 0.5\alpha + \alpha^2, & \beta_0 &= -\alpha\beta - \alpha^2. \end{aligned} \qquad (4.4.23)$$

We now have the characteristic polynomial:

$$\sum_{j=0}^{3} (\alpha_j + 2\nu\Omega\gamma_j + \Omega^2\beta_j)r^j = 0, \qquad (4.4.24)$$

where the α_j are given in eqns (4.4.20) and the γ_j and β_j are given in eqns (4.3.23). We can examine the conditions for stability by making the Routh–Hurwitz transformation (Appendix 1) $r = (1+z)/(1-z)$ to give

$$a_0 z^3 + a_1 z^2 + a_2 z + a_3 = 0. \qquad (4.4.25)$$

The transformed equation gives:

$$\begin{aligned} &(\alpha_3 + 2\nu\Omega\gamma_3 + \Omega^2\beta_3)(1+z)^3 \\ &\quad + (\alpha_2 + 2\nu\Omega\gamma_2 + \Omega^2\beta_2)(1+z)(1-z^2) \\ &\quad + (\alpha_1 + 2\nu\Omega\gamma_1 + \Omega^2\beta_1)(1-z)(1-z^2) \\ &\quad + (\alpha_0 + 2\nu\Omega\gamma_0 + \Omega^2\beta_0)(1-z)^3 = 0. \end{aligned} \qquad (4.4.26)$$

Collecting the like powers of z then gives

$$\begin{aligned}
a_0 &= \alpha_3 - \alpha_2 + \alpha_1 - \alpha_0 + 2\nu\Omega(\gamma_3 - \gamma_2 + \gamma_1 - \gamma_0) \\
&\quad + \Omega^2(\beta_3 - \beta_2 + \beta_1 - \beta_0) \\
a_1 &= 3\alpha_3 - \alpha_2 - \alpha_1 + 3\alpha_0 + 2\nu\Omega(3\gamma_3 - \gamma_2 - \gamma_1 + 3\gamma_0) \\
&\quad + \Omega^2(3\beta_3 - \beta_2 - \beta_1 + 3\beta_0) \\
a_2 &= 3\alpha_3 + \alpha_2 - \alpha_1 - 3\alpha_0 + 2\nu\Omega(3\gamma_3 + \gamma_2 - \gamma_1 - 3\gamma_0) \\
&\quad + \Omega^2(3\beta_3 + \beta_2 - \beta_1 - 3\beta_0) \\
a_3 &= \sum_{j=0}^{3} \alpha_j + 2\nu\Omega \sum_{j=0}^{3} \gamma_j + \Omega^2 \sum_{j=0}^{3} \beta_j.
\end{aligned} \tag{4.4.27}$$

We then have the stability conditions from Appendix 1:

$$a_0 > 0, \qquad a_j \geq 0, \qquad j = 1, 2, 3 \text{ and } a_1 a_2 \geq a_0 a_3. \tag{4.4.28}$$

Substituting from eqns (4.4.27), we have

$$4 + 2\nu\Omega(-4\alpha - 8\alpha^2) + \Omega^2(-1 + 4\beta + 8\alpha\beta + 4\alpha^2) > 0 \tag{4.4.29}$$

$$4 + 4\nu\Omega + \Omega^2(4\beta - 1 - 4\alpha^2) \geq 0, \qquad 4\nu\Omega + \Omega^2 \geq 0, \qquad \Omega^2 \geq 0 \tag{4.4.30}$$

and the last of the inequalities (4.4.28) gives

$$\begin{aligned}
&\{4 + 4\nu\Omega + \Omega^2(4\beta - 1 - 4\alpha^2)\}(4\nu\Omega + \Omega^2) \\
&\qquad \geq \Omega^2\{4 - 8\nu\Omega(\alpha + 2\alpha^2) + \Omega^2(4\beta - 1 + 8\alpha\beta + 4\alpha^2)\}.
\end{aligned}$$

Multiplying this out we have

$$-\alpha\Omega^3(\beta + \alpha) + \nu\Omega^2(\alpha + 2\beta) + 2\Omega\nu^2 + 2\nu \geq 0. \tag{4.4.31}$$

Inequalities (4.4.30) are satisfied if $\beta \geq 0.25 + \alpha^2$. From inequality (4.4.29) we see that α must be negative so it is easier to put $\alpha = -q$ while we manoeuvre here, then we have from (4.4.29) that it is satisfied if

$$q(1 - 2q) \geq 0, \quad \text{and} \quad 4\beta - 1 - 8q\beta + 4q^2 > 0, \quad \text{i.e.} \quad 0 \leq q < 0.5 \tag{4.4.32}$$

and

$$\beta > 0.25(1 - 4q^2)/(1 - 2q) = 0.25 + 0.5q. \tag{4.4.33}$$

Substituting $q = -\alpha$ in inequality (4.4.31) gives

$$q\Omega^3(\beta - q) + \nu\Omega^2(2\beta - q) + 2\Omega\nu^2 + 2\nu \geq 0. \tag{4.4.34}$$

Inequality (4.4.34) is then satisfied if

$$\beta > q. \tag{4.4.35}$$

From the above inequalities we conclude that for stability we need

$$0 \geq \alpha > -0.5 \quad \text{and} \quad \beta \geq 0.25 - 0.5\alpha. \tag{4.4.36}$$

Hilber *et al.* (1977) give numerical results for the roots of the characteristic polynomial when $\Omega \rightarrow$ infinity, i.e. for

$$\beta_3 r^3 + \beta_2 r^2 + \beta_1 r + \beta_0 = 0, \tag{4.4.37}$$

with $\beta = (0.5 + \gamma)^2/4$, the value of β which with $\alpha = 0$ guarantees complex roots for the whole range of frequencies. Evidently with $\alpha < 0$ the roots become real as Ω increases so that the algorithm is A-stable but not C-stable by the definitions in Chapter 2. The characteristic polynomial here is a cubic. When there are two complex roots these are approximating the solution of the differential equation; the third root is spurious. Hilber *et al.* conclude that the value of α must be greater than $-\frac{1}{3}$ for the third root in eqn (4.4.37) not to dominate. It is evident from the numerical results in Hilber *et al.* and in Adams and Wood that the values of α of interest are negative and numerically small, i.e. $0 > \alpha > -0.3$. From Fig. 4.4.1 from Adams and Wood (1983) and the numerical results given in Hilber *et al.* (1977) Fig. 1, we can see that we can obtain a value of the spectral radius as $\Omega = \omega \Delta t$ tends to infinity equal to about 0.8, for example by taking $\alpha = -0.1$. But the numerical results in Hilber *et al.* show that it is no good trying to make this spectral radius less than 0.5 because then the spurious root takes over.

With the damping ratio $\nu = 0$ and $B_1 = B_3$ for the second order method as above we can use eqns (4.2.29) and (4.2.30) to give the fractional period increase and the increase in damping for smaller values of Ω. From eqns (4.4.20) we have

$$C_3 = \frac{1}{24} \sum_{j=0}^{3} j^4 \alpha_j = 25/12, \qquad D_3 = \frac{1}{120} \sum_{j=0}^{3} j^5 \alpha_j = 3/2 \tag{4.4.38}$$

and from eqns (4.4.23) we have

$$\begin{aligned} C_1 &= \tfrac{1}{2} \sum_{j=0}^{3} j^2 \beta_j = 2 + \beta - \alpha^2, \\ D_1 &= \tfrac{1}{6} \sum_{j=0}^{3} j^3 \beta_j = 2\beta + \alpha\beta - \alpha^2 + 4/3. \end{aligned} \tag{4.4.39}$$

Hence substituting from eqns (4.4.38) and (4.4.39) into eqn (4.2.29), we have the phase error given by

$$-\epsilon = \frac{\omega\Omega^2}{2}(\beta - \alpha^2 - 1/12) + O(\Delta t^4). \tag{4.4.40}$$

For $\beta > 0.25 - 0.5\alpha$ the bracket on the right of eqn (4.4.40), $\beta - \alpha^2 - 1/12 > 1/6 - 0.5\alpha - \alpha^2$, so that for α small and negative there is always a period increase. This is another example of period increase with unconditional stability. Substituting from eqns (4.4.38) and (4.4.39) into eqn (4.2.31), we have the algorithmic damping given by

$$\delta = \frac{\omega\Omega^3}{4}\{-4(\tfrac{1}{12} - \beta + \alpha^2) + 2(\tfrac{1}{6} - 2\beta - \alpha\beta + \alpha^2)\} + O(\Delta t^5)$$

$$= -\omega\Omega^3\alpha(\beta + \alpha)/2 + O(\Delta t^5). \tag{4.4.41}$$

Thus the algorithmic damping is $O(\Delta t^3)$ here and is positive for $0 > \alpha > -0.3$.

Figures 4.4.1 and 4.4.2 from Adams and Wood (1983) show

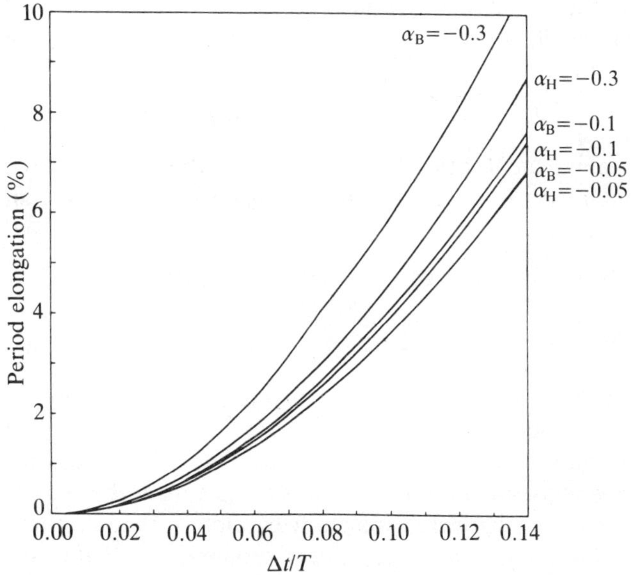

Fig. 4.4.2. Percentage period elongation plotted against $\Delta t/T$ for $\alpha_H = -0.05$, -0.1, -0.3, and $\alpha_B = -0.05$, -0.1, -0.3.

the spectral radius of the amplification matrices for the HHT and Bossak alpha methods (i.e. the modulus of the dominant root of the characteristic polynomial) and the percentage period elongation, plotted against $\Delta t/T$ where $T = 2\pi/\omega$ for the numerical solution of eqn (4.4.9) with zero natural damping.

Aboudi (1976), dealing with two-dimensional wave propagation in a non-linear elastic half-space, takes the equivalent of the central difference algorithm (Chapter 2) given by the Newmark algorithm with $\beta = 0$ and $\gamma = 0.5$ and brings in extra parameters θ_1 and $\theta_2 = 1 - \theta_1$. The corresponding treatment of

$$\ddot{x} + \omega^2 x = 0 \tag{4.4.42}$$

is by the explicit scheme

$$x_{n+1} = 2x_n - x_{n-1} - \Omega^2(\theta_2 x_n + \theta_1 x_{n-1}). \tag{4.4.43}$$

Comparing eqn (4.4.43) with eqn (4.4.2) and remembering that the acceleration term is to be replaced by the central difference expression, the HHT equivalent is

$$x_{n+1} = 2x_n - x_{n-1} - \Omega^2\{(1+\alpha)x_n - \alpha x_{n-1}\}, \tag{4.4.44}$$

so that the Aboudi algorithm corresponds to HHT with $\theta_2 = (1+\alpha)$ and $\theta_1 = -\alpha$. We obtain the stability polynomial for the Aboudi algorithm by putting $x_{n+1} = r x_n$ which produces

$$r^2 + r(-2 + \theta_2\Omega^2) + 1 + \theta_1\Omega^2 = 0. \tag{4.4.45}$$

Then making the Routh–Hurwitz transformation $r = (1+z)/(1-z)$ we obtain $a_0 z^2 + a_1 z + a_2 = 0$ and the Routh–Hurwitz conditions for stability from Appendix 1:

$$a_0 > 0, \qquad a_1 \geq 0, \qquad a_2 \geq 0$$

give

$$\Omega^2(\theta_1 + \theta_2) \geq 0, \qquad 4 - \Omega^2(\theta_2 - \theta_1) > 0, \qquad \theta_1 \leq 0. \tag{4.4.46}$$

The corresponding conditions with the HHT α parameter then are $\alpha > 0$ (whereas with the unconditionally stable scheme above α is negative) and $\Omega^2 < 4/(1+2\alpha)$. Since $\Omega = \omega\Delta t$ this last condition is the bound on the time step size for stability. Aboudi claims that this algorithm makes it possible to handle the steep solutions which appear in non-linear problems. He quotes results obtained with $\theta_1 = -1/9$, $\theta_2 = 1 - \theta_1$, i.e. $\alpha = 1/9$ for (a) propagating shock waves in a half-space where there are two displacement components and (b) propagating circularly polarized waves in a half-space where there are three displacement components.

Looking in more detail at the Aboudi algorithm, with zero natural damping and $\beta=0$, $\gamma=0.5$, eqns (4.4.20) give the relevant parameters as:

$$\begin{aligned} &\alpha_2=1, \quad &\alpha_1=-2, \quad &\alpha_0=1 \\ &\beta_2=0, \quad &\beta_1=1-\theta_1, \quad &\beta_0=\theta_1. \end{aligned} \tag{4.4.47}$$

Hence from eqns (4.2.10) we have

$$B_1=\sum_{j=0}^{2} j\beta_j=1-\theta_1, \quad \text{and} \quad B_3=\sum_{j=0}^{2} j^3\alpha_j/3!=1. \tag{4.4.48}$$

Thus this is not a second order method with $\theta_1 \neq 0$ because the condition $B_1=B_3$ which is necessary for the $O(\Delta t)$ term in eqn (4.2.11) to vanish is not satisfied.

Further from eqns (4.2.15) we have:

$$\begin{aligned} C_1&=\sum_{j=0}^{2} j^2\beta_j/2=(1-\theta_1)/2, \\ C_3&=\sum_{j=0}^{2} j^4\alpha_j/4!=7/12. \end{aligned} \tag{4.4.49}$$

Hence substituting in eqn (4.2.28) we have the phase error for the Aboudi algorithm given by

$$\epsilon=\frac{\omega^3\Delta t^2}{8}\{\theta_1(-\theta_1-4)+4(1/12+\theta_1/2)\},$$

i.e.

$$\epsilon=\frac{\omega^3\Delta t^2}{8}(-\theta_1^2-2\theta_1+1/3). \tag{4.4.50}$$

Since the expression $-\theta_1^2-2\theta_1+1/3$ is greater than zero for small negative values of θ_1 including the value $\theta_1=-1/9$ used by Aboudi we again have a period decrease ($\epsilon>0$, eqn (4.2.18)), with this explicit method. Equation (4.2.26) shows that the algorithmic damping is given by

$$\delta=-\theta_1\omega^2\Delta t/2, \quad \text{to} \quad O(\Delta t) \tag{4.4.51}$$

which again shows that we must have a negative value of θ_1 (corresponding to positive α) for stability.

4.5. Generalized Newmark GNpj

A generalized form of Newmark's method was introduced by Katona in the NUMETA 85 conference January 1985. It was published later that year in Katona and Zienkiewicz (1985) as Part 3 of a series of papers on 'A unified set of single step algorithms'. Parts 1 and 2 of this series (Zienkiewicz *et al.* 1984; Wood 1984b, 1985) describe a family of methods called there SSpj where these symbols stand for Single-Step p-stage methods using the function and derivatives up to and including the $(p-1)$th at each time level with p parameters at choice and to be applied to the numerical solution of differential equations of order j. The SSpj algorithms are obtained by a weighted residual method. They are closely allied to the generalized Newmark methods and will be described in §4.6.

This generalized form of Newmark's method was originally called the 'beta-m' method because it used m parameters $\beta_0, \beta_1, \ldots, \beta_{m-1}$, the particular parameter β being favoured because it was one of those used by Newmark himself. With the notation already in use for the SSpj algorithms, however, it is more convenient here to refer to the generalized Newmark method as the 'beta-p' or GNpj algorithm.

To introduce the generalized Newmark method, we rewrite the Newmark equations (3.2.1) and (3.2.2) now using $\overset{(1)}{\boldsymbol{x}}_n$, $\overset{(2)}{\boldsymbol{x}}_n$ to represent the approximations to $\dot{\boldsymbol{x}}(n\Delta t)$, $\ddot{\boldsymbol{x}}(n\Delta t)$ respectively:

$$\boldsymbol{x}_{n+1} = \boldsymbol{x}_n + \Delta t \overset{(1)}{\boldsymbol{x}}_n + \frac{\Delta t^2}{2} \overset{(2)}{\boldsymbol{x}}_n + \beta_0 \frac{\Delta t^2}{2} \Delta \overset{(2)}{\boldsymbol{x}}_n \tag{4.5.1}$$

and

$$\overset{(1)}{\boldsymbol{x}}_{n+1} = \overset{(1)}{\boldsymbol{x}}_n + \Delta t \overset{(2)}{\boldsymbol{x}}_n + \beta_1 \Delta t \Delta \overset{(2)}{\boldsymbol{x}}_n, \tag{4.5.2}$$

where $\Delta \overset{(2)}{\boldsymbol{x}}_n = \overset{(2)}{\boldsymbol{x}}_{n+1} - \overset{(2)}{\boldsymbol{x}}_n$ and β_0 and β_1 are replacing the usual Newmark parameters 2β and γ respectively. Equations (4.5.1) and (4.5.2) can be regarded as Taylor series expansions with 'remainder' terms

$$\beta_0 \frac{\Delta t^2}{2} \Delta \overset{(2)}{\boldsymbol{x}}_n = \frac{\Delta t^3}{3!} \frac{\mathrm{d}\boldsymbol{x}^3}{\mathrm{d}t^3} (n\Delta t + \xi_1), \tag{4.5.3}$$

$$\beta_1 \Delta t \Delta \overset{(2)}{\boldsymbol{x}}_n = \frac{\Delta t^2}{2!} \frac{\mathrm{d}\boldsymbol{x}^3}{\mathrm{d}t^3} (n\Delta t + \xi_2). \tag{4.5.4}$$

For example, if the displacement function $x(t)$ is actually a fourth degree polynomial:

$$x(t)=a_0+a_1t+a_2t^2+a_3t^3+a_4t^4, \tag{4.5.5}$$

this gives

$$\frac{\mathrm{d}^2x}{\mathrm{d}t^2}=2a_2+6a_3t+12a_4t^2, \qquad \frac{\mathrm{d}^3x}{\mathrm{d}t^3}=6a_3+24a_4t. \tag{4.5.6}$$

Then, putting $t=n\Delta t$, the Taylor series expansion (4.5.1) with the remainder term (4.5.3) is exact if

$$\beta_0\{6a_3\Delta t+12a_4(2t\Delta t+\Delta t^2)\}/2=\Delta t\{6a_3+24a_4(t+\xi_1)\}/6, \tag{4.5.7}$$

which is satisfied by

$$\beta_0=1/3 \quad \text{and} \quad \xi_1=\Delta t/2. \tag{4.5.8}$$

The Taylor series expansion (4.5.2) with remainder term (4.5.4) is exact if

$$\beta_1\Delta t\{6a_3\Delta t+12a_4(2t\Delta t+\Delta t^2)\}=\Delta t^2\{6a_3+24a_4(t+\xi_2)\}/2, \tag{4.5.9}$$

which is satisfied by

$$\beta_1=0.5 \quad \text{and} \quad \xi_2=\Delta t/2. \tag{4.5.10}$$

The standard Newmark algorithm can be expressed as

$$\overset{(q)}{\boldsymbol{x}}_{n+1}=\boldsymbol{a}_{q,n}+b_q\Delta\overset{(2)}{\boldsymbol{x}}_n, \qquad q=0,\,1,\,2, \tag{4.5.11}$$

where the vector

$$\boldsymbol{a}_{q,n}=\sum_{j=q}^{2}\overset{(j)}{\boldsymbol{x}}_n\Delta t^{j-q}/(j-q)! \tag{4.5.12}$$

is known from the values at the beginning of the time step and

$$b_q=\beta_q\Delta t^{2-q}/(2-q)!, \quad \text{with} \quad \beta_2\equiv1. \tag{4.5.13}$$

The quantities given by eqns (4.5.11) are then substituted into the dynamic vibration equation at time $t=(n+1)\Delta t$:

$$\boldsymbol{M}\overset{(2)}{\boldsymbol{x}}_{n+1}+\boldsymbol{C}\overset{(1)}{\boldsymbol{x}}_{n+1}+\boldsymbol{K}\boldsymbol{x}_{n+1}=\boldsymbol{f}_{n+1} \tag{4.5.14}$$

to provide an equation for the increment in the acceleration:

$$(b_2\boldsymbol{M} + b_1\boldsymbol{C} + b_0\boldsymbol{K})\Delta \overset{(2)}{\boldsymbol{x}}_n = \boldsymbol{f}_{n+1} - (\boldsymbol{M}\boldsymbol{a}_{2,n} + \boldsymbol{C}\boldsymbol{a}_{1,n} + \boldsymbol{K}\boldsymbol{a}_{0,n}). \tag{4.5.15}$$

Equation (4.5.15) is solved for the $\Delta \overset{(2)}{\boldsymbol{x}}_n$ and this is then substituted into eqns (4.5.11) in order to update the eqns for $\boldsymbol{x}_{n+1}$ and $\overset{(1)}{\boldsymbol{x}}_{n+1}$ ready for the next time step. This forms the GN22 algorithm. It corresponds to the Newmark method in Chapter 3 considered as a one-step three-stage algorithm.

The Generalized Newmark GNp2 algorithm follows the same pattern by taking

$$\overset{(q)}{\boldsymbol{x}}_{n+1} = \boldsymbol{a}_{q,n} + b_q \Delta \overset{(p)}{\boldsymbol{x}}_n, \qquad q = 0.\ 1, \ldots, p, \tag{4.5.16}$$

where

$$\boldsymbol{a}_{q,n} = \sum_{j=q}^{p} \overset{(j)}{\boldsymbol{x}}_n \Delta t^{j-q}/(j-q)! \tag{4.5.17}$$

and

$$b_q = \beta_q \Delta t^{p-q}/(p-q)!, \qquad \beta_p \equiv 1. \tag{4.5.18}$$

The quantities in eqns (4.5.16) are substituted into eqn (4.5.14) to give the equation for the increment in the pth derivative:

$$(b_2\boldsymbol{M} + b_1\boldsymbol{C} + b_0\boldsymbol{K})\Delta \overset{(p)}{\boldsymbol{x}}_n = \boldsymbol{f}_{n+1} - (\boldsymbol{M}\boldsymbol{a}_{2,n} + \boldsymbol{C}\boldsymbol{a}_{1,n} + \boldsymbol{K}\boldsymbol{a}_{0,n}). \tag{4.5.19}$$

Equation (4.5.19) is solved for the increment $\Delta \overset{(p)}{\boldsymbol{x}}_n$ and this is substituted into eqns (4.5.16) in order to update the values of $\boldsymbol{x}_{n+1}, \overset{(1)}{\boldsymbol{x}}_{n+1}, \ldots, \overset{(p)}{\boldsymbol{x}}_{n+1}$ ready for the next time step. The Generalized Newmark algorithm GNp2 was originally given in Katona and Zienkiewicz (1985) in terms of calculating increments in the pth derivative but from the point of view of the analysis here it is simpler to write the equations in the form:

$$\overset{(q)}{\boldsymbol{x}}_{n+1} = \hat{\boldsymbol{a}}_{q,n} + b_q \overset{(p)}{\boldsymbol{x}}_{n+1}, \qquad q = 0, 1, \ldots, p-1, \tag{4.5.20}$$

where

$$\hat{\boldsymbol{a}}_{q,n} = \boldsymbol{a}_{q,n} - b_q \overset{(p)}{\boldsymbol{x}}_n, \tag{4.5.21}$$

and b_q is given by eqn (4.5.18) as before. Then substituting into eqn (4.5.14) gives an equation now for the new value of the pth derivative:

$$(b_2\boldsymbol{M} + b_1\boldsymbol{C} + b_0\boldsymbol{K})\overset{(p)}{\boldsymbol{x}}_{n+1} = \boldsymbol{f}_{n+1} - (\boldsymbol{M}\hat{\boldsymbol{a}}_{2,n} + \boldsymbol{C}\hat{\boldsymbol{a}}_{1,n} + \boldsymbol{K}\hat{\boldsymbol{a}}_{0,n}). \tag{4.5.22}$$

The new value of the pth derivative is obtained by inverting the same matrix as before in eqn (4.5.19) and the new values of $\overset{(j)}{\boldsymbol{x}}_{n+1}$, $j = 0, 1, \ldots, p-1$ are obtained by substituting for the pth derivative in eqns (4.5.20). This generalized Newmark algorithm has the advantage of all single-step algorithms that it is easy to change the time step size. If $b_0 = 0$, i.e. $\beta_0 = 0$ and the matrices $\boldsymbol{M}$ and $\boldsymbol{C}$ are lumped then eqn (4.5.22) gives a simple explicit algorithm but this will be only conditionally stable. The GN*pj* algorithm is closely allied to the SS*pj* family of algorithms discussed in the next section, hence its accuracy and stability properties will be included there.

To simplify the algebra in the analysis and comparison of the GN*pj* and SS*pj* algorithms we apply them to the scalar equation

$$m\ddot{x} + \mu\dot{x} + kx = 0, \tag{4.5.23}$$

which can be regarded as having been obtained from the modal decomposition of the homogeneous form of the dynamic vibration equation as discussed in Chapter 1. The equation corresponding to (4.5.22) is then

$$(b_2 m + b_1\mu + b_0 k)\overset{(p)}{x}_{n+1} = -(m\hat{a}_{2,n} + \mu\hat{a}_{1,n} + k\hat{a}_{0,n}). \tag{4.5.24}$$

We can then write the system of $(p+1)$ equations formed from the p equations (4.5.20) together with (4.5.24) for the GNp2 algorithm applied to the numerical solution of eqn (4.5.23) as

$$\boldsymbol{X}_{n+1}^{(G)} = \boldsymbol{A}^{(G)}\boldsymbol{X}_n^{(G)}, \tag{4.5.25}$$

where the vector

$$\boldsymbol{X}_n^{(G)} = \left[x_n,\ \Delta t\overset{(1)}{x}_n, \ldots,\ \Delta t^p\overset{(p)}{x}_n\right]^{\mathrm{T}} \tag{4.5.26}$$

and the amplification matrix $\boldsymbol{A}^{(G)}$ is of order $p+1$.

We can now form the linear combination:

$$\sum_{j=0}^{p+1} q_j \boldsymbol{X}_{n+j} = \left(\sum_{j=0}^{p+1} q_j \boldsymbol{A}^{(G)^j}\right)\boldsymbol{X}_n. \tag{4.5.27}$$

Then if the q_j in eqn (4.5.27) are the coefficients in the characteristic polynomial of the matrix $\boldsymbol{A}^{(G)}$ we have

$$\sum_{j=0}^{p+1} q_j \boldsymbol{X}_{n+j} = \boldsymbol{0} \tag{4.5.28}$$

by the Cayley–Hamilton theorem (the matrix satisfies its own characteristic polynomial). Thus $x_n, \overset{(1)}{x}_n, \ldots, \overset{(p)}{x}_n$ all satisfy the same recurrence relation. It makes no difference if $\boldsymbol{X}_n$ is written as $\left[x_n, \overset{(1)}{x}_n, \ldots, \overset{(p)}{x}_n\right]$ or as in eqn (4.5.26) since the powers of Δt will cancel. Hence, given sufficiently accurate starting values, $x_n, \overset{(1)}{x}_n, \ldots, \overset{(p)}{x}_n$ are all obtained to the same order of accuracy. This result applies to any consistent one-step p-stage algorithm; it applies also to the SS*pj* methods introduced in the next section and to the even more general algorithms of Thomas and Gladwell (1988), discussed in §4.10.

As we saw with the standard Newmark algorithm given by $p = 2$, the amplification matrix $\boldsymbol{A}^{(G)}$ must have a zero eigenvalue so that the corresponding multistep form is a p-step, not a $(p+1)$-step.

We can see that the amplification matrix $\boldsymbol{A}^{(G)}$ in the GN*p*2 algorithm is a generalization of the GN22 amplification matrix given in eqns (3.2.9a). The entries in the matrix are linear combinations of the terms m, $\Delta t\mu$, and $\Delta t^2 k$. The coefficients in the characteristic polynomial of the amplification matrix which are the coefficients of the corresponding p-step recurrence will also be linear combinations of these terms. This is also true of the SS*pj* family of algorithms introduced in Zienkiewicz *et al.* (1984) and we shall show in the next section that there is a linear relationship between these two sets of algorithms.

4.6. The SS*pj* family of algorithms

The SS*pj* family of algorithms for the numerical solution of the second order equation

$$\boldsymbol{M}\ddot{\boldsymbol{x}} + \boldsymbol{C}\dot{\boldsymbol{x}} + \boldsymbol{K}\boldsymbol{x} = \boldsymbol{f}, \tag{4.6.1}$$

or equally well for the first order equation

$$\boldsymbol{C}\dot{\boldsymbol{x}} + \boldsymbol{K}\boldsymbol{x} = \boldsymbol{f} \tag{4.6.2}$$

by putting $\boldsymbol{M} \equiv 0$ in the algorithm, was inspired by supposing that the unknown function $x(t)$ is treated like the finite element approximation in space discretization by being represented in the time step by a polynomial. Putting $\tau = t - n\Delta t$, we suppose that in the time step $n\Delta t \le t \le (n+1)\Delta t$, i.e. $0 \le \tau \le \Delta t$, the function $\boldsymbol{x}(t)$ is represented by

$$\boldsymbol{x}(\tau) = \boldsymbol{x}_n + \tau \overset{(1)}{\boldsymbol{x}}_n + \tau^2 \overset{(2)}{\boldsymbol{x}}_n/2 + \cdots \boldsymbol{\alpha}_n^{(p)}\tau^p/p!, \tag{4.6.3}$$

where we are writing $\overset{(j)}{\boldsymbol{x}}_n$ to represent $\mathrm{d}^j x_n/\mathrm{d}t^j$ as in §4.5. We assume that we know $\boldsymbol{x}_n, \overset{(1)}{\boldsymbol{x}}_n, \ldots, \overset{(p-1)}{\boldsymbol{x}}_n$ at the start of the time step and we have to calculate the unknown vector $\boldsymbol{\alpha}_n^{(p)}$ from which the new values $\boldsymbol{x}_{n+1}, \overset{(1)}{\boldsymbol{x}}_{n+1}, \ldots, \overset{(p-1)}{\boldsymbol{x}}_{n+1}$ are to be determined by substituting $\tau = \Delta t$ in eqn (4.6.3) and its derivatives:

$$\begin{aligned}
\boldsymbol{x}_{n+1} &= \sum_{q=0}^{p-1} \overset{(q)}{\boldsymbol{x}}_n \frac{\Delta t^q}{q!} + \boldsymbol{\alpha}_n^{(p)} \frac{\Delta t^p}{p!} = \hat{\boldsymbol{x}}_{n+1} + \boldsymbol{\alpha}_n^{(p)} \frac{\Delta t^p}{p!} \\
\overset{(1)}{\boldsymbol{x}}_{n+1} &= \sum_{q=1}^{p-1} \overset{(p)}{\boldsymbol{x}}_n \frac{\Delta t^{q-1}}{(q-1)!} + \boldsymbol{\alpha}_n^{(p)} \frac{\Delta t^{p-1}}{(p-1)!} \\
&= \overset{(\hat{1})}{\boldsymbol{x}}_{n+1} + \boldsymbol{\alpha}_n^{(p)} \frac{\Delta t^{p-1}}{(p-1)!}
\end{aligned} \tag{4.6.4}$$

etc.

The vector $\boldsymbol{\alpha}_n^{(p)}$ is determined by substituting for $\boldsymbol{x}(\tau)$ and its derivatives as given by the approximation (4.6.3) into the weighted residual equation:

$$\int_0^{\Delta t} W(\tau)(\boldsymbol{M}\ddot{\boldsymbol{x}} + \boldsymbol{C}\dot{\boldsymbol{x}} + \boldsymbol{K}\boldsymbol{x} - \boldsymbol{f})\,\mathrm{d}\tau = \boldsymbol{0}. \tag{4.6.5}$$

The $\boldsymbol{\alpha}_n^{(p)}$ may be regarded as an estimate of the pth derivative averaged over the time step.

We define a set of $p+1$ parameters θ_q, $q = 0, 1, \ldots, p$ by

$$\theta_0 = 1 \quad \text{and} \quad \theta_q = \left(\int_0^{\Delta t} W(\tau)\tau^q\,\mathrm{d}\tau\right) \Big/ \left(\Delta t^q \int_0^{\Delta t} W(\tau)\,\mathrm{d}\tau\right). \tag{4.6.6}$$

Then from eqn (4.6.3) we have, putting $(\int_0^{\Delta t} W(\tau)\,\mathrm{d}\tau)^{-1} = R$,

$$R\int_0^{\Delta t} W(\tau)\boldsymbol{x}\,\mathrm{d}\tau = \sum_{q=0}^{p-1} \overset{(q)}{\boldsymbol{x}}_n \frac{\Delta t^q}{q!}\,\theta_q + \boldsymbol{\alpha}_n^{(p)} \frac{\Delta t^p}{p!}\,\theta_p,$$

$$R\int_0^{\Delta t} W(\tau)\dot{\boldsymbol{x}}\,\mathrm{d}\tau = \sum_{q=0}^{p-1} \overset{(q)}{\boldsymbol{x}}_n \frac{\Delta t^{q-1}}{(q-1)!}\,\theta_{q-1} + \boldsymbol{\alpha}_n^{(p)} \frac{\Delta t^{p-1}}{(p-1)!}\,\theta_{p-1},$$

$$R\int_0^{\Delta t} W(\tau)\ddot{\boldsymbol{x}}\,\mathrm{d}\tau = \sum_{q=2}^{p-1} \overset{(q)}{\boldsymbol{x}}_n \frac{\Delta t^{q-2}}{(q-2)!}\,\theta_{q-2} + \boldsymbol{\alpha}_n^{(p)} \frac{\Delta t^{p-2}}{(p-2)!}\,\theta_{p-2},$$

$$R\int_0^{\Delta t} W(\tau)\boldsymbol{f}\,\mathrm{d}\tau = \bar{\boldsymbol{f}}. \tag{4.6.7}$$

Substituting into eqn (4.6.5) and cancelling the factor R we have

$$\boldsymbol{M}\left(\tilde{\ddot{\boldsymbol{x}}}_{n+1} + \boldsymbol{\alpha}_n^{(p)} \frac{\Delta t^{p-2}}{(p-2)!}\,\theta_{p-2}\right) + \boldsymbol{C}\left(\tilde{\dot{\boldsymbol{x}}}_{n+1} + \boldsymbol{\alpha}_n^{(p)} \frac{\Delta t^{p-1}}{(p-1)!}\,\theta_{p-1}\right) + \boldsymbol{K}\left(\tilde{\boldsymbol{x}}_{n+1} + \boldsymbol{\alpha}_n^{(p)} \frac{\Delta t^{p}}{p!}\,\theta_{p}\right) - \bar{\boldsymbol{f}} = \boldsymbol{0} \tag{4.6.8}$$

where the quantities

$$\begin{aligned}\tilde{\boldsymbol{x}}_{n+1} &= \sum_{q=0}^{p-1} \overset{(q)}{\boldsymbol{x}}_n \Delta t^q \theta_q / q!\\ \tilde{\dot{\boldsymbol{x}}}_{n+1} &= \sum_{q=1}^{p-1} \overset{(q)}{\boldsymbol{x}}_n \Delta t^{q-1} \theta_{q-1} / (q-1)!\\ \tilde{\ddot{\boldsymbol{x}}}_{n+1} &= \sum_{q=2}^{p-1} \overset{(q)}{\boldsymbol{x}}_n \Delta t^{q-2} \theta_{q-2} / (q-2)!\end{aligned} \tag{4.6.9}$$

are known from the start of the time step and can be regarded as predictor values. The value of $\boldsymbol{\alpha}_n^{(p)}$ is then found from the equation:

$$\left\{\frac{\Delta t^{p-2}}{(p-2)!}\theta_{p-2}\boldsymbol{M}+\frac{\Delta t^{p-1}}{(p-1)!}\theta_{p-1}\boldsymbol{C}+\frac{\Delta t^{p}}{p!}\theta_{p}\boldsymbol{K}\right\}\boldsymbol{\alpha}_n^{(p)}$$
$$=\bar{\boldsymbol{f}}-\boldsymbol{M}\ddot{\tilde{\boldsymbol{x}}}_{n+1}-\boldsymbol{C}\dot{\tilde{\boldsymbol{x}}}_{n+1}-\boldsymbol{K}\tilde{\boldsymbol{x}}_{n+1}. \quad (4.6.10)$$

The steps of the algorithm are very similar to those for Newmark, i.e.

1. Evaluate the predictor values $\tilde{\boldsymbol{x}}_{n+1}, \dot{\tilde{\boldsymbol{x}}}_{n+1}, \ddot{\tilde{\boldsymbol{x}}}_{n+1}$ from eqns (4.6.9) using the known values at the start of the time step.
2. Calculate $\boldsymbol{\alpha}_n^{(p)}$ from eqn (4.6.10).
3. Substitute for $\boldsymbol{\alpha}_n^{(p)}$ into eqns (4.6.4) to obtain the corrector values of $\boldsymbol{x}_{n+1}, \dot{\boldsymbol{x}}_{n+1}$, etc.

   Repeat for next time step.

It is of course vital to have sufficiently accurate starting values for the first time step. We expect with the second order differential equation (4.6.1) to have given values for the displacement $\boldsymbol{x}(0)$ and the initial velocity $\dot{\boldsymbol{x}}(0)$. It is usual to suppose that any higher derivatives required with $p>2$ can be obtained from the differential equation and its derivatives. In Zienkiewicz *et al.* (1984) it is suggested that it is always possible to set the initial displacement, velocity, and any other derivatives required to start the algorithm, equal to zero before the imposition of the forcing function $\boldsymbol{f}(t)$. This procedure prevents the initial spurious 'kick' in the numerical solution also known as 'overshoot' which will be discussed in §4.9, but its use depends on whether it is appropriate for the particular physical problem being solved.

As with the GN*pj* family of algorithms, the SS*pj* algorithms are automatically consistent because of the way they are constructed. The accuracy and stability properties depend on the values chosen for the parameters $\theta_q, q=1, 2,\ldots, p$.

Enright (1986) recommends a type of error control, which he uses with Runge–Kutta algorithms, depending on constructing an interpolating polynomial and requiring a measure of the residuals at points in the time step to be less than some tolerance. The SS*pj* algorithms have the polynomials all ready for this form of error control but the author of this book does not know of any experiments with this.

Before studying some particular examples of SSpj algorithms which include many well-known engineering algorithms developed separately, we look in the next section at the relationship between the SSpj and the GNpj families.

4.7. Matching the SSpj and *GNpj* algorithms (Wood 1985)

In §(4.5) we expressed the GNpj algorithm when applied to the homogeneous equation

$$m\ddot{x} + \mu\dot{x} + kx = 0 \tag{4.7.1}$$

in the form

$$\boldsymbol{X}_{n+1}^{(G)} = \boldsymbol{A}^{(G)}\boldsymbol{X}_n^{(G)} \tag{4.7.2}$$

where

$$\boldsymbol{X}_n^{(G)} = \left[x_n,\ \Delta t\dot{x}_n, \ldots,\ \Delta t_n^p \overset{(p)}{\boldsymbol{x}_n}\right]^{\mathrm{T}} \tag{4.7.3}$$

and the amplification matrix $\boldsymbol{A}^{(G)}$ is of order $(p+1)$. We note that the GNpj algorithm has the disadvantage when compared with the SSpj algorithm of having to hold the quantity $\overset{(p)}{\boldsymbol{x}_n}$ from one time step to the next. The SSpj algorithm naturally lends itself to an expression like eqn (4.7.2) but with an amplification matrix of order p. For the application of the SSpj algorithm to the numerical solution of eqn (4.7.1) this system of p equations is formed from the eqns (4.6.4) for x_{n+1}, $\overset{(1)}{x}_{n+1}, \ldots,\ \overset{(p-1)}{x}_{n+1}$ with the value of $\alpha_n^{(p)}$ having been substituted from

$$\left\{\frac{\Delta t^{p-2}}{(p-2)!}\,\theta_{p-2}m + \frac{\Delta t^{p-1}}{(p-1)!}\,\theta_{p-1}\mu + \frac{\Delta t^p}{p!}\,\theta_p k\right\}\alpha_n^{(p)}$$
$$= -m\tilde{\ddot{x}}_{n+1} - \mu\tilde{\dot{x}}_{n+1} - k\tilde{x}_{n+1} \tag{4.7.4}$$

and eqns (4.6.8).

In order to match the two algorithms, we write the SSpj equations in a different form from that corresponding to eqn (4.7.2). We first write eqn (4.6.10), substituting from eqns (4.6.9), and then eqns (4.6.4) putting $x_{n+1} = rx_n$, $\overset{(1)}{x}_{n+1} =$

$r\overset{(1)}{x}_n, \ldots, \overset{(p-1)}{x}_{n+1} = r\overset{(p-1)}{x}_n$, which gives the system of equations:

$$\boldsymbol{S}\boldsymbol{Y}_n = \boldsymbol{0}, \tag{4.7.5}$$

where

$$\boldsymbol{Y}_n^{\mathrm{T}} = \left[x_n, \Delta t\overset{(1)}{x}_n, \ldots, \Delta t^{p-1}\overset{(p-1)}{x}_n, \Delta t^p \alpha_n^p\right]$$

and $\boldsymbol{S}$ is the $(p+1)$ order matrix:

$$\boldsymbol{S} = \begin{bmatrix} b_0 & b_1 & b_2 & . & b_{p-1} & b_p \\ 1-r & 1 & 1/2 & . & \dfrac{1}{(p-1)!} & \dfrac{1}{p!} \\ 0 & 1-r & 1 & . & . & . \\ . & . & . & . & . & . \\ 0 & . & . & 1-r & 1 & 1/2! \\ 0 & 0 & . & 0 & 1-r & 1 \end{bmatrix}, \tag{4.7.6}$$

where $b_0 = \theta_0 k \Delta t^2$, $b_1 = \theta_0 \mu \Delta t + \theta_1 k \Delta t^2$, and

$$b_q = \frac{m}{(q-2)!}\theta_{p-2} + \frac{\mu \Delta t}{(q-1)!}\theta_{q-1} + \frac{k\Delta t^2}{q!}\theta_q, \qquad q = 2, 3, \ldots, p \tag{4.7.7}$$

with $\theta_0 \equiv 1$.

We next take the GN*pj* algorithm given by eqn (4.5.19) together with eqns (4.5.16). To simplify the algebra in the process of matching the SS*pj* and GN*pj* algorithms we replace the beta parameters in the GN*pj* algorithm, $\beta_0, \beta_1, \ldots, \beta_p$ by $\gamma_p, \gamma_{p-1}, \ldots, \gamma_1, \gamma_0$, respectively, with $\gamma_0 = 1$ corresponding to $\beta_p = 1$ (gamma is the other Newmark parameter). We next put

$$x_{n+1} = rx_n, \qquad \overset{(1)}{x}_{n+1} = r\overset{(1)}{x}_n, \ldots, \overset{(p)}{x}_{n+1} = r\overset{(p)}{x}_n$$

and this gives the system of homogeneous equations:

$$\boldsymbol{G}\boldsymbol{X}_n^{(G)} = \boldsymbol{0}, \tag{4.7.8}$$

where the vector $\boldsymbol{X}_n^{(G)}$ is the same as in eqn (4.7.3) and the matrix $\boldsymbol{G}$ is given by

$$\boldsymbol{G} =$$

$$\begin{bmatrix} a_0 & a_1 & a_2 & \cdot & a_{p-1} & a_p \\ 1-r & 1 & 1/2 & \cdot & \dfrac{1}{(p-1)!} & \dfrac{1}{p!}\{1+\gamma_p(r-1)\} \\ 0 & 1-r & 1 & \cdot & \dfrac{1}{(p-2)!} & \dfrac{1}{(p-1)!}\{1+\gamma_{p-1}(r-1)\} \\ \cdot & \cdot & \cdot & \cdot & \cdot & \cdot \\ 0 & 0 & \cdot & 1-r & 1 & \dfrac{1}{2!}\{1+\gamma_2(r-1)\} \\ 0 & 0 & \cdot & 0 & 1-r & 1+\gamma_1(r-1) \end{bmatrix},$$

where $a_0 = k\Delta t^2$, $a_1 = \mu\Delta t + k\Delta t^2$,

$$a_q = \frac{m}{(q-2)!} + \frac{\mu\Delta t}{(q-1)!} + \frac{k\Delta t^2}{q!}, \qquad q = 2, 3, \dots, (p-1)$$

and

$$\begin{aligned} a_p = {} & \frac{m}{(p-2)!}\{1+\gamma_{p-2}(r-1)\} \\ & + \frac{\mu\Delta t}{(p-1)!}\{1+\gamma_{p-1}(r-1)\} \\ & + \frac{k\Delta t^2}{p!}\{1+\gamma_p(r-1)\}. \end{aligned} \tag{4.7.9}$$

Now, since the system of eqns (4.7.6) only gives a non-trivial solution for $\boldsymbol{Y}_n$ if the determinant of the matrix $\boldsymbol{S}$, written $\det \boldsymbol{S}$, is zero, the equation in r given by this condition must be the stability polynomial of the SS*pj* algorithm. The condition $\det \boldsymbol{G} = 0$ similarly gives the stability polynomial of the GN*pj* algorithm. We express $\det \boldsymbol{S}$ in the form:

$$\det \boldsymbol{S} \equiv mS_1 + \mu\Delta t S_2 + k\Delta t^2 S_3. \tag{4.7.10}$$

Then from eqns (4.7.6) and (4.7.7) we have

$$S_3 = \begin{vmatrix} 1 & \theta_1 & \dfrac{\theta_2}{2!} & . & . & \dfrac{\theta_{p-1}}{(p-1)!} & \dfrac{\theta_p}{p!} \\ 1-r & 1 & \dfrac{1}{2!} & . & . & \dfrac{1}{(p-1)!} & \dfrac{1}{p!} \\ 0 & 1-r & 1 & . & . & . & . \\ 0 & 0 & . & . & 1-r & 1 & \dfrac{1}{2!} \\ 0 & 0 & . & . & 0 & 1-r & 1 \end{vmatrix}. \qquad (4.7.11)$$

We now make the operation:

row 2 replaced by row 2 + $(r-1)$ times row 1

which leaves the value of the determinant S_3 unaltered and gives unity in the leading position as the only non-zero entry in the first column. Thus we have $S_3 = \Delta_p$ where Δ_p is the pth order determinant given by:

$$\Delta_p = \begin{vmatrix} c_1 & c_2 & . & . & c_{p-1} & c_p \\ 1-r & 1 & . & . & \dfrac{1}{(p-2)!} & \dfrac{1}{(p-1)!} \\ 0 & 1-r & . & . & . & . \\ . & . & . & . & . & . \\ 0 & 0 & . & 1-r & 1 & \dfrac{1}{2!} \\ 0 & 0 & . & 0 & 1-r & 1 \end{vmatrix}, \qquad (4.7.12)$$

where

$$c_q = \frac{1}{q!}\{1 + \theta_q(r-1)\}, \qquad q = 1, 2, \ldots, p. \qquad (4.7.13)$$

We now express the determinant of the matrix $\boldsymbol{G}$ in eqn (4.7.9) in the form:

$$\det \boldsymbol{G} \equiv mG_1 + \mu\Delta t G_2 + k\Delta t^2 G_3. \qquad (4.7.14)$$

We thus have:

$$G_3 = \begin{vmatrix} 1 & 1 & \frac{1}{2!} & \cdot & \cdot & \frac{1}{(p-1)!} & \frac{1}{p!}\{1+\gamma_p(r-1)\} \\ 1-r & 1 & \frac{1}{2!} & \cdot & \cdot & \frac{1}{(p-1)!} & \frac{1}{p!}\{1+\gamma_p(r-1)\} \\ 0 & 1-r & 1 & \cdot & \cdot & \frac{1}{(p-2)!} & \frac{1}{(p-1)!}\{1+\gamma_{p-1}(r-1)\} \\ \cdot & \cdot & \cdot & \cdot & \cdot & \cdot & \cdot \\ 0 & 0 & \cdot & \cdot & 1-r & 1 & \frac{1}{2!}\{1+\gamma_2(r-1)\} \\ 0 & 0 & \cdot & \cdot & 0 & 1-r & 1+\gamma_1(r-1) \end{vmatrix}. \tag{4.7.15}$$

Subtracting row 2 from row 1 in the determinant G_3 leaves only one non-zero entry in the first column and hence we have:

$$G_3 = r\begin{vmatrix} 1 & \frac{1}{2!} & \cdot & \cdot & \frac{1}{(p-1)!} & \frac{1}{p!}\{1+\gamma_p(r-1)\} \\ 1-r & 1 & \cdot & \cdot & \frac{1}{(p-2)!} & \frac{1}{(p-1)!}\{1+\gamma_{p-1}(r-1)\} \\ 0 & 1-r & \cdot & \cdot & \cdot & \cdot \\ \cdot & \cdot & \cdot & \cdot & \cdot & \cdot \\ 0 & 0 & \cdot & 1-r & 1 & \frac{1}{2!}\{1+\gamma_2(r-1)\} \\ 0 & 0 & \cdot & 0 & 1-r & 1+\gamma_1(r-1) \end{vmatrix} \tag{4.7.16}$$

Comparing eqns (4.7.12) and (4.7.16) we can see that the two determinants will multiply out to give the same result hence we have:

$$G_3 = r\Delta_p, \tag{4.7.17}$$

where Δ_p is given by eqn (4.7.12), if we take

$$\theta_j = \gamma_j, \qquad j = 1, 2, \ldots, p. \tag{4.7.18}$$

We next separate out the coefficient of $\mu\Delta t$ in det S, i.e. from eqn (4.7.6) we have:

$$S_2 = \begin{vmatrix} 0 & 1 & \theta_1 & \cdot & \cdot & \cdot & \dfrac{\theta_{p-1}}{(p-1)!} \\ 1-r & 1 & \dfrac{1}{2!} & \cdot & \cdot & \cdot & \dfrac{1}{p!} \\ 0 & 1-r & 1 & \cdot & \cdot & \cdot & \dfrac{1}{(p-1)!} \\ \cdot & \cdot & \cdot & \cdot & \cdot & \cdot & \cdot \\ 0 & 0 & \cdot & \cdot & 1-r & 1 & \dfrac{1}{2!} \\ 0 & 0 & \cdot & \cdot & 0 & 1-r & 1 \end{vmatrix}. \tag{4.7.19}$$

There is only one non-zero term in the first column of the determinant in eqn (4.7.19) and hence we have

$$S_2 = (r-1)\begin{vmatrix} 1 & \theta_1 & \dfrac{\theta_2}{2!} & \cdot & \cdot & \dfrac{\theta_{p-1}}{(p-1)!} \\ 1-r & 1 & \dfrac{1}{2!} & \cdot & \cdot & \dfrac{1}{(p-1)!} \\ 0 & 1-r & 1 & \cdot & \cdot & \cdot \\ \cdot & \cdot & \cdot & \cdot & \cdot & \cdot \\ 0 & 0 & \cdot & 1-r & 1 & \dfrac{1}{2!} \\ 0 & 0 & \cdot & 0 & 1-r & 1 \end{vmatrix}.$$

Then, again replacing

row 2 by row 2 + $(r-1)$ times row 1

we have

$$S_2 = (r-1)\Delta_{p-1} \tag{4.7.20}$$

from eqn (4.7.12).

Performing operations on the determinant G_2 obtained from the coefficient of $\mu\Delta t$ in eqn (4.7.9) similar to the operations used on G_3 gives

$$G_2 = r(r-1)\Delta_{p-1} \tag{4.7.21}$$

with conditions (4.7.18).

Similarly we have

$$S_1 = (r-1)^2\Delta_{p-2} \tag{4.7.22}$$

and

$$G_1 = r(r-1)^2\Delta_{p-2} \tag{4.7.23}$$

again with conditions (4.7.18).

Hence, apart from the trivial factor r in the GNpj polynomial, the SSpj and GNpj algorithms have the same stability polynomial provided we identify the parameters in the Katona and Zienkiewicz (1985) 'beta-m' notation as:

$$\theta_0 = 1 = \beta_p, \qquad \theta_j = \beta_{p-j}, \qquad j = 1, 2, \ldots, p. \tag{4.7.24}$$

The extra factor r in the GNpj polynomial is there because this algorithm requires the approximate value of the pth derivative to be retained from one time step to the next as can be seen from eqn (4.5.18). With conditions (4.7.24) the two algorithms then have equivalent stability conditions and correspond to the same p-step method. We can see this by considering the general p-step algorithm

$$\sum_{j=0}^{p} (m\alpha_j + \mu\Delta t\gamma_j + k\Delta t^2\beta_j)x_{n+j} = 0 \tag{4.7.25}$$

as applied to the numerical solution of eqn (4.7.1), using the notation as in §4.2. The corresponding stability polynomial is

$$\sum_{j=0}^{p} (m\alpha_j + \mu\Delta t\gamma_j + k\Delta t^2\beta_j)r^j = 0. \tag{4.7.26}$$

Comparing the equation giving the stability polynomial of SS*p*2, i.e. from eqn (4.7.10),

$$mS_1 + \mu\Delta t S_2 + k\Delta t^2 S_3 = 0 \tag{4.7.27}$$

with eqn (4.7.26) we see that we can equate coefficients to give

$$S_1 = \sum_{j=0}^{p} \alpha_j r^j, \qquad S_2 = \sum_{j=0}^{p} \gamma_j r^j, \qquad S_3 = \sum_{j=0}^{p} \beta_j r^j. \tag{4.7.28}$$

Thus the coefficients of the powers of r in the expansions of the determinants S_1, S_2, S_3 give the coefficients in the p-step algorithm corresponding to SS*p*2. Since we have shown above that with the conditions (4.7.24) GN*p*2 has the same stability polynomial (apart from the trivial r factor) as SS*p*2, we can now see that the two single-step algorithms correspond to the same p-step algorithm.

By setting $m = 0$ we obtain the corresponding results for SS*p*1 and GN*p*1 applied to the first order equation

$$\mu\dot{x} + kx = 0$$

and this will be discussed later in this chapter.

With the close relation between the SS*pj* and GN*pj* families of algorithms demonstrated above, it is not surprising that the papers Zienkiewicz *et al.* (1984) and Katona and Zienkiewicz (1985) each claim the inclusion of the same examples of well-known algorithms derived previously. We shall discuss these in the next section but first we make some general observations:

1. The idea of forming an algorithm for the numerical solution of the dynamic vibration equation by inserting a polynomial approximation for the displacement into a weighted residual equation was originally put forward by Zienkiewicz (1976) in a paper entitled 'A new look at Newmark, Houbolt and other time-stepping formulae. A weighted residual approach'. The method is used there to derive (a) a two-step and (b) a three-step recurrence in the displacements only. The polynomial approximation in each case is expressed in terms of Lagrange 'shape functions' like those used in the finite element spatial discretization. For example, for the two-step recurrence, it is supposed that in the interval

$$-1 \leq \xi = \tau/\Delta t \leq 1, \tag{4.7.29}$$

we have

$$x = \sum_{i=n-1}^{n+1} N_i x_i, \tag{4.7.30}$$

where

$$\begin{aligned} N_{n+1} &= \xi(1+\xi)/2, \quad \Delta t\dot{N}_{n+1} = (1+2\xi)/2, \quad \Delta t^2\ddot{N}_{n+1} = 1, \\ N_n &= (1-\xi^2), \quad \Delta t\dot{N}_n = -2\xi, \quad \Delta t^2\ddot{N}_n = -2 \\ N_{n-1} &= -\xi(1-\xi)/2, \quad \Delta t\dot{N}_{n-1} = (-1+2\xi)/2, \quad \Delta t^2\ddot{N}_{n-1} = 1 \cdots . \end{aligned} \tag{4.7.31}$$

These are substituted into the weighted residual equation

$$\int_{-1}^{1} W(\xi) \sum_{i=n-1}^{n+1} (m\ddot{N}_i + \mu\dot{N}_i + kN_i)x_i \, \mathrm{d}\xi = \int_{-1}^{1} \sum_{i=n-1}^{n+1} f_i N_i \, \mathrm{d}\xi, \tag{4.7.32}$$

where $W(\xi)$ is the 'weight', and supposing that the forcing function $f(t)$ is interpolated from its 'nodal' values in the same ways as the displacement. The parameters referred to in Katona and Zienkiewicz (1985) Table IV as Z_j are then given by

$$Z_j = \int_{-1}^{1} W(\xi)\xi^j \, \mathrm{d}\xi \Big/ \int_{-1}^{1} W(\xi) \, \mathrm{d}\xi. \tag{4.7.33}$$

These parameters Z_j as given by eqn (4.7.33) are then related to the standard Newmark parameters β and γ by the equations

$$Z_1 + Z_2 = 2\beta, \qquad Z_1 + 0.5 = \gamma. \tag{4.7.34}$$

It is important to note that this particular relationship depends on the interval of ξ values in eqns (4.7.29) onwards being taken as $[-1, 1]$ and not $[0, 2]$ or $[-2, 0]$. We point this out here because it can be confusing. In Zienkiewicz (1976) the weighted integral equation is given with the integral over the equivalent of $0 \leq \xi \leq 2$, but the parameters are defined with integrals over $-1 \leq \xi \leq 1$ and the apparently indefinite integrals in Table IV in Katona and Zienkiewicz (1985) must be over $[-1, 1]$. If instead we take

$$x = \sum_{n}^{n+2} N_i x_i \tag{4.7.35}$$

with

$$N_{n+2} = \xi(\xi-1)/2, \quad \Delta t\dot{N}_{n+2} = (2\xi-1)/2, \quad \Delta t^2\ddot{N}_{n+2} = 1$$
$$N_{n+1} = -\xi(\xi-2), \quad \Delta t\dot{N}_{n+1} = -2\xi+2, \quad \Delta t^2\ddot{N}_{n+1} = -2,$$
$$N_n = (\xi-1)(\xi-2)/2, \quad \Delta t\dot{N}_n = (2\xi-3))/2, \quad \Delta t^2\ddot{N}_n = 1 \tag{4.7.36}$$

then substituting into

$$\int_0^2 W(\xi) \sum_n^{n+2} (m\ddot{N}_i + \mu\dot{N}_i + kN_i)x_i \, \mathrm{d}\xi = 0$$

and supposing that $Z'_j = \int_0^2 W\xi^j \, \mathrm{d}\xi / \int_0^2 W \, \mathrm{d}\xi$, we have

$$Z'_1 - 0.5 = \gamma, \qquad Z'_2 - Z'_1 = 2\beta, \tag{4.7.37}$$

not the same relationship as in eqns (4.7.34). Similarly the particular relations given in Katona and Zienkiewicz (1985) Table V for the Zienkiewicz (1976) cubic approximation parameters and those for the beta-3 (GN32) algorithm depend on the intervals for ξ being taken as $[0, 3]$ rather than $[-1, 2]$, say or $[-2, 1]$. In Katona and Zienkiewicz (1985) Table VI, the corresponding interval of integration is again not specified. Different intervals will give different values to the parameters but they will of course be linearly related.

Wood (1984a) includes a discussion of various groupings of the independent equations involved in these weighted residual methods. For instance, the quadratic variation in time can be expressed in terms of three quantities which can be chosen as v_{n+1}, x_n, v_n, or x_{n+1}, x_n, v_n, $\left(\text{using } v_n = \overset{(1)}{x_n}\right)$. The weighted residual equation then gives a relation between these three quantities. The statement of the quadratic variation can then be used to give x_{n+1} in terms of v_{n+1}, x_n, v_n or v_{n+1} in terms of x_{n+1}, x_n, v_n. Similarly the cubic variation can be expressed in terms of x_{n+1}, x_n, v_n and either x_{n-1} or v_{n-1}. For the purpose of the discussion in this text, we stay with the simple relations (4.6.4) obtained in connection with the SS*pj* algorithm using the weighted residual equation (4.6.5). As we are approximating only in one particular time step, the integration interval can only be $[0, \Delta t]$ here.

2. Another aspect of the SSpj algorithm which seems worth mentioning is as follows. Because of the association of ideas with finite elements in space through the representation of the displacement by a polynomial of degree p in the time step 'element' and the way that the algorithm is obtained from a weighted residual equation, intuition may suggest that the resultant errors in the displacement, velocity, acceleration, etc. are $O(\Delta t^{p+1})$, $O(\Delta t^{p})$, $O(\Delta t^{p-1})$,... respectively. But time-stepping schemes do not work like this. We have seen in §4.5 that with single-step algorithms of the SSpj or GNpj type, the function and all derivatives up to and including the pth are approximated to the same order of accuracy, given sufficiently accurate starting values. The GNpj algorithm includes the approximation for the pth derivative; the SSpj algorithm is using an averaged value for the pth derivative but the derivative can be recovered from the differential equation supposing this is sufficiently differentiable. Tables 4.7.1 and 4.7.2 from Wood (1987b) show this effect for SS22 and SS32 respectively. The

Table 4.7.1. Tables for SS22 and SS32 showing ratio of errors as time step is halved, from Wood (1987b) Further Results... Tables 1 and 2.

(a) SS22: $\theta_1 = \theta_2 = 0.5$, $v = 0.1$. Errors at $t = 5$.

Δt	Displacement		Velocity	
0.5	3.04×10^{-3}		9.26×10^{-3}	
0.25	8.71×10^{-4}	3.5	2.31×10^{-3}	4.0
0.125	2.25×10^{-4}	3.9	5.77×10^{-4}	4.0
0.062 5	5.67×10^{-5}	4.0	1.44×10^{-4}	4.0
0.031 25	1.42×10^{-5}	4.0	3.60×10^{-5}	4.0
0.015 625	3.55×10^{-6}	4.0	9.01×10^{-6}	4.0

(B) SS22: $\theta_1 = 0.6$, $\theta_2 = 0.605$, $v = 0.5$. Errors at $t = 5$.

Δt	Displacement		Velocity	
0.5	2.14×10^{-3}		2.12×10^{-3}	
0.25	1.04×10^{-3}	2.0	7.42×10^{-4}	2.8
0.125	5.24×10^{-4}	2.0	2.94×10^{-4}	2.5
0.062 5	2.64×10^{-4}	2.0	1.28×10^{-4}	2.3
0.031 25	1.32×10^{-4}	2.0	5.92×10^{-5}	2.2
0.015 625	6.63×10^{-5}	2.0	2.84×10^{-5}	2.1

Table 4.7.2. Tables for SS22 and SS32 showing ratio of errors as time step is halved, from Wood (1987b) Further Results... Tables 1 and 2.

(a) SS32: $\theta_1 = 1.4$, $\theta_2 = 1.96$, $\theta_3 = 2.744$, $\nu = 0.5$. Errors at $t = 5$.

Δt	Displacement		Velocity		Acceleration	
0.5	2.90×10^{-3}		3.91×10^{-3}		4.02×10^{-3}	
0.25	9.01×10^{-4}	3.2	7.34×10^{-4}	5.3	1.04×10^{-3}	3.9
0.125	2.35×10^{-4}	3.8	1.58×10^{-4}	4.6	2.56×10^{-4}	4.1
0.062 5	6.26×10^{-5}	3.8	3.69×10^{-5}	4.3	6.31×10^{-5}	4.0
0.031 25	1.50×10^{-5}	4.2	8.93×10^{-6}	4.1	1.57×10^{-5}	4.0
0.015 625	3.76×10^{-6}	4.0	2.20×10^{-6}	4.0	3.90×10^{-6}	4.0

(b) SS32: $\theta_1 = 1.05$, $\theta_2 = 1.1$, $\theta_3 = 1.14$, $\nu = 0.1$. Errors at $t = 5$.

Δt	Displacement		Velocity		Acceleration	
0.5	3.37×10^{-3}		5.9×10^{-3}		4.33×10^{-3}	
0.25	1.04×10^{-3}	3.7	1.47×10^{-3}	4.0	1.16×10^{-3}	3.7
0.125	2.67×10^{-4}	3.9	3.65×10^{-4}	4.0	2.98×10^{-4}	3.9
0.062 5	6.76×10^{-5}	3.9	9.10×10^{-5}	4.0	7.53×10^{-5}	4.0
0.031 25	1.70×10^{-5}	4.0	2.27×10^{-5}	4.0	1.89×10^{-5}	4.0
0.015 625	4.25×10^{-6}	4.0	5.67×10^{-6}	4.0	4.74×10^{-6}	4.0

numbers alongside the columns of the errors give the ratios of the errors as the time step is halved. We have to take into account the way the error accumulates as the algorithm is repeated over successive time steps. In general the cumulative error is of the same order as the truncation error as defined in eqn (2.2.4), i.e. $O(\Delta t^{p+1-j})$ for SS*pj* and GN*pj*. Details of the accuracy of the SS*pj* algorithms for particular values of the parameter *p* will be discussed in the next section.

3. We emphasise that the GN*pj* and SS*pj* algorithms can be matched exactly when in their equivalent *p*-step forms by taking $\theta_j = \beta_{p-j}$, $j = 1, 2, \ldots, p$, $\theta_0 = \beta_p = 1$. Given the same starting conditions they then produce exactly the same results. Treated as single-step algorithms, however, they cannot in general be made to give the same results by matching the parameters like this. In Chapter 7 we shall show that for non-linear equations they can also have quite different stability properties.

4.8. Particular examples of the SSp2 algorithm

We now consider particular examples of the SSp2 algorithm applied to the numerical solution of the scalar equation

$$m\ddot{x} + \mu\dot{x} + kx = 0. \tag{4.8.1}$$

We recall that if there is a forcing function $f(t)$ on the right-hand side of eqn (4.8.1) then accuracy is maintained by approximating $f(t)$ in the same way as the kx term as shown in §2.4. This means that if the SSp2 algorithm applied to eqn (4.8.1) has an equivalent p-step form

$$\sum_{j=0}^{p} (\alpha_j m + \Delta t\gamma_j\mu + \Delta t^2\beta_j k)x_{n+j} = 0 \tag{4.8.2}$$

the non-homogeneous equation should have the forcing function replaced by

$$\bar{f} = \sum_{j=0}^{p} \beta_j f((n+j)\Delta t). \tag{4.8.3}$$

SS22

Table 4.8.1 shows the coefficients in the two-step equivalent of SS22 as in eqn (4.8.2) with the parameters θ_q defined as in eqn (4.6.6.).

In Chapter 3 we expressed the Newmark algorithm as a one-step algorithm in $\boldsymbol{Y}_n = [x_n, \Delta t\dot{x}_n]^{\mathrm{T}}$ with the coefficients in the amplification matrix given by eqns (3.2.13). The result of writing the SS22 algorithm out in the same way for comparison gives an

Table 4.8.1. The SS22 coefficients for the two-step form.

j	α_j	γ_j	β_j
0	1	$\theta_1 - 1$	$(-2\theta_1 + \theta_2 + 1)/2$
1	-2	$-2\theta_1 + 1$	$(2\theta_1 - 2\theta_2 + 1)/2$
2	1	θ_1	$\theta_2/2$

amplification matrix $\hat{A} = \{\hat{A}_{ij}\}$ where

$$\begin{aligned}\hat{A}_{11} &= \{m + \theta_1 \Delta t\mu + 0.5(\theta_2 - 1)\Delta t^2 k\}/D \\ \hat{A}_{12} &= \{m + (\theta_1 - 0.5)\Delta t\mu + 0.5(\theta_2 - \theta_1)\Delta t^2 k\}/D \\ \hat{A}_{21} &= -\Delta t^2 k/D \\ \hat{A}_{22} &= \{m + (\theta_1 - 1)\Delta t\mu + (0.5\theta_2 - \theta_1)\Delta t^2 k\}/D,\end{aligned} \tag{4.8.4}$$

where

$$D = \{m + \theta_1 \Delta t\mu + 0.5\theta_2 \Delta t^2 k\}. \tag{4.8.5}$$

Comparing eqns (3.2.9b) and (4.8.5) and the sets of eqns (3.2.13) and (4.8.4), we see that the two algorithms can only be the same if $2\beta = \gamma = \theta_1 = \theta_2$. Only then can the two schemes given the same results. For $2\beta \neq \gamma$, $\theta_1 \neq \theta_2$, the results will be different.

Table 4.8.1 gives the coefficients in the stability polynomial of the SS22 algorithm:

$$\sum_{j=0}^{2} (\alpha_j m + \gamma_j \Delta t\mu + \beta_j \Delta t^2 k) r^j = 0. \tag{4.8.6}$$

From eqns (4.2.9) we see that the method is consistent and the parameter $A = 1$.

Further from the definitions in eqns (4.2.10) and (4.2.15) we have

$$B_1 = B_2 = \theta_1 + 0.5, \qquad B_3 = 1 \tag{4.8.7}$$

and

$$C_1 = (\theta_1 + \theta_2)/2 + 1/4, \qquad C_2 = \theta_1 + 1/6, \qquad C_3 = 7/12. \tag{4.8.8}$$

Hence in general with $\mu \neq 0$ from eqns (4.2.21) and (4.2.22) the amplitude and phase errors are each $O(\Delta t)$. If $\mu = 0$ from eqns (4.2.24) and (4.2.25) we have the amplitude error $O(\Delta t)$ and the phase error $O(\Delta t^2)$.

With $\theta_1 = 0.5$ which makes $B_1 = B_2 = B_3$, we have

$$C_1 = (1 + \theta_2)/2, \qquad C_2 = 2/3, \qquad C_3 = 7/12. \tag{4.8.9}$$

Hence if $\mu = 0$ we can see that the product of the roots of the stability polynomial in eqn (4.8.6) is equal to

$$\{m + \tfrac{1}{2}(-2\theta_1 + \theta_2 + 1)\Delta t^2 k\}/(m + \tfrac{1}{2}\theta_2 \Delta t^2 k),$$

which, with $\theta_1 = 0.5$ is unity for any value of the time step and so

Table 4.8.2. Parameters for methods included in SS22.

SS22	Newmark	Constant acceleration	Linear acceleration	Numerov	Central difference
$\theta_2 =$	2β	1/2	1/3	1/6	0
$\theta_1 =$	γ	1/2	1/2	1/2	1/2

there is zero amplitude error; the phase error from eqn (4.2.29) is $O(\Delta t^2)$. If, further we have $C_1 = C_3$, i.e. $\theta_2 = 1/6$ then the phase error is $O(\Delta t^4)$ from eqn (4.2.29). This method given by $\theta_1 = 1/2$, $\theta_2 = 1/6$, for the differential equation with zero natural damping is called the Numerov or Fox–Goodwin (1949) method. Table 4.8.2 shows the parameters for some methods included in SS22.

If we take $\theta_1 = 0.5$ when $\mu \neq 0$ from eqns (4.2.21) and (4.2.22) we have amplitude and phase errors each of $O(\Delta t^2)$.

As shown in Appendix 1, we obtain the stability conditions for the algorithm by applying the Routh–Hurwitz transformation $r = (1+z)/(1-z)$ to the stability polynomial in eqn (4.8.6) with the coefficients given in Table 4.8.1, The Routh–Hurwitz conditions then give

$$\begin{aligned} &2m + (2\theta_1 - 1)\Delta t\mu + (\theta_2 - \theta_1)\Delta t^2 k > 0 \\ &\Delta t^2 k \geq 0 \\ &\Delta t\mu + (\theta_1 - 0.5)\Delta t^2 k \geq 0. \end{aligned} \tag{4.8.10}$$

It is understood here that $m, k > 0$ and $\mu \geq 0$, hence the conditions on the parameters θ_1, θ_2 for unconditional stability are

$$\theta_2 \geq \theta_1 \geq 0.5. \tag{4.8.11}$$

From the discussion in §2.3 we see that (4.8.11) is the condition for the SS22 algorithm to be *A* stable (Dahlquist 1963). Further results from Dahlquist 1963 which are relevant here are:

1. An explicit linear multistep method cannot be *A*-stable.

2. The order of the error of an implicit *A*-stable method cannot be greater than two.

3. The A-stable linear multistep method with second order error and the smallest error constant is the trapezium rule. (The error constant referred to here is the number multiplying the $O(\Delta t^2)$ term in the error.) The trapezium rule applied to the numerical integration of the first order equation

$$\dot{z} = -Bz \tag{4.8.12}$$

over one time step gives:

$$z_{n+1} - z_n = -\Delta t B(z_{n+1} + z_n)/2. \tag{4.8.13}$$

If eqn (4.8.12) is the first order form of the dynamic vibration equation (4.8.1), writing $\dot{x} = v$, we have

$$z = [x, v]^{\mathrm{T}}, \qquad B = \begin{bmatrix} 0 & -1 \\ \omega^2 & 2\nu\omega \end{bmatrix}, \tag{4.8.14}$$

where

$$\omega^2 = k/m, \qquad 2\nu\omega = \mu/m.$$

Writing $z_{n+1} = rz_n$ in eqn (4.8.13) gives

$$r(I + 0.5\Delta t B)z_n = (I - 0.5\Delta t B)z_n. \tag{4.8.15}$$

Then the condition for non-trivial z_n gives the characteristic polynomial of the algorithm, i.e. the vanishing of the determinant

$$\begin{vmatrix} r-1 & -(r+1)\Delta t/2 \\ (r+1)\omega^2\Delta t/2 & r-1+\nu\omega\Delta t(r+1) \end{vmatrix}. \tag{4.8.16}$$

Multiplying out this determinant gives the polynomial

$$r^2(1 + \nu\omega\Delta t + \omega^2\Delta t^2/4) + r(-2 + \omega^2\Delta t^2/2) \\ + 1 - \nu\omega\Delta t + \omega^2\Delta t^2/4 = 0 \tag{4.8.17}$$

or

$$r^2(m + 0.5\Delta t\mu + 0.25\Delta t^2 k) + r(-2m + 0.5\Delta t^2 k) \\ + m - 0.5\Delta t\mu + 0.25\Delta t^2 k = 0. \tag{4.8.18}$$

Comparing the coefficients in the polynomial (4.8.18) with those given in Table 4.8.1 we see that the SS22 algorithm which corresponds to the trapezium rule is given by the parameters $\theta_1 = \theta_2 = 0.5$ (Newmark $2\beta = \gamma = 0.5$). We see from the stability condition (4.8.11) that the stability for these values of the parameters is 'marginal', i.e. they lie at the end of the stability range.

Table 4.8.3. SS22 accuracy and stability.

	$\mu=0$			$\mu\neq 0$		
Parameters	ϵ	δ	Stability	ϵ	δ	Stability
θ_1, θ_2	$O(\Delta t^2)$	$O(\Delta t)$	$\theta_2 \geq \theta_1 \geq 1/2$	$O(\Delta t)$	$O(\Delta t)$	$\theta_2 \geq \theta_1 \geq 1/2$
$\theta_1 = 1/2$	$O(\Delta t^2)$	0	$\theta_2 \geq 1/2$	$O(\Delta t^2)$	$O(\Delta t^2)$	$\theta_2 \geq 1/2$
$\left.\begin{array}{l}\theta_1 = 1/2\\ \theta_2 = 1/6\end{array}\right\}$	$O(\Delta t^4)$	0	$\omega^2\Delta t^2 < 6$	$O(\Delta t^2)$	$O(\Delta t^2)$	$\omega^2\Delta t^2 < 6$

Table 4.8.3 gives a summary of the accuracy and stability results for the SS22 algorithm.

SS32

Table 4.8.4 gives the coefficients for the SS32 equivalent three-step algorithm which are also those of the stability polynomial

$$\sum_{j=0}^{3} (\alpha_j m + \gamma_j \Delta t \mu + \beta_j \Delta t^2 k) r^j = 0. \tag{4.8.19}$$

From eqns (4.2.9) the algorithm is consistent as expected and the parameter $A = 1/\theta_1$. Further from eqns (4.2.10) we have

$$B_1 = B_2 = B_3 = A(1+\theta_1). \tag{4.8.20}$$

Hence from eqns (4.2.21) and (4.2.22), the SS32 algorithm for $\mu \neq 0$ gives error of at least $O(\Delta t^2)$ for both amplitude and phase for any values of the θ parameters.

From eqn (4.2.15) we have

$$C_1 = C_2 = A(2\theta_1 + \theta_2 + 4/3)/2, \qquad C_3 = A(3\theta_1 + 7/6)/2. \tag{4.8.21}$$

Table 4.8.4. SS32 coefficients for the three-step form.

j	α_j	γ_j	β_j
0	$A(-\theta_1+1)$	$A(2\theta_1-\theta_2-1)/2$	$A(-3\theta_1+3\theta_2-\theta_3+1)/6$
1	$A(3\theta_1-2)$	$A(-4\theta_1+3\theta_2)/2$	$A(-6\theta_2+3\theta_3+4)/6$
2	$A(-3\theta_1+1)$	$A(2\theta_1-3\theta_2+1)/2$	$A(3\theta_1+3\theta_2-3\theta_3+1)/6$
3	1	$A\theta_2/2$	$A\theta_3/6$

where $A = 1/\theta_1$.

From eqn (4.2.21) we see that with $\mu \neq 0$ we require $C_1 = C_2 = C_3$ for the phase error to be $O(\Delta t^3)$. This means that we must have

$$\theta_2 - \theta_1 = -1/6. \tag{4.8.22}$$

This also makes the amplitude error $O(\Delta t^3)$ from eqn (4.2.22). From eqns (4.2.16) taking $\theta_1 = \frac{1}{2}$, $\theta_2 = \frac{1}{3}$, and $\theta_3 = \frac{1}{4}$ makes $D_1 = D_2 = D_3$ as well which means that the $O(\Delta t^3)$ terms also disappear from eqn (4.2.20) and the amplitude and phase errors are each $O(\Delta t^4)$, but we show later that then we cannot satisfy the stability conditions with $\mu \neq 0$.

If $\mu = 0$, i.e. no natural damping, then in general from eqns (4.2.21) and (4.2.22) ($\mu = 0$ implies $\alpha = \pi/2$ hence $\cos\alpha = 0$, $\sin 4\alpha = \sin 2\alpha = 0$), we have that the phase error ϵ is $O(\Delta t^2)$ and the amplitude error δ is $O(\Delta t^3)$. With $\theta_2 = \theta_1 - 1/6$ to make $C_1 = C_3$ we have from eqns (4.2.24) and (4.2.25) that the phase error ϵ is $O(\Delta t^4)$ and the amplitude error δ is $O(\Delta t^3)$. If we also have $D_1 = D_3$ then ϵ is $O(\Delta t^4)$ and δ is $O(\Delta t^5)$; this requires $\theta_3 = \theta_1/2$ as well as $\theta_2 = \theta_1 - 1/6$. These results are summarized in Table 4.8.5.

Table 4.8.6 shows the SS32 θ parameter values which give the equivalents of some other well known algorithms devised by engineers. Houbolt (1950) in a paper on a recurrent matrix solution for the dynamic response of elastic aircraft uses the idea of making a cubic curve pass through four successive values of

Table 4.8.5. SS32 accuracy and stability.

	$\mu = 0$			$\mu \neq 0$		
Parameters	ϵ	δ	Stability	ϵ	δ	Stability
a, b, c*	$O(\Delta t^2)$	$O(\Delta t^3)$	$a > 0$, $b \geq 0$, $0 \leq c \leq 3ab$	$O(\Delta t^2)$	$O(\Delta t^2)$	See text
$b = -1/3$	$O(\Delta t^4)$	$O(\Delta t^3)$	$a > 0$, $c \leq -a$, $\omega^2\Delta t^2 < 6a/\lvert c\rvert$	$O(\Delta t^3)$	$O(\Delta t^3)$	See text
$b = -1/3$ $a + c = 0$	$O(\Delta t^4)$	$O(\Delta t^5)$	$\omega\Delta t^2 < 6$			

* $a = \theta_1 - 0.5$, $b = \theta_2 - \theta_1 - 1/6$, $c = \theta_3 - 1.5\theta_2 + 0.25$

Table 4.8.6. Other algorithms included in SS32.

	Wilson-θ	Houbolt	Bossak $\mu=0$	Bossak $\mu\neq 0$	HHT $\mu=0$	HHT $\mu\neq 0$
θ_3	θ^3	6	$6\beta_B$	$1-3\alpha_B$	$6\beta_H(1+\alpha_H)$	$(1-3\alpha_H)(1+\alpha_H)$
θ_2	θ^2	$\frac{11}{3}$	$2\beta_B-\alpha_B+\frac{2}{3}$	$1-2\alpha_B$	$2\beta_H-2\alpha_H^2+\frac{2}{3}$	$1-\alpha_H-2\alpha_H^2$
θ_1	θ	2	$1-\alpha_B$	$1-\alpha_B$	1	1

the displacement. This is then differentiated to obtain the velocity and displacement and these are substituted into the equation of motion. In the Zienkiewicz (1976) paper, this is shown to be a particular case of supposing a cubic variation and then substituting into the weighted residual equation. The SS32 single-step equivalent to Houbolt is given by $\theta_1=2$, $\theta_2=11/3$, $\theta_3=6$.

The Wilson-θ method introduced in Bathe and Wilson (1973) is a single-step algorithm which assumes a cubic variation of the displacement of the form

$$x=x_n+\tau\dot{x}_n+\frac{\tau^2}{2}\ddot{x}_n+\frac{\tau^3}{6\Delta t}(\ddot{x}_{n+1}-\ddot{x}_n), \tag{4.8.23}$$

where $0\leq\tau\leq\theta\Delta t$.

Then the displacement and derivatives are substituted from eqn (4.8.23) into the dynamic vibration equation collocated at time $t=(n+\theta)\Delta t$, i.e.

$$m\ddot{x}_{t+\theta\Delta t}+\mu\dot{x}_{t+\theta\Delta t}+kx_{t+\theta\Delta t}=f_{t+\theta\Delta t}. \tag{4.8.24}$$

This algorithm corresponds to SS32 with parameters given by $\theta_1=\theta$, $\theta_2=\theta^2$, $\theta_3=\theta^3$. Bathe and Wilson (1973) show that for unconditional stability the Wilson-θ algorithm must have $\theta\geq 1.37$.

Figures 4.8.1 and 4.8.2 (from Wood 1977) show the maximum modulus root of the stability polynomial and the percentage period elongation respectively plotted against $\Delta t/T$ (where $T=2\pi/\omega$ is the natural period of the motion with no natural damping) for various unconditionally stable examples of SS32. These examples all show the period elongation expected with unconditional stability. Table 4.8.7 gives the corresponding θ values.

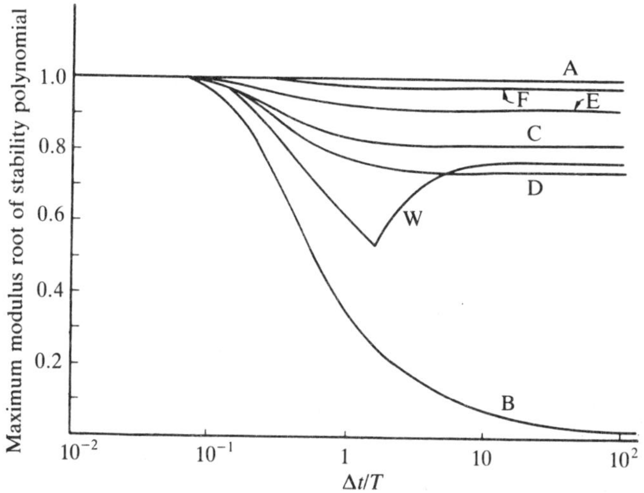

Fig. 4.8.1. Maximum modulus root of stability polynomial plotted against $\Delta t/T$ for the θ values given in Table 4.8.5.

Example A shows unconditional stability with zero algorithmic damping and minimum period elongation, B is Houbolt which has all roots of the stability polynomial zero for infinite $\omega\Delta t$, C, D, E, F are experiments to indicate the parameter values which give a maximum modulus root at infinity between 0.7 and unity. They also show that keeping down the percentage period elongation means that the maximum modulus root of the stability polynomial stays near unity. Example W is the Wilson $\theta = 1.4$ algorithm.

With the parameters for curves A to F, roots of the cubic stability polynomial are always two complex and one real. With the Wilson $\theta = 1.4$ parameters the discriminant of the cubic vanishes for $\Delta t/T = 2.4$ ($\omega\Delta t = 15$) approximately so that there are three real roots for the higher values of $\Delta t/T$ and this produces the different shape of the curve representing the maximum modulus.

Figure 4.8.3 (from Wood 1986) shows the percentage period increase and decrease for various examples of the SS22 and SS32 algorithms with zero natural damping. In this figure, the curves are labelled U for unconditionally stable and E for explicit,

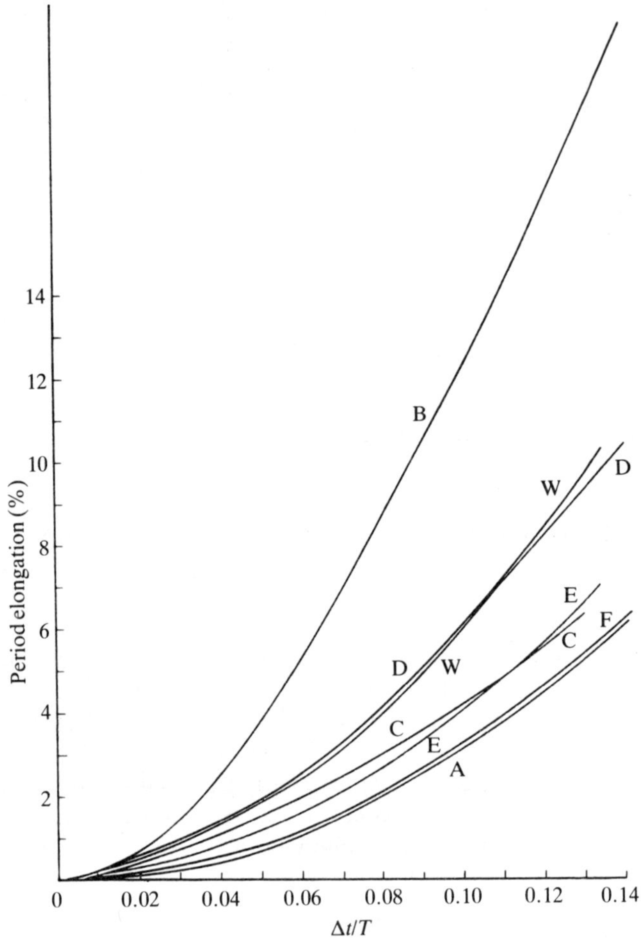

Fig. 4.8.2. Percentage period elongation plotted against $\Delta t/T$ for the θ values given in Table 4.8.5.

hence conditionally stable and 3UW is the SS32 corresponding to Wilson $\theta = 1.4$, 3UH is the SS32 corresponding to Houbolt, 2U1 is the SS22 trapezium rule equivalent with $\theta_1 = \theta_2 = 0.5$, 3U1 is a marginal case of the SS32 algorithm with $\theta_1 = 0.5$, $\theta_2 = 2/3$, $\theta_3 = 0.75$ which gives zero algorithmic damping; the stability polynomial is the same as that for 2U1 with the extra factor

Table 4.8.7. The θ values for the curves in Figs 4.8.1 and 4.8.2.

	θ_1	θ_2	θ_3	
A	1.5	1.6667	2.25	
B	2	3.6667	6	Houbolt
C	5.5	6.6667	9.85	
D	2	2.6667	4	
E	2	2.3167	3.325	
F	50.5	51.6667	78.25	
H	1.4	1.98	2.744	Wilson $\theta = 1.4$

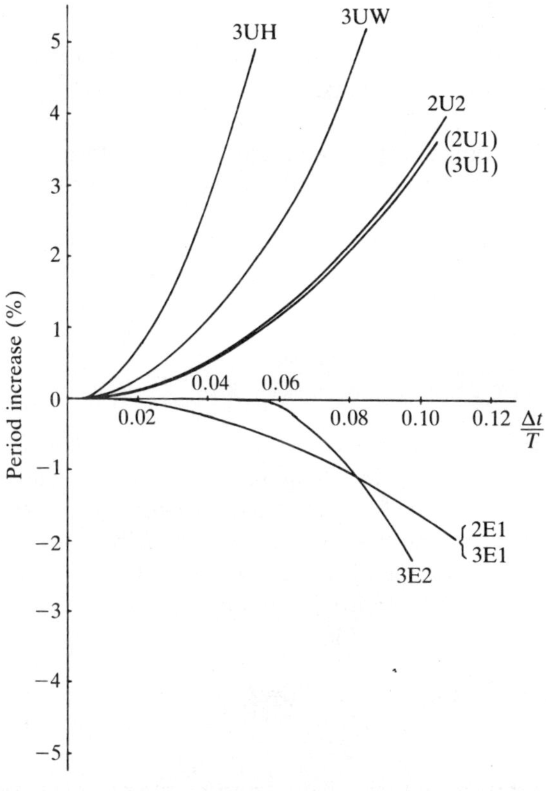

Fig. 4.8.3. Percentage period increase and decrease for some unconditionally stable (U) algorithms, and some explicit (E) algorithms respectively.

$(r+1)$. In Fig. 4.8.3 the curves for 2U1 and 3U1 are the same. 2U2 is the SS22 algorithm with $\theta_1 = 0.6$, $\theta_2 = 0.605$ to give a small amount of algorithmic damping. Figure 4.8.3 also shows results from explicit algorithms which all give a period decrease: 2E1 is the SS22 algorithm with $\theta_1 = 0.5$, $\theta_2 = 0$; 2E1 is SS32 with $\theta_1 = 0.5$, $\theta_2 = 1/6$, $\theta_3 = 0$, which is a marginal case referred to later because it is an example of an algorithm which is conditionally stable with zero natural damping but unconditionally unstable when natural damping is present.

To some extent here we can also fit in the Bossak and Hilber–Hughes–Taylor algorithms described under the heading of Newmark variations in § 4.4. Taking the Bossak algorithm first, this is a three-stage, three-parameter algorithm like SS32. The three parameters are referred to here as $\alpha_B, \beta_B, \gamma_B$. But SS32 is automatically a second order algorithm and as shown in §4.4 the Bossak algorithm is only second order if we take $\gamma_B = 0.5 - \alpha_B$ which reduces the number of free parameters to two.

We then find that provided $\mu = 0$, that is, no natural damping, we can match the coefficients in the SS32 algorithm by taking

$$\theta_3 = 6\beta_B, \qquad \theta_2 = 2\beta_B - \alpha_B + 2/3, \qquad \theta_1 = 1 - \alpha_B. \quad (4.8.25)$$

But if $\mu \neq 0$ we can only match the coefficients by making $\beta = (1 - 3\alpha_B)/6$ and taking

$$\theta_3 = 1 - 3\alpha_B, \qquad \theta_2 = 1 - 2\alpha_B, \qquad \theta_1 = 1 - \alpha_B. \quad (4.8.26)$$

The case is similar with the HHT (Hilber–Hughes–Taylor 1977) algorithm discussed in detail in §4.4. Using a subscript H to distinguish the HHT coefficients, in order to make the HHT algorithm comparable with the automatically second order SS32 algorithm, as we found in §4.4 we have to take $\gamma_H = 0.5 - \alpha_H$. Then if there is no natural damping, that is, no γ_j terms, we can match coefficients by taking

$$\theta_3 = 6\beta_H(1 + \alpha_H), \qquad \theta_2 = 2\beta_H - 2\alpha_H^2 + 2/3, \qquad \theta_1 = 1. \quad (4.8.27)$$

But if $\mu \neq 0$ so that we also have to match the γ_j terms, we must have $\beta_H = (1 - 3\alpha_H)/6$ so that there is now only the one free parameter and we have to match the coefficients by taking

$$\theta_3 = (1 - 3\alpha_H)(1 + \alpha_H), \qquad \theta_2 = 1 - \alpha_H - 2\alpha_H^2, \qquad \theta_1 = 1. \tag{4.8.28}$$

Table 4.8.6 summarizes these results.

We next look at the stability conditions for the SS32 algorithm. In order to use the Routh–Hurwitz conditions as described in Appendix 2 we make the substitution $r = (1+z)/(1-z)$ in the characteristic polynomial (4.8.19) to give the cubic equation

$$a_0 z^3 + a_1 z^2 + a_2 z + a_3 = 0. \tag{4.8.29}$$

The Routh–Hurwitz criterion for the roots of the characteristic polynomial (4.8.29) to have modulus less than or equal to unity then means that we have to satisfy:

$$a_0 > 0,\ a_1 \geq 0, \begin{vmatrix} a_1 & a_3 \\ a_0 & a_2 \end{vmatrix} \geq 0, \begin{vmatrix} a_1 & a_3 & 0 \\ a_0 & a_2 & 0 \\ 0 & a_1 & a_3 \end{vmatrix} \geq 0. \tag{4.8.30}$$

Conditions (4.8.30) reduce to

$$a_0 > 0,\ a_1 \geq 0,\ a_2 \geq 0,\ a_3 \geq 0, \quad \text{and} \quad a_1 a_2 - a_0 a_3 \geq 0. \tag{4.8.31}$$

Making the convenient substitutions

$$a = \theta_1 - 1/2, \quad b = \theta_2 - \theta_1 - 1/6, \quad c = \theta_3 - 3\theta_2/2 + 1/4 \tag{4.8.32}$$

the stability conditions (4.8.31) are

$$12am + (6b+1)\Delta t\mu + 2c\Delta t^2 k > 0 \tag{4.8.33a}$$

$$2m + 2a\Delta t\mu + b\Delta t^2 k \geq 0 \tag{4.8.33b}$$

$$\Delta t\mu + a\Delta t^2 k \geq 0 \tag{4.8.33c}$$

$$\Delta t^2 k \geq 0 \tag{4.8.33d}$$

and

$$12m\mu + 12a\Delta t\mu^2 + (12a^2 - 1)\Delta t^2 \mu k + 2(3ab - c)\Delta t^3 k^2 \geq 0. \tag{4.8.34}$$

For the inequalities (4.8.33) to be satisfied for all values of the time step Δt, we require

$$a \geq 0,\ b \geq 0,\ c \geq 0 \tag{4.8.35}$$

whether the natural damping term μ is zero or not. When the natural damping μ is zero it is simple to satisfy inequality (4.8.34) as well by taking

$$3ab - c \geq 0. \tag{4.8.36}$$

But when the natural damping term is not zero the situation is more complicated.

Authors frequently seem to assume when testing time-stepping algorithms for the numerical solution of the dynamic vibration equation that it is only necessary to use the undamped case

$$m\ddot{x} + kx = 0. \tag{4.8.37}$$

Intuition appears to be suggesting that any algorithm which is stable with the undamped equation (4.8.37) must be stable with some non-zero natural damping to help. This is certainly true with the generalized Newmark and the SS22 algorithms as we can see from the stability conditions in eqn (4.8.10), for example. It makes no difference here whether μ is zero or not. With the SS32 algorithm, however, we have a situation where we can have unconditional stability for the numerical solution of the undamped vibration equation but only conditional stability when natural damping is present, i.e. the natural damping does not necessarily help with the stability of the algorithm (Wood 1987a). In order to study the effect of the inequality (4.8.34), we regard it as a cubic in $\omega\Delta t$, where $\omega^2 = k/m$ as before, and we also substitute $\mu = 2v\omega m$ (where v is the fraction of critical damping with $0 \leq v < 1$), so that inequality (4.8.34) becomes:

$$\begin{aligned} f(\omega\Delta t) = (3ab - c)(\omega\Delta t)^3 - (1 - 12a^2)v(\omega\Delta t)^2 \\ + 24av^2(\omega\Delta t) + 12v \geq 0. \end{aligned} \tag{4.8.38}$$

If the damping ratio $v \neq 0$ and we have $(3ab - c) > 0$, we have $f(0) > 0$ and $f(\infty) > 0$. We want to examine the possibility that the cubic curve given by $y = f(\omega\Delta t)$ is as shown in Fig. 4.8.4, i.e. the curve dips below the axis for some range of positive values of $\omega\Delta t$. We can see that if $(12a^2 - 1) > 0$ then all the coefficients in the cubic $f(\omega\Delta t)$ are positive. The conditions for $f(\omega\Delta t) \geq 0$ for all positive values of $\omega\Delta t$ may also be expressed as $3ab - c \geq 0$ and either

(i) the cubic has no real turning points; or
(ii) the cubic has real turning points where $\omega\Delta t = \delta_1, \delta_2$, $\delta_2 > \delta_1$, and $f(\delta_2) > 0$.

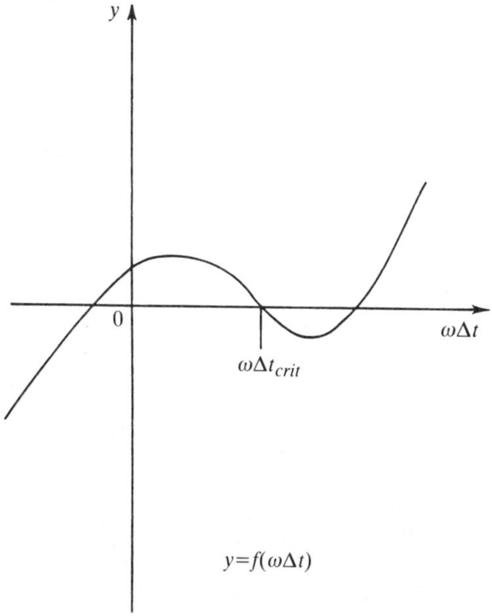

Fig. 4.8.4. Example of a damping cubic curve

After some algebra we thus arrive at the complete conditions for unconditional stability of the SS32 algorithm with the damping ratio $\nu \neq 0$, i.e. the conditions for the satisfaction of the inequalities (4.8.33) and (4.8.34): $a > 0$, $b \geq 0$, $c \geq 0$ and either

(a) $12a^2 \geq 1$ and $3ab - c \geq 0$; or

(b) $0 < 12a^2 \leq 1$ and $72a(3ab - c) \geq \min\{A, B\}$

where

$$A = (12a^2 - 1)^2, \tag{4.8.39}$$

and

$$\begin{aligned} 0 \leq B = {} & 16a^2\nu^2\{3(12a^2 - 1) + 32a^2\nu^2\} \\ & + 8a\nu(16a^2\nu^2 - 12a^2 + 1)^{3/2}, \end{aligned} \tag{4.8.40}$$

If $\nu = 0$, this gives the same results as in Table 4.8.5, $\mu = 0$.

Examples

1. If we take $a = 0.1$ then $12a^2 < 1$ and condition (b) applies. If we have $\nu = 0.1$, then from eqns (4.8.39) and (4.8.40)

$A = 0.774$ and $B = 0.07$; hence we want to make $3ab - c \geq 0.01$ approximately. With $c \geq 0$ we must make $0.3b \geq 0.01$. Thus we can choose $b = 0.1$, $c = 0.01$ and then the θ parameters are $\theta_1 = 0.6$, $\theta_2 = 0.867$, $\theta_3 = 1.06$.

2. If we take $a = 1.5$ then condition (a) applies and we require $4.5b \geq c \geq 0$. If we now take $b = 1.5$ and $c = 0.75$, we have the Houbolt (1950) algorithm with $\theta_1 = 2$, $\theta_2 = 11/3$, $\theta_3 = 6$.

From the second and fourth terms in $f(\omega\Delta t)$, we see that the curve is certainly above the axis for

$$12 > (1 - 12a^2)(\omega\Delta t)^2,$$

i.e.

$$(\omega\Delta t)^2 < 12/(1 - 12a^2), \quad \text{if } (1 - 12a^2) > 0. \tag{4.8.41}$$

But then, $12 < 12/(1 - 12a^2)$, because we have $0 < (1 - 12a^2) < 1$, hence we certainly have $f(\omega\Delta t) > 0$ for $\omega\Delta t < \sqrt{12} \cong 3.464$.

By using some trial and error we find that we can make the value of $\omega\Delta t$ where $f(\omega\Delta t)$ changes from positive to negative with increasing $\omega\Delta t$ approximately equal to $\sqrt{12}$ for $v = 0.1$ (10 per cent natural damping) with $a = 0.01$, $b = 0.1$, and $c = 0.0027$. These values do not satisfy the stability conditions (a) or (b). They correspond to $\theta_1 = 0.51$, $\theta_2 = 0.77666$, $\theta_3 = 0.9177$ and they make $f(3.5) = -0.00227$, for example. With $v = 0$ these values of the parameters give unconditional stability because they satisfy the inequalities (4.8.33) and (4.8.34); a, b, and c are each positive and $3ab - c = 0.003 - 0.0027 = 0.0003$. Figures 4.8.5(a) and (b) and 4.8.6(a) and (b) (from Wood 1987a) illustrate these results. The value of $\omega\Delta t = 5$ is chosen to make the instability more obviously visible. The results with such a large value of $\omega\Delta t$ are of course very inaccurate; these experiments are designed to show the growing solution, i.e. the instability, which is present when there is natural damping but absent when the natural damping is zero. Results are shown for (a) zero and (b) sinusoidal forcing function. It is interesting to note that all the algorithms of SS32 or the corresponding three-step form mentioned in the literature all happen to have the parameter a such that $a > 1/\sqrt{12} \cong 0.289$ so that the results demonstrated here have never appeared. With Houbolt (Houbolt 1950), $a = 1.5$, with the Wilson-θ algorithm (Wilson 1968), since $\theta > 1.366$, and $a = \theta_1 - 0.5 = \theta - 0.5 > 0.866$ and with the Hilber–Hughes–Taylor (1977) algorithm $a = 0.5$.

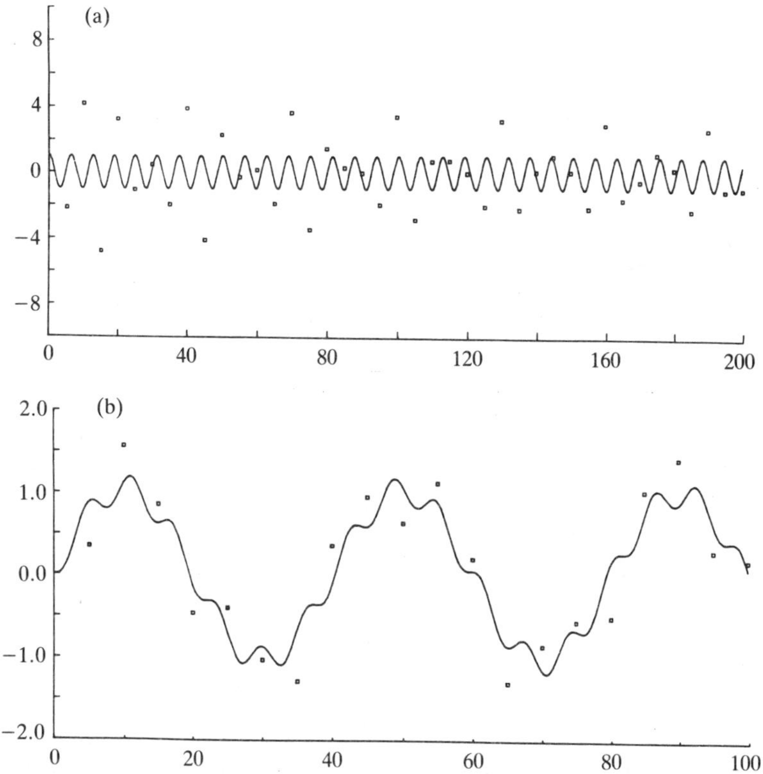

Fig. 4.8.5. SS32 with $\theta_1 = 0.51$, $\theta_2 = 0.77666$, $\theta_3 = 0.9177$, $\Delta t = 5$, $\nu = 0$, showing inaccurate but stable solution with (a) $\ddot{x} + x = 0$ and (b) $\ddot{x} + x = \sin(\pi t/20)$.

Figure 4.8.7 illustrates SS32 with the value of θ_1 taken to give $12a^2 = 1$ to make inequality (4.8.34) simple to satisfy. We can have an explicit scheme with $\theta_3 = 0$ and with $O(\Delta t^3)$ error by taking $\theta_2 - \theta_1 = -1/6$ from eqn (4.8.22), i.e. $b = -1/3$ and $c = -(1 + \sqrt{3})/4$ from eqn (4.8.32). The stability conditions for this explicit scheme can then be deduced from inequalities (4.8.33a) and (4.8.33b) which give

$$(1 + \sqrt{3})\omega^2\Delta t^2 + 4\nu\omega\Delta t - 4\sqrt{3} < 0$$

and

$$\omega^2\Delta t^2 - 2\sqrt{3}\nu\omega\Delta t - 6 \leq 0.$$

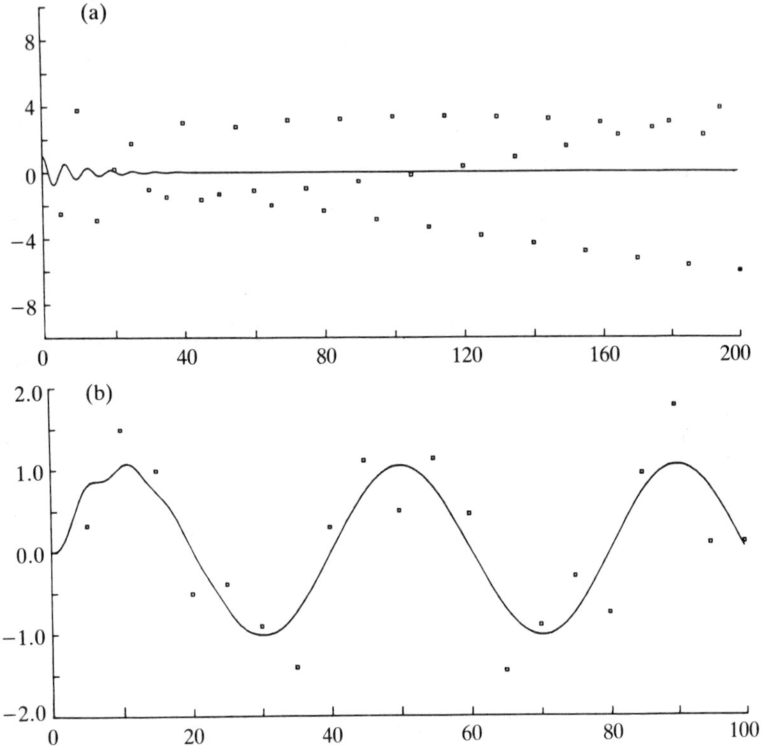

Fig. 4.8.6. Shows instability with the same values of θ_1, θ_2, θ_3, and Δt as in Fig. 4.8.5 but with $\nu = 0.1$, for (a) $\ddot{x} + 0.2\dot{x} + x = 0$, and (b) $\ddot{x} + 0.2\dot{x} + x = \sin(\pi t/20)$.

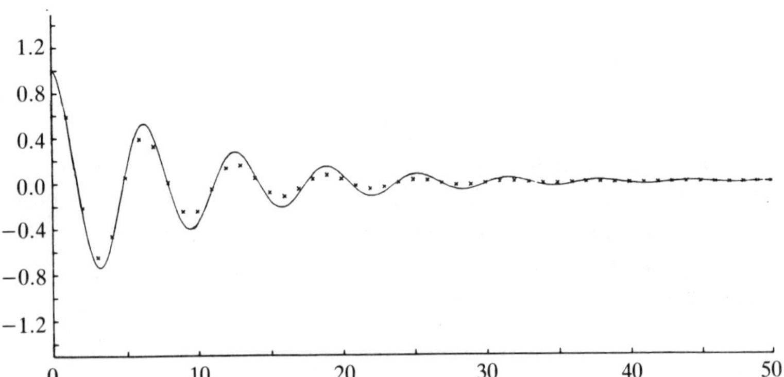

Fig. 4.8.7. The explicit scheme SS32 with $\theta_1 = 0.7887$, $\theta_2 = 0.6220$, $\theta_3 = 0$, $\nu = 0.1$ is still stable for $\Delta t = 1$.

These inequalities are satisfied for $\omega\Delta t \leq 1$ for any value of $\nu < 1$. Figure 4.8.7 (from Wood 1986) shows this SS32 explicit algorithm used on $\ddot{x} + 0.2\dot{x} + x = 0$, $x(0) = 1$, $\dot{x}(0) = 0$ with $\nu = 0.1$ and $\Delta t = 1$ to show that it is stable with this comparatively large time step.

We can also illustrate the unexpected effect that the presence or absence of natural damping has on the stability with the marginal case 3E1 illustrated in Fig. 4.8.3. 3E1 is an explicit example of SS32 with $\theta_1 = 0.5$, $\theta_2 = 1/6$, $\theta_3 = 0$. It is marginal because these values of the parameters make $a = 0$, $b = -0.5$, and $c = 0$ and thus violate stability condition (4.8.33a). However, when the natural damping parameter $\mu = 0$, inequality (4.8.33b) gives $\omega\Delta t \leq 2$ and stable results can be obtained (Wood 1986). With $\mu \neq 0$ the left-hand side of inequality (4.8.33a) becomes equal to $-2\mu\Delta t$ and the stability conditions cannot be satisfied; the algorithm becomes unconditionally unstable. Figure 4.8.8 (from Wood 1986) illustrates this effect with a time step which would give stability with the undamped problem.

We conclude from all this that using SS32 with $\mu \neq 0$ it is simplest to take $a > 0.289$ together with $b \geq 0$ and $c \geq 0$. The

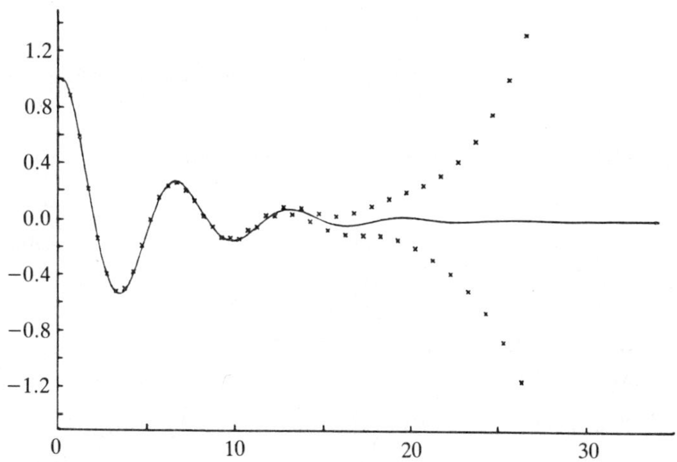

Fig. 4.8.8. SS32 with $\theta_1 = 0.5$, $\theta_2 = 0.16667$, $\theta_3 = 0$, $\Delta t = 0.5$ is unconditionally unstable with $\nu = 0.2$

corresponding conditions on the θ-parameters are then

$$\theta_1 = a + 0.5 > 0.789, \qquad \theta_2 \geq \theta_1 + 1/6 > 0.956, \qquad \theta_3 \geq 3\theta_2/2 - 0.25 > 1.184. \tag{4.8.42}$$

From eqn (4.8.22) for an $O(\Delta t^3)$ error we have to take $b = -1/3$, i.e. $\theta_2 = \theta_1 - 1/6$ which violates the conditions for unconditional stability and thus verifies the Dahlquist (1963) result that it is not possible to have an A-stable algorithm with better than $O(\Delta t^2)$ error.

If we do take $b = -1/3$ to obtain $O(\Delta t^3)$ error, the stability conditions (4.8.33) and (4.8.34) become

$$\begin{aligned} 12am - \Delta t\mu + 2c\Delta t^2 k &> 0 \\ 2m + 2a\Delta t\mu - \tfrac{1}{3}\Delta t^2 k &\geq 0 \\ \mu + a\Delta tk &\geq 0 \\ \Delta t^2 k &\geq 0 \end{aligned} \tag{4.8.43}$$

and

$$12m\mu + 12a\Delta t\mu^2 + (12a^2 - 1)\Delta t^2\mu k - 2(a + c)\Delta t^3 k^2 \geq 0.$$

If the natural damping term $\mu = 0$ we can satisfy inequalities (4.8.43) by taking $a > 0$, $a + c \leq 0$, and $\omega^2\Delta t^2 < 6a/|c|$, i.e. $\theta_1 > 0.5$, $\theta_2 = \theta_1 - 1/6$, $\theta_3 < \theta_1/2$ and

$$\omega^2\Delta t^2 < 6(\theta_1 - 0.5)/|\theta_3 - 1.5\theta_1 + 0.5|. \tag{4.8.44}$$

If the natural damping term $\mu \neq 0$, we can satisfy inequalities (4.8.43) by taking $a > 1/\sqrt{12}$, $a + c = 0$, and $\omega^2\Delta t^2 < 6$, i.e.

$$\theta_1 > 0.789, \quad \theta_2 = \theta_1 - 1/6, \quad \theta_3 = \theta_1/2 \quad \text{and} \quad \omega^2\Delta t^2 < 6. \tag{4.8.45}$$

From eqns (4.2.20) for the error to be $O(\Delta t^4)$ we must have $a = c = 0$ as well as $b = -\frac{1}{3}$. If the natural damping term $\mu \neq 0$ the inequalities (4.8.43) are then incompatible. If $\mu = 0$ we can have amplitude and phase error $O(\Delta t^4)$ with $b = -\frac{1}{3}$, $a > 0$, $c = -a$, and $\omega^2\Delta t^2 < 6$, i.e.

$$\theta_1 > 0.5, \quad \theta_2 = \theta_1 - 1/6, \quad \theta_3 = \theta_1/2, \quad \text{and} \quad \omega^2\Delta t^2 < 6. \tag{4.8.46}$$

Table 4.8.5 summarizes these results.

Table 4.8.8. Coefficients for SS42.

j	α_j	γ_j	β_j
0	$a(-2\theta_1+\theta_2+1)$	$\frac{a}{3}(3\theta_1-3\theta_2+\theta_3-1)$	$\frac{a}{12}(-4\theta_1+6\theta_2-4\theta_3+\theta_4+1)$
1	$a(6\theta_1-4\theta_2-1)$	$\frac{a}{3}(-3\theta_1+9\theta_2-4\theta_3-3)$	$\frac{a}{12}(-12\theta_1-6\theta_2+12\theta_3-4\theta_4+11)$
2	$a(-6\theta_1+6\theta_2-1)$	$\frac{a}{3}(-3\theta_1-9\theta_2+6\theta_3+3)$	$\frac{a}{12}(12\theta_1-6\theta_2-12\theta_3+6\theta_4+11)$
3	$a(2\theta_1-4\theta_2+1)$	$\frac{a}{3}(3\theta_1+3\theta_2-4\theta_3+1)$	$\frac{a}{12}(4\theta_1+6\theta_2+4\theta_3-4\theta_4+1)$
4	1	$a\theta_3/3$	$a\theta_4/12$

where $a=1/\theta_1$

SS42

Table 4.8.8 gives the coefficients for the stability polynomial of the algorithm SS42. These are also the coefficients for the equivalent four-step scheme. The usefulness of SS42 is limited because each SSp2 has automatically an error which is $O(\Delta t^{p-1})$, hence SS42 has an error of $O(\Delta t^3)$ and thus by the Dahlquist (1963) theorem cannot be unconditionally stable. In §4.10, however, we refer to a wider class of algorithms introduced by Thomas and Gladwell (1988) which includes the SSp2 group. Some suggestions are made by Thomas and Gladwell about using these higher order methods with $p>3$.

4.9. The algorithm SS32B

In this section, we discuss an algorithm introduced in Penry and Wood (1985). This algorithm is equivalent to SS31 applied to the first order form of the dynamic vibration equation but is referred to by the name SS32B to distinguish this particular application from SS31 applied to the semi-discretized parabolic equation.

Hilber and Hughes (1978) give a list of desirable properties for an algorithm for the numerical integration of the system of structural dynamic equations:

$$\boldsymbol{M}\ddot{\boldsymbol{x}}+\boldsymbol{C}\dot{\boldsymbol{x}}+\boldsymbol{K}\boldsymbol{x}=\boldsymbol{f}(t). \tag{4.9.1}$$

These desirable properties are:

(1) unconditional stability for linear problems;

(2) only one set of implicit equations to be solved at each time step;

(3) second order accuracy;

(4) controllable algorithmic damping of the higher frequency modes;

(5) must be self-starting. For example, two-step algorithms require some procedure for generating x_1 from the values of $x(0)$ and $\dot{x}(0)$ which are normally given with the second order differential equation. Special consideration must be given to the order of the procedure used for generating x_1. For instance, an Euler forward difference approximation will ruin the accuracy of the two-step equivalent of SS22 with $\theta_1 = \theta_2 = 0.5$ which is the second order trapezium rule. A single-step algorithm avoids this difficulty.

All these properties (1) to (5) above hold for the algorithms of the GN32 and SS32 groups (which include the Hilber–Hughes–Taylor algorithm as a particular case when this is second order). In the Hilber and Hughes (1978) paper, the authors effectively add another desirable property to the list above:

(6) no tendency for pathological overshooting.

This phenomenon of 'overshooting' is discussed in Goudreau and Taylor (1972) who give results for the response of a simple one-degree-of-freedom oscillator to an initial displacement. They point out that when large time steps are used the Wilson-θ algorithm significantly overshoots the exact solution in the first few time steps, although eventually damping out the solution. They give results for the Wilson–θ with $\theta = 1.4$ and also with the value $\theta = 1.5$ which is used in Farhoomand (1970) which is better but still has inconveniently large overshooting. Goudreau and Taylor (1972) also study the transmission of a triangular pulse in an elastic slab. They obtain a converged result with a sufficiently small value of the time step and then compare the performance of the Wilson-θ algorithm with various versions of the Newmark algorithm:

1. $2\beta = \gamma = 0.5$ (trapezium rule equivalent).
2. $\gamma = 0.5$, $\beta = 1/12$ (Fox–Goodwin 1949) This has higher order accuracy (Table 4.8.2) but is only conditionally stable.

3. $\gamma = 0.55$, $\beta = 0.276$, which is unconditionally stable with slight damping.

4. $\gamma = 0.55$, $\beta = 1/12$, which has slight damping and is conditionally stable.

All the methods show some tendency to produce spurious oscillations at the points of discontinuity of the solution but the Wilson-θ method does seem to be particularly jumpy in this example. Goudreau and Taylor recommend using method (3), which is unconditionally stable with a small amount of damping to eliminate the inaccurate higher frequency modes. They call the use of algorithmic damping here to reduce the oscillations at the points of discontinuity a 'data smoothing technique' and liken it to the use of artificial viscosity to reduce oscillations around shocks (Richtmyer and Morton 1967).

Hilber and Hughes (1978) further discuss the phenomenon of overshoot. They recommend specifically computing the result of the application of an algorithm over one time step for the full range of values expected for $\omega\Delta t$ and comparing the computed x_1 and $\overset{(1)}{x}_1$ with the maximum absolute values of the exact x and $\dot{x}$ to detect any tendency to overshoot. See Exercise 4.6 for some numerical tests on Wilson $\theta = 1.4$ and Houbolt indicating that overshooting on the first time step does only occur with the Δt larger than the standard one-tenth of period.

Hilber and Hughes (1978) compare

(1) Newmark with $2\beta = \gamma = 0.5$ (trapezium rule equivalent);

(2) Newmark with $\gamma = 0.6$, $\beta = 0.3025$;

(3) the Hilber–Hughes–Taylor algorithm discussed in §4.3 with $\alpha = -0.1$;

(4) Wilson-θ with $\theta = 1.4$;

and two collocation methods which are a mixture of Newmark and Wilson-θ. Hilber and Hughes decide that the HHT algorithm (3) gives a good performance with respect to overshooting as well as for algorithmic damping and relative period error.

Figures 4.9.1 to 4.9.4 illustrate the overshoot effect with the Bossak α-method (Wood *et al.* 1980). Figure 4.9.1 shows the effect of using the Bossak method on the solution of $\ddot{x} + \omega^2 x = 0$

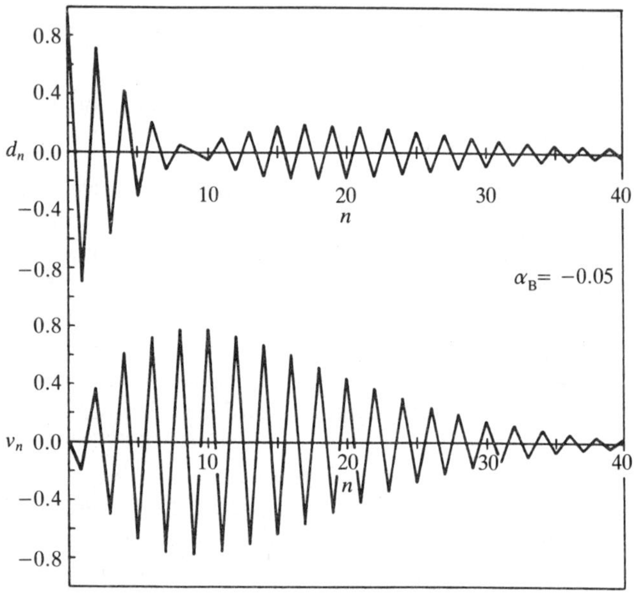

Fig. 4.9.1. Overshoot with Bossak $\alpha_B = -0.05$ with starting vector $[1, 0, -\Omega^2]^T$ and $\Delta t = 20\pi$.

with $\alpha_B = -0.05$ and the unit displacement start giving $[1, 0, -\omega^2 \Delta t^2]^T$ for the starting vector. With $\omega = 1$ the time step $\Delta t = 20\pi$ has the same large value as is used by Hilber and Hughes (1978), i.e. $\Delta t = 10T$ where T is the natural period. The reasoning is that with convergent algorithms there can be no overshoot when $\omega\Delta t \rightarrow 0$, hence the effect is investigated with large values of $\omega\Delta t$. The corresponding HHT method with $\alpha_H = -0.05$ differs from the results in Fig. 4.9.1 by less than 1%. Figures 4.9.2a and b show the Bossak and HHT methods with the same starting vector and time step as for Fig. 4.9.1 and α_H, α_B each equal to -0.3. The two sets of results are very close and there is no overshoot in the displacement in either case. Figure 4.9.3 shows results with unit velocity and zero displacement start for $\alpha_B = -0.1$. There is no overshoot. The results with $\alpha_H = -0.1$ are very nearly the same.

One motive for the investigation in Penry and Wood (1985) was to extend the study of overshoot effects. The idea put forward in Zienkiewicz (1978) of always starting from a known

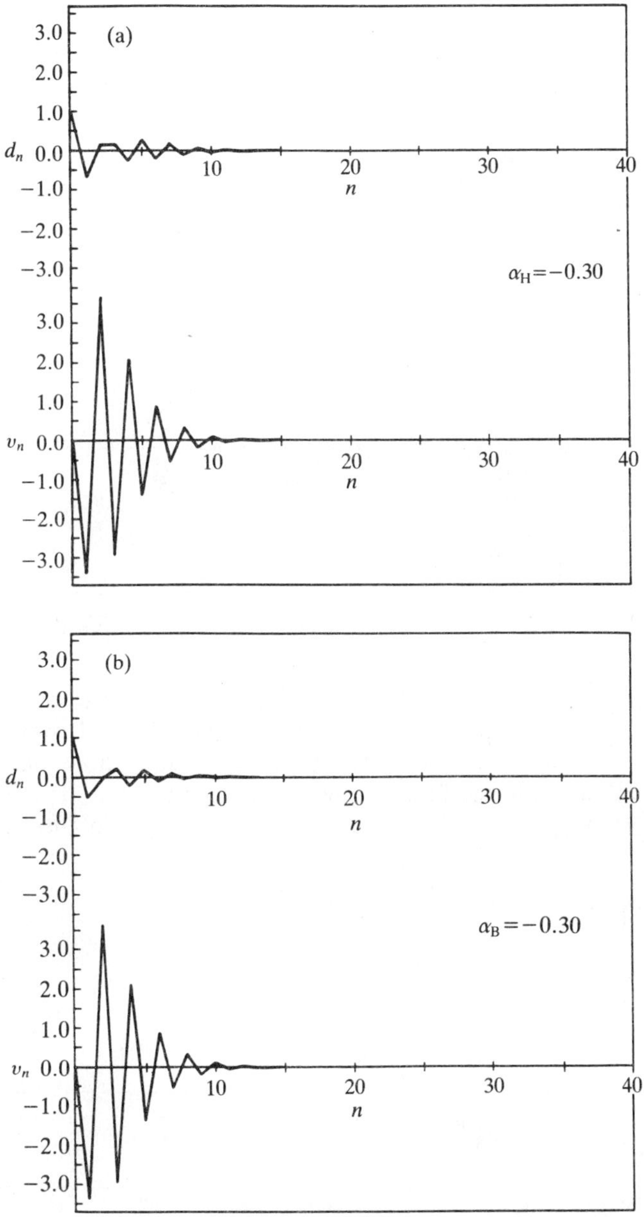

Fig. 4.9.2a,b. Overshoot with starting vector $[1, 0, -\Omega_e^2]^T$, $\Delta t = 20\pi$ and (a) $\alpha_H = -0.3$, (b) $\alpha_B = -0.3$.

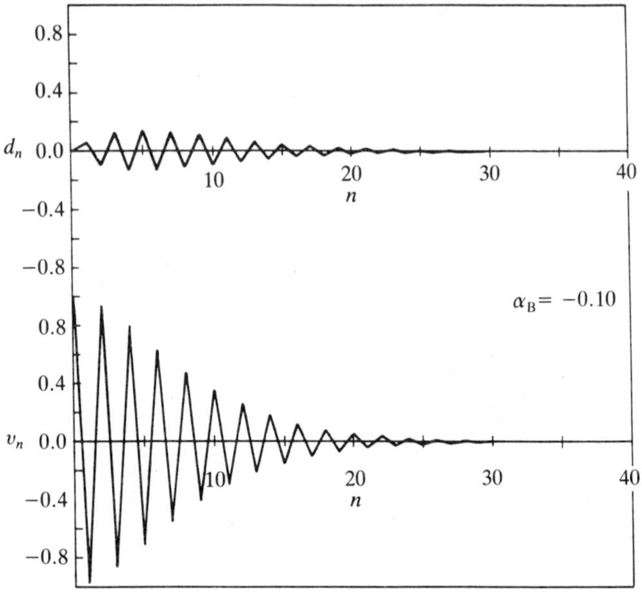

Fig. 4.9.3. Overshoot with $\alpha_B = -0.1$, starting vector $[0, \Delta t, 0]^T$ and $\Delta t = 20\pi$.

steady state solution for $t < 0$ and smoothing out discontinuities may, when appropriate, cut out the Wilson-θ type of overshoot effect in the first few time steps, but there may also be discontinuities in the forcing function later in the run. The Penry and Wood paper examines the effect of these with the SS32B

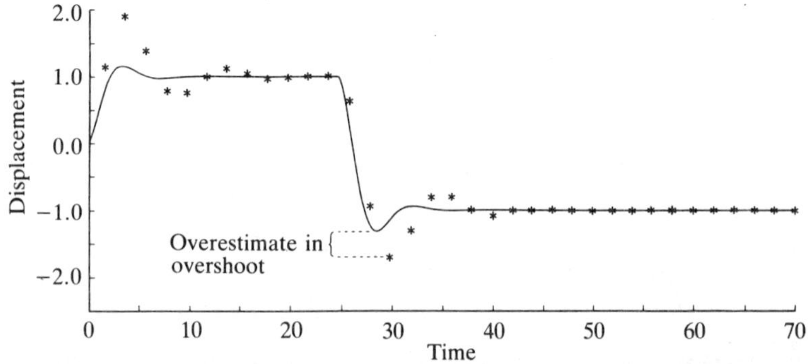

Fig. 4.9.4. Example of overestimate in overshoot. SSpj = SS32. $\theta_1 = 1.40000$, $\theta_2 = 1.96000$, $\theta_3 = 2.74400$, $\Delta t = 2.0000$, NU = 0.5000.

algorithm and compares its performance with that of other algorithms.

Another motivation for investigating the SS32B algorithm is as follows. We know from Dahlquist (1963) that it is not possible to have A-stability with better than $O(\Delta t^2)$ error. But with $A(\alpha)$-stability (see §2.3) it is possible to have $O(\Delta t^3)$ error. The question is how much damping is needed to give $A(\alpha)$-stability with results which are competitive with the A-stable SS22 and SS32 algorithms.

As shown in Chapter 1 if we start with the typical system of structural dynamic equations as in (4.9.1) obtained from the finite element space discretization of a structure, with the matrices $\boldsymbol{M}$ and $\boldsymbol{K}$ symmetric and positive definite and Rayleigh damping, we can make a modal decomposition and show that the exact solution of the system of equations (4.9.1) is the particular integral plus a complementary function given by

$$\sum_{j=1}^{N} a_j \exp(\lambda_j t)\boldsymbol{u}_j,$$

where λ_j satisfies the equation

$$m_j\lambda_j^2 + \mu_j\lambda_j + k_j = 0, \quad j = 1, 2, \ldots, N \tag{4.9.2}$$

and m_j, μ_j, and k_j are given by

$$\boldsymbol{M}\boldsymbol{u}_j = m_j\boldsymbol{u}_j, \qquad \boldsymbol{C}\boldsymbol{u}_j = \mu_j\boldsymbol{u}_j, \qquad \boldsymbol{K}\boldsymbol{u}_j = k_j\boldsymbol{u}_j. \tag{4.9.3}$$

We can reduce the second order system of equations (4.9.1) to a first order system by making the substitution $\boldsymbol{v} = \dot{\boldsymbol{x}}$ and reforming the equations as

$$\dot{\boldsymbol{z}} = -\boldsymbol{B}\boldsymbol{z} + \boldsymbol{F}, \tag{4.9.4}$$

where $\boldsymbol{z}^{\mathrm{T}} = [\boldsymbol{x}, \boldsymbol{v}]$, $\boldsymbol{F}^{\mathrm{T}} = [\boldsymbol{0}, \boldsymbol{M}^{-1}\boldsymbol{f}]$, and

$$\boldsymbol{B} = \begin{bmatrix} \boldsymbol{0} & -\boldsymbol{I} \\ \boldsymbol{M}^{-1}\boldsymbol{K} & \boldsymbol{M}^{-1}\boldsymbol{C} \end{bmatrix}. \tag{4.9.5}$$

The matrix $-\boldsymbol{B}$ has $2N$ eigenvalues given by the $2N$ total roots of the N quadratics in eqn (4.9.2). Supposing the damping in each mode of the system is less than critical, this means that the matrix $-\boldsymbol{B}$ has $2N$ eigenvalues in complex conjugate pairs given by

$$\lambda_j = \omega_j \exp\{\mathrm{i}(\pi \pm \alpha_j)\}, \tag{4.9.6}$$

where $i=\sqrt{-1}$, $\omega_j^2=k_j/m_j$, and $\cos\alpha_j=\nu_j$, where ν_j is the fraction of critical damping in the jth mode, i.e. $\mu_j=2\nu_j\omega_j/m_j$ as before.

These $2N$ eigenvalues of the matrix $-\boldsymbol{B}$ lie in the Argand diagram shown in Fig. 2.3 in the wedge given by $\arg(-\lambda)\le\alpha$ where $\alpha=\max_j\{\alpha_j\}$.

For simplicity we demonstrate the algorithm SS32B on the scalar equation (dropping subscripts)

$$m\ddot{x}+\mu\dot{x}+kx=f. \tag{4.9.7}$$

We again use the notation $\overset{(p)}{x}$ to denote the approximation to the pth derivative of x with respect to time.

The SSp2B algorithm

1. Calculate

$$\begin{aligned}
\tilde{x}_n &= x_n+\theta_1\Delta t\,\overset{(1)}{x}_n+\cdots+\theta_{p-1}\frac{\Delta t^{p-1}}{(p-1)!}\overset{(p-1)}{x}_n\\
\tilde{\tilde{x}}_n &= \overset{(1)}{x}_n+\theta_1\Delta t\,\overset{(2)}{x}_n+\cdots\\
&\quad+\theta_{p-2}\frac{\Delta t^{p-2}}{(p-2)!}\overset{(p-1)}{x}_n,\ p>1,\quad \text{if } p=1,\ \tilde{\tilde{x}}_n=0\\
\tilde{v}_n &= v_n+\theta_1\Delta t\,\overset{(1)}{v}_n+\cdots+\theta_{p-1}\frac{\Delta t^{p-1}}{(p-1)!}\overset{(p-1)}{v}_n\\
\tilde{\tilde{v}}_n &= \overset{(1)}{v}_n+\theta_1\Delta t\,\overset{(2)}{v}_n+\cdots\\
&\quad+\theta_{p-2}\frac{\Delta t^{p-1}}{(p-2)!}\overset{(p-1)}{v}_n,\ p>1,\quad \text{if } p=1,\ \tilde{\tilde{v}}_n=0.
\end{aligned} \tag{4.9.8}$$

2. Calculate β_n^p from the equation

$$\begin{aligned}
&\left\{\theta_{p-1}\frac{\Delta t^{p-1}}{(p-1)!}m+\theta_p\frac{\Delta t^p}{p!}\mu+\frac{\theta_p^2\Delta t^{p+1}}{\theta_{p-1}p(p!)}k\right\}\beta_n^p\\
&\qquad=-m\tilde{\tilde{v}}_n-\mu\tilde{v}_n-k\left\{\tilde{x}_n+\frac{\theta_p\Delta t}{p\theta_{p-1}}(\tilde{v}_n-\tilde{\tilde{x}}_n)\right\}+\bar{f},
\end{aligned} \tag{4.9.9}$$

where $\bar{f}$ is the weighted residual form of the forcing function $f(t)$.

3. Calculate α_n^p from the equation

$$\alpha_n^p = \left\{\frac{\theta_{p-1}\Delta t^{p-1}}{(p-1)!}\right\}^{-1}\left\{\theta_p \frac{\Delta t^p}{p!}\beta_n^p + (\tilde{v}_n - \tilde{\dot{x}}_n)\right\}. \quad (4.9.10)$$

4. Obtain x_{n+1}, $\overset{(1)}{x}_{n+1}$, v_{n+1}, $\overset{(1)}{v}_{n+1}$, etc. by substituting for α_n^p and β_n^p from (2) and (3) into the equations

$$x_{n+1} = x_n + \Delta t \overset{(1)}{x}_n + \cdots + \frac{\Delta t^{p-1}}{(p-1)!}\overset{(p-1)}{x_n} + \frac{\Delta t^p}{p!}\alpha_n^p$$

$$\overset{(1)}{x}_{n+1} = \overset{(1)}{x}_n + \Delta t \overset{(2)}{x}_n + \cdots + \frac{\Delta t^{p-1}}{(p-1)!}\alpha_n^p, \quad \text{etc.}$$

$$v_{n+1} = v_n + \Delta t \overset{(1)}{v}_n + \cdots + \frac{\Delta t^{p-1}}{(p-1)!}\overset{(p-1)}{v_n} + \frac{\Delta t^p}{p!}\beta_n^p$$

$$\overset{(1)}{v}_{n+1} = \overset{(1)}{v}_n + \Delta t \overset{(2)}{v}_2 + \cdots + \frac{\Delta t^{p-1}}{(p-1)!}\beta_n^p, \quad \text{etc.} \quad (4.9.11)$$

5. Repeat for next time step.

When we are using the algorithm on the system of equations (4.9.1), the matrix to be inverted at each time step is

$$\theta_{p-1}\frac{1}{(p-1)!}\boldsymbol{M} + \theta_p \frac{\Delta t}{p!}\boldsymbol{C} + \frac{\theta_p^2 \Delta t^2}{\theta_{p-1} p(p!)}\boldsymbol{K}$$

which is of the same order as that in the SS*pj* algorithm. The extra equations in step (3) are scalar equations which do not require a matrix inversion.

In Penry and Wood (1985) the algorithms SS22, SS32 and SS32B are used for the numerical solution of the representative equation

$$\ddot{x} + 2v\dot{x} + x = f(t), \quad (4.9.12)$$

where (i) $f(t)$ is the step function:

$$\begin{aligned} f(t) &= 0, \ t < 0 \\ f(t) &= 1, \ 0 \le t < 25 \\ f(t) &= -1, \ t \ge 25, \end{aligned} \quad (4.9.13)$$

and (ii) $f(t)$ is the periodic function

$$f(t) = \sin(\pi t/20). \quad (4.9.14)$$

The initial conditions are taken as $x(0)=0$, $\dot{x}(0)=0$. The exact solutions are obtained using the Green's function which is given, for example, in Symon (1977).

The step function (i) is chosen in order to look at the overshoot effect, not only at the start but also at the sudden change in the forcing function from +1 to −1 at time $t=25$. Figure 4.9.4 illustrates the overshoot effect with the SS32 parameters which give the equivalent of the Wilson $\theta=1.4$ Algorithm (Wilson 1968; Bathe and Wilson 1973) used on Problem (i). Figure 4.9.5 shows that we may also get a negative overshoot or 'undershoot'. This result is with the SS32 parameters which give the equivalent of the Houbolt algorithm (Houbolt 1950).

Tables 4.9.1 to 4.9.12 show numerical results obtained by using the algorithms SS22, SS32, and SS32B for the numerical solution of eqn (4.9.7) with the forcing functions (i) and (ii). The error is given at time $t=5$ as a typical point and measured as $|x_n - x(5)|$, where $x(5)$ is the exact solution and x_n is the numerical solution at $n\Delta t=5$. This error is also expressed as a percentage of the exact solution. With the step function (i) the tables also give the overshoot or undershoot at $t=25$ measured as shown in Figs. 4.9.2 and 4.9.3 and this is also expressed as a percentage of the exact minimum value of $x(t)$ after the jump in the forcing function.

The forcing function is approximated by taking the weighted

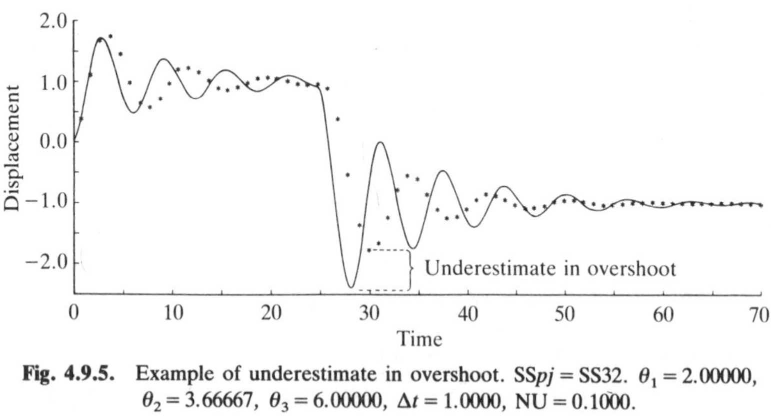

Fig. 4.9.5. Example of underestimate in overshoot. SS*pj* = SS32. $\theta_1=2.00000$, $\theta_2=3.66667$, $\theta_3=6.00000$, $\Delta t=1.0000$, NU = 0.1000.

average value

$$\bar{f} = \theta_1 f_{n+1} + (1 - \theta_1) f_n \tag{4.9.15}$$

as recommended in Zienkiewicz *et al.* (1984). The results show that this is adequate for first and second order methods but not for the third order SS32B algorithm.

SS22

1. $\theta_1 = \theta_2 = 0.5$ (equivalent to Newmark with $2\beta = \gamma = 0.5$).

As shown in §4.8 this is equivalent to using the trapezium rule and by the result in Dahlquist (1963) it is the most accurate second order method; it gives no algorithmic damping.

Tables 4.9.1 and 4.9.2 show numerical results for the step and sine functions respectively with the value of the time step Δt progressively halved. The second order accuracy shows in the way the ratio of the errors at time $t = 5$ tends to the value 4 as

Tables 4.9.1 to 4.9.12. Numerical results from SS*pj* used on $\ddot{x} + 2v\dot{x} + x = f$ with $f = 0$, sine and step functions.

Scheme: SS22; parameters: $\theta_1 = 0.5$, $\theta_2 = 0.5$; forcing term: $f = +0$, $t < 0$; $f = +1$, $0 \le t \le 25$; $f = -1$, $t > 25$.

Damping: $v = 0.5$

Δt	Overshoot (% error)	Error at $t = 5.0$ (% error)
0.5	2.34×10^{-2} (1.77)	8.23×10^{-3} (7.66×10^{-1})
0.25	6.33×10^{-3} (4.78×10^{-1})	1.97×10^{-3} (1.83×10^{-1})
0.125	1.54×10^{-3} (1.16×10^{-1})	4.87×10^{-4} (4.54×10^{-2})
0.062 5	3.85×10^{-4} (2.90×10^{-2})	1.22×10^{-4} (1.13×10^{-2})
0.031 25	9.62×10^{-5} (7.25×10^{-3})	3.04×10^{-5} (2.83×10^{-3})
0.015 625	2.41×10^{-5} (1.81×10^{-3})	7.59×10^{-6} (7.06×10^{-4})

Damping: $v = 0.1$

Δt	Overshoot (% error)	Error at $t = 5.0$ (% error)
0.5	-1.83×10^{-2} (7.68×10^{-1})	5.70×10^{-2} (6.33)
0.25	-7.14×10^{-3} (2.98×10^{-1})	1.42×10^{-2} (1.58)
0.125	-1.22×10^{-3} (5.10×10^{-2})	3.55×10^{-3} (3.94×10^{-1})
0.062 5	-3.01×10^{-4} (1.25×10^{-2})	8.88×10^{-4} (9.85×10^{-2})
0.031 25	-7.70×10^{-5} (3.44×10^{-3})	2.22×10^{-4} (2.46×10^{-2})
0.015 625	-1.97×10^{-5} (8.20×10^{-4})	5.55×10^{-5} (6.16×10^{-3})

Table 4.9.2.

Scheme: SS22; parameters: $\theta_1 = 0.5$, $\theta_2 = 0.5$; forcing term: $f = \sin(\pi t/20)$.

Damping: $\nu = 0.5$

Δt	Error at $t = 5.0$	Error (%)
0.5	7.83×10^{-5}	1.32×10^{-2}
0.25	2.86×10^{-5}	4.82×10^{-3}
0.125	7.69×10^{-6}	1.29×10^{-3}
0.062 5	1.96×10^{-6}	3.29×10^{-4}
0.031 25	4.91×10^{-7}	8.25×10^{-5}
0.015 625	1.23×10^{-7}	2.07×10^{-5}

Damping: $\nu = 0.1$

Δt	Error at $t = 5.0$	Error (%)
0.5	3.04×10^{-3}	3.80×10^{-1}
0.25	8.71×10^{-4}	1.09×10^{-1}
0.125	2.25×10^{-4}	2.81×10^{-2}
0.062 5	5.67×10^{-5}	7.09×10^{-3}
0.031 25	1.42×10^{-5}	1.78×10^{-3}
0.015 625	3.55×10^{-6}	4.44×10^{-4}

the time step is decreased. The ratios of the overshoots for natural damping ratio $\nu = 0.5$ and the undershoots for $\nu = 0.1$ also tend to a value approximately equal to 4. The approximation for $\bar{f}$ given by eqn (4.9.15) is adequate here.

2. $\theta_1 = 0.6$, $\theta_2 = 0.605$.

These parameters are chosen to give a small amount of algorithmic damping and maintain unconditional stability ($\theta_2 \geq \theta_1 \geq 0.5$). Tables 4.9.3 and 4.9.4 show the numerical results for the step and sine functions respectively. The algorithm is now first order accurate and the overshoot and undershoot and the error at time $t = 5$ each show a ratio approximately equal to 2 as the time step is progressively halved.

SS32

This algorithm requires for the starting conditions the value of the acceleration as well as the given initial conditions of the

Table 4.9.3.

Scheme: SS22; parameters: $\theta_1 = 0.6$, $\theta_2 = 0.605$; forcing term: $f = 0$, $t < 0$; $f = +1$, $0 \le t \le 25$; $f = -1$, $t > 25$.

Damping: $\nu = 0.5$

Δt	Overshoot (% error)	Error at $t = 5.0$ (% error)
0.5	-1.62×10^{-2} (1.24)	1.70×10^{-2} (1.6)
0.25	-7.01×10^{-3} (5.37×10^{-1})	6.69×10^{-3} (6.23×10^{-1})
0.125	-3.55×10^{-3} (2.68×10^{-1})	2.90×10^{-3} (2.70×10^{-1})
0.062 5	-2.12×10^{-3} (1.60×10^{-1})	1.34×10^{-3} (1.25×10^{-1})
0.031 25	-1.15×10^{-3} (8.66×10^{-2})	6.46×10^{-4} (6.01×10^{-2})
0.015 625	-5.968×10^{-4} (4.50×10^{-2})	3.16×10^{-4} (2.94×10^{-2})

Damping: $\nu = 0.1$

Δt	Overshoot (% error)	Error at $t = 5.0$ (% error)
0.5	6.89×10^{-2} (2.88)	9.54×10^{-2} (10.59)
0.25	2.85×10^{-2} (1.19)	3.68×10^{-2} (4.09)
0.125	1.74×10^{-2} (7.23×10^{-1})	1.54×10^{-2} (1.71)
0.062 5	8.86×10^{-3} (3.69×10^{-1})	6.91×10^{-3} (7.67×10^{-1})
0.031 25	4.40×10^{-3} (1.83×10^{-1})	3.25×10^{-3} (3.61×10^{-1})
0.015 625	2.23×10^{-3} (9.28×10^{-2})	1.58×10^{-3} (1.75×10^{-1})

displacement and velocity. The differential equation is used to generate the acceleration from

$$\ddot{x}(0) = f(0) - 2\nu\dot{x}(0) - x(0). \qquad (4.9.16)$$

The Penry and Wood (1985) numerical experiments show that with the step function (4.9.13) which is discontinuous at $t = 0$, it is necessary to use

$$\ddot{x}(0) = f(0+) - 2\nu\dot{x}(0) - x(0),$$

thus using the value of $f(t)$ which applies to the time step over which we are integrating.

All the SS32 algorithms are automatically of second order accuracy.

1. $\theta_1 = 2$, $\theta_2 = 11/3$, $\theta_3 = 6$

This is the SS32 equivalent of the algorithm introduced by Houbolt (1950) which was a three-step in the displacements only.

Table 4.9.4.

Scheme: SS22; parameters: $\theta_1 = 0.6$, $\theta_2 = 0.605$; forcing term: $f = \sin(\pi t/20)$.

Damping: $\nu = 0.5$

Δt	Error at $t = 5.0$	Error (%)
0.5	2.14×10^{-3}	3.60×10^{-1}
0.25	1.04×10^{-3}	1.75×10^{-1}
0.125	5.24×10^{-4}	8.81×10^{-2}
0.062 5	2.64×10^{-4}	4.43×10^{-2}
0.031 25	1.32×10^{-4}	2.23×10^{-2}
0.015 625	6.63×10^{-5}	1.11×10^{-2}

Damping: $\nu = 0.1$

Δt	Error at $t = 5.0$	Error (%)
0.5	7.74×10^{-3}	9.68×10^{-1}
0.25	4.67×10^{-3}	5.85×10^{-1}
0.125	2.59×10^{-3}	3.24×10^{-1}
0.062 5	1.36×10^{-3}	1.71×10^{-1}
0.031 25	6.98×10^{-4}	8.74×10^{-2}
0.015 625	3.54×10^{-4}	4.43×10^{-2}

As shown in §4.8 this algorithm is unconditionally stable. Tables 4.9.5 and 4.9.6 show the numerical results of the Houbolt SS32 algorithm applied to the step and sine functions respectively. It is interesting to see that the undershoot with the lower natural damping $\nu = 0.1$ which was used in Fig. 4.9.5 to illustrate undershoot becomes an overshoot with the higher natural damping $\nu = 0.5$. Also the undershoot with $\nu = 0.1$ is no longer in such a regular ratio as with the previous results but the error at time $t = 5$ shows a ratio approximately equal to 4 as the time step is halved as expected. From Table 4.8.4 the θ parameters for the SS32 Houbolt equivalent make the coefficients $\beta_2 = \beta_1 = \beta_0 = 0$, this means that as $\omega \Delta t$ becomes very large the roots of the characteristic polynomial are all zero so that this algorithm is as heavily damped as is possible.

2. $\theta_1 = 1.4$, $\theta_2 = 1.96$, $\theta_3 = 2.744$.

This is the SS32 equivalent of the Wilson-θ algorithm with

Table 4.9.5.

Scheme: SS32; parameters: $\theta_1 = 2.0$, $\theta_2 = 11/3$, $\theta_3 = 6.0$ (single step Houbolt equivalent); forcing term: $f = 0$, $t < 0$; $f = +1$, $0 \le t \le 25$; $f = -1$, $t > 25$.

Damping: $\nu = 0.5$

Δt	Overshoot (% error)	Error at $t = 5.0$ (% error)
0.5	1.80×10^{-2} (1.35)	5.81×10^{-2} (5.41)
0.25	1.71×10^{-2} (1.28)	7.87×10^{-3} (7.32×10^{-1})
0.125	5.10×10^{-3} (3.83×10^{-1})	1.20×10^{-3} (1.12×10^{-1})
0.062 5	1.35×10^{-3} (1.01×10^{-1})	2.22×10^{-4} (2.07×10^{-2})
0.031 25	3.46×10^{-4} (2.29×10^{-2})	4.69×10^{-5} (4.36×10^{-3})
0.015 625	8.99×10^{-5} (6.76×10^{-3})	1.07×10^{-5} (9.96×10^{-4})

Damping: $\nu = 0.1$

Δt	Overshoot (% error)	Error at $t = 5.0$ (% error)
0.5	-1.98×10^{-1} (8.31)	2.26×10^{-1} (25.1)
0.25	-7.14×10^{-2} (3.10)	6.93×10^{-2} (7.69)
0.125	-2.03×10^{-2} (8.52×10^{-1})	1.79×10^{-2} (1.98)
0.062 5	-5.20×10^{-3} (2.18×10^{-1})	4.43×10^{-3} (4.92×10^{-1})
0.031 25	-1.68×10^{-3} (7.05×10^{-2})	1.10×10^{-3} (1.22×10^{-1})
0.015 625	-5.76×10^{-4} (2.42×10^{-2})	2.72×10^{-4} (3.02×10^{-2})

$\theta = 1.4$ ($\theta_1 = \theta$, $\theta_2 = \theta^2$, $\theta_3 = \theta^3$). The Wilson-θ algorithm is discussed in §4.8. Tables 4.9.7 and 4.9.8 show the numerical results with the step and sine function respectively. We observe that the Wilson-θ overshoot becomes an undershoot at the jump in the step function with the lower natural damping. The overshoot and the error at time $t = 5$ show a ratio approximately equal to 4 as the time step is halved but not the undershoot.

3. $\theta_1 = 1.05$, $\theta_2 = 1.1$, $\theta_3 = 1.15$.

These parameters are chosen to give the SS32 Bossak–Newmark equivalent (Wood *et al.* 1980). The original Bossak–Newmark, as explained in §4.4, is in general only first order accurate. With the parameters chosen as in Table 4.8.5 the Bossak–Newmark algorithm is second order accurate and the parameters here are taken with the Bossak $\alpha_B = -0.05$. As with the Hilber–Hughes–Taylor Newmark variation (Hilber *et al.* 1977) discussed in §4.4,

Table 4.9.6.

Scheme: SS32; parameters: $\theta_1 = 2.0$, $\theta_2 = 11/3$, $\theta_3 = 6.0$ (single step Houbolt equivalent); forcing term: $f = \sin(\pi t/20)$.

Damping: $\nu = 0.5$

Δt	Error at $t = 5.0$	Error (%)
0.5	2.48×10^{-3}	4.18×10^{-1}
0.25	1.94×10^{-3}	3.26×10^{-1}
0.125	5.84×10^{-4}	9.82×10^{-2}
0.062 5	1.54×10^{-4}	2.59×10^{-2}
0.031 25	3.92×10^{-5}	6.59×10^{-3}
0.015 625	9.89×10^{-6}	1.66×10^{-3}

Damping: $\nu = 0.1$

Δt	Error at $t = 5.0$	Error (%)
0.5	5.97×10^{-3}	7.47×10^{-1}
0.25	5.57×10^{-3}	6.97×10^{-1}
0.125	1.99×10^{-3}	2.49×10^{-1}
0.062 5	5.66×10^{-4}	7.09×10^{-2}
0.031 25	1.49×10^{-4}	1.87×10^{-2}
0.015 625	3.82×10^{-5}	4.78×10^{-3}

the useful values of the extra parameter α are numerically small and negative. The Bossak–Newmark algorithm is conditionally stable; the stability conditions are given by inequalities (4.8.33) and (4.8.34). From eqns (4.8.32) we have

$$a = 0.55, \qquad b = -0.1167, \qquad c = -0.25.$$

Hence $12a^2 - 1 > 0$ and $3ab - c > 0$ and therefore inequality (4.8.34) is satisfied. The conditions on $\omega\Delta t$ from (4.8.33a) and (4.8.33b) give

$$\omega\Delta t < 3.7 \text{ approximately, for } \nu = 0.1$$

and

$$\omega\Delta t < 3.9 \text{ approximately, for } \nu = 0.5.$$

With the 'rule of thumb' of at least 10 time steps per period $\omega\Delta t < 2\pi/10 = 0.63$ approximately, so with this the Bossak–Newmark algorithm is safe. Tables 4.9.9 and 4.9.10 show the

Table 4.9.7.

Scheme: SS32; parameters: $\theta_1 = 1.4$, $\theta_2 = 1.96$, $\theta_3 = 2.744$ (single step Wilson-$\theta = 1.4$ equivalent); forcing term: $f = 0$, $t < 0$; $f = +1$, $0 \le t \le 25$; $f = -1$, $t > 25$.

Δt	Overshoot (% error)	Error at $t = 5.0$ (% error)
Damping: $\nu = 0.5$ (overestimate at overshoot)		
0.5	3.04×10^{-2} (2.30)	1.28×10^{-2} (1.19)
0.25	1.06×10^{-2} (8.02×10^{-1})	1.87×10^{-3} (1.74×10^{-1})
0.125	2.80×10^{-3} (2.12×10^{-1})	3.44×10^{-4} (3.20×10^{-2})
0.062 5	7.07×10^{-4} (5.35×10^{-2})	7.31×10^{-5} (6.80×10^{-3})
0.031 25	1.78×10^{-4} (1.35×10^{-2})	1.68×10^{-5} (1.56×10^{-3})
0.015 625	4.52×10^{-5} (3.42×10^{-3})	4.02×10^{-6} (3.74×10^{-4})
Damping: $\nu = 0.1$ (underestimate at overshoot)		
0.5	-7.80×10^{-2} (3.27)	1.06×10^{-1} (11.8)
0.25	-2.06×10^{-2} (8.65×10^{-1})	2.79×10^{-2} (3.10)
0.125	-6.04×10^{-3} (2.53×10^{-1})	6.97×10^{-3} (7.74×10^{-1})
0.062 5	-1.81×10^{-3} (7.60×10^{-2})	1.73×10^{-3} (1.92×10^{-1})
0.031 25	-6.99×10^{-4} (2.93×10^{-2})	4.31×10^{-3} (4.78×10^{-2})
0.015 625	-2.73×10^{-4} (1.14×10^{-2})	1.09×10^{-3} (1.32×10^{-2})

numerical results. It is interesting to see how well this algorithm performs with an overshoot much smaller in magnitude than those with Houbolt as well as with the Wilson-θ and better than the SS22 trapezium rule equivalent for damping ratio $\nu = 0.5$ though not with $\nu = 0.1$.

SS32B

The SS32B algorithm given in eqns (4.9.8) to (4.9.11) requires starting values

$$x_0, \overset{(1)}{x}_0, \overset{(2)}{x}_0$$

and

$$v_0, \overset{(1)}{v}_0, \overset{(2)}{v}_0.$$

We take

$$x_0 = x(0), \quad \overset{(1)}{x}_0 = \dot{x}(0), \quad \overset{(2)}{x}_0 = \ddot{x}(0)$$

Table 4.9.8.

Scheme: SS32; parameters: $\theta_1=1.4$, $\theta_2=1.96$, $\theta_3=2.744$ (single step Wilson-$\theta=1.4$ equivalent); forcing term: $f=\sin(\pi t/20)$

Damping: $\nu=0.5$

Δt	Error at $t=5.0$	Error (%)
0.5	2.90×10^{-3}	5.01×10^{-1}
0.25	9.01×10^{-4}	1.52×10^{-1}
0.125	2.35×10^{-4}	3.95×10^{-2}
0.062 5	6.26×10^{-5}	1.05×10^{-2}
0.031 25	1.50×10^{-5}	2.52×10^{-3}
0.015 625	3.76×10^{-6}	6.32×10^{-4}

Damping $\nu=0.1$

Δt	Error at $t=5.0$	Error (%)
0.5	8.68×10^{-3}	1.09
0.25	3.15×10^{-3}	3.94×10^{-1}
0.125	8.92×10^{-4}	1.12×10^{-1}
0.062 5	2.34×10^{-4}	2.93×10^{-2}
0.031 25	5.98×10^{-5}	7.48×10^{-3}
0.015 625	1.51×10^{-5}	1.89×10^{-3}

and

$$v_0=\dot{x}(0),\quad \overset{(1)}{v_0}=\ddot{x}(0),\quad \overset{(2)}{v_0}=\dddot{x}(0),$$

where $\ddot{x}(0)$ is obtained as in SS32 from eqn (4.9.16) and the third derivative $\dddot{x}(0)$ is generated by supposing we can differentiate the differential equation to produce

$$\dddot{x}(0)=\frac{d^3x(0)}{dt^3}=\dot{f}(0)-2\nu\ddot{x}(0)-\dot{x}(0). \tag{4.9.17}$$

With the step function (i) the value taken for $\dot{f}(0)$ is zero.

The SS32B algorithm is equivalent to using SS31 on the first order form of the second order equation (4.9.12). SS31 is the algorithm obtained from the eqns (4.6.4) to (4.6.10) by putting $M\equiv0$, the matrix referred to as C then becomes the mass matrix. The θ parameters used for the SS32B numerical results shown in Tables 4.9.11 and 4.9.12 are the same as with the SS32

Table 4.9.9.

Scheme: SS32; parameters: $\theta_1 = 1.05$, $\theta_2 = 1.1$, $\theta_3 = 1.15$ Bossak–Newmark equivalent); forcing term: $f = 0$, $t < 0$; $f = +1$, $0 \le t \le 25$; $f = -1$, $t > 25$.

Damping: $\nu = 0.5$

Δt	Overshoot (% error)	Error at $t = 5.0$ (% error)
0.5	1.72×10^{-3} (1.30×10^{-1})	1.55×10^{-3} (1.44×10^{-1})
0.25	6.23×10^{-3} (4.71×10^{-1})	3.36×10^{-4} (3.13×10^{-2})
0.125	5.39×10^{-4} (4.08×10^{-2})	7.86×10^{-5} (7.31×10^{-3})
0.062 5	1.44×10^{-4} (1.09×10^{-2})	1.90×10^{-5} (1.77×10^{-3})
0.031 25	4.17×10^{-5} (3.15×10^{-3})	4.67×10^{-6} (4.35×10^{-2})
0.015 625	1.34×10^{-5} (1.01×10^{-3})	1.16×10^{-6} (1.08×10^{-4})

Damping: $\nu = 0.1$

Δt	Overshoot (% error)	Error at $t = 5.0$ (% error)
0.5	-8.65×10^{-4} (3.63×10^{-2})	3.28×10^{-2} (3.64)
0.25	-1.57×10^{-3} (6.59×10^{-2})	8.23×10^{-3} (9.13×10^{-1})
0.125	-1.84×10^{-3} (7.72×10^{-2})	2.05×10^{-3} (2.28×10^{-1})
0.062 5	-3.40×10^{-4} (1.43×10^{-2})	5.12×10^{-4} (5.68×10^{-2})
0.031 25	-2.78×10^{-4} (1.17×10^{-2})	1.28×10^{-4} (1.42×10^{-2})
0.015 625	-1.19×10^{-4} (4.99×10^{-3})	3.19×10^{-5} (3.54×10^{-3})

(1) Houbolt algorithm. These give the single-step equivalent of the Gear three-step algorithm (Gear 1969), designed for use with a first order differential equation. The parameters are the same as with the Houbolt algorithm because Gear also designed his series of p-step algorithms, $p = 2, 3,\ldots$ always to have zero roots for what corresponds here to $\omega\Delta t$ tending to infinity.

Tables 4.9.11 and 4.9.12 show the numerical results of using the SS32B algorithm on eqn (4.9.12) with the step function (i) and sine function (ii) respectively. With a zero forcing function this algorithm gives third order accuracy as expected, i.e. the errors decrease by a factor approximately equal to 8 as the time step is halved. With the step function, Table 4.9.11 shows the error at time $t = 5$ also decreasing by a factor approximately equal to 8 which shows that, as expected, taking the weighted average as in eqn (4.9.15) is adequate for a forcing function which is constant for $0 \le t \le 25$. With the sine function, however,

Table 4.9.10.

Scheme: SS32; parameters: $\theta_1 = 1.05$, $\theta_2 = 1.1$, $\theta_3 = 1.15$ (Bossak–Newmark equivalent); forcing term: $f = \sin(\pi t/20)$.

Damping: $\nu = 0.5$

Δt	Error at $t = 5.0$	Error (%)
0.5	9.23×10^{-4}	1.55×10^{-1}
0.25	2.32×10^{-4}	3.91×10^{-2}
0.125	5.83×10^{-5}	9.30×10^{-3}
0.062 5	1.46×10^{-5}	2.46×10^{-3}
0.031 25	3.65×10^{-6}	7.63×10^{-4}
0.015 625	9.14×10^{-7}	1.54×10^{-4}

Damping: $\nu = 0.1$

Δt	Error at $t = 5.0$	Error (%)
0.5	3.87×10^{-3}	4.85×10^{-1}
0.25	1.04×10^{-3}	1.31×10^{-1}
0.125	2.67×10^{-4}	3.35×10^{-2}
0.062 5	6.76×10^{-5}	8.46×10^{-3}
0.031 25	1.70×10^{-5}	2.12×10^{-3}
0.015 625	4.25×10^{-6}	5.32×10^{-4}

the ratio of the errors is approximately 4 because the approximation (4.9.15) is not adequate here. This demonstrates the importance of a sufficiently accurate approximation of the forcing function and shows that the formula (4.9.15) recommended in Zienkiewicz *et al.* (1984) will not do for higher order methods. The consistency conditions for the p-step scheme discussed in §2.4 necessarily depend on the approximation of the forcing function as well as of the rest of the differential equation. It is suggested there that it is reasonable to take such a scheme applied to the differential equation

$$m\ddot{x} + \mu\dot{x} + kx = f(t)$$

as

$$\sum_{j=0}^{p} (\alpha_j m + \gamma_j \Delta t \mu + \beta_j \Delta t^2 k) x_{n+j} = \Delta t^2 \sum_{j=0}^{p} \beta_j f(\overline{n+j}\Delta t),$$

i.e. $f(t)$ is approximated in the same pattern as the kx term.

Table 4.9.11.

Scheme: SS32B; parameters: $\theta_1 = 2.0$, $\theta_2 = 11/3$, $\theta_3 = 6.0$ (Single step equivalent to Gear three-step); forcing term: $f = 0$, $t < 0$; $f = +1$, $0 \le t \le 25$; $f = -1$, $t > 25$.

Damping: $\nu = 0.5$

Δt	Overshoot (% error)	Error at $t = 5.0$ (% error)
0.5	3.75×10^{-2} (2.83)	1.06×10^{-2} (9.85×10^{-1})
0.25	4.95×10^{-3} (3.74×10^{-1})	1.62×10^{-3} (1.51×10^{-1})
0.125	6.42×10^{-4} (4.86×10^{-2})	2.11×10^{-4} (1.96×10^{-2})
0.062 5	2.23×10^{-4} (1.69×10^{-2})	2.67×10^{-5} (2.48×10^{-3})
0.031 25	5.80×10^{-5} (4.39×10^{-3})	3.35×10^{-6} (3.11×10^{-4})
0.015 625	1.23×10^{-5} (9.30×10^{-4})	4.19×10^{-7} (3.90×10^{-5})

Damping: $\nu = 0.1$

Δt	Overshoot (% error)	Error at $t = 5.0$ (% error)
0.5	-5.30×10^{-2} (2.22)	9.56×10^{-3} (1.06)
0.25	-2.16×10^{-2} (9.06×10^{-1})	2.87×10^{-3} (3.18×10^{-1})
0.125	-3.68×10^{-3} (1.54×10^{-1})	5.87×10^{-4} (6.51×10^{-2})
0.062 5	-1.44×10^{-3} (6.04×10^{-2})	8.66×10^{-5} (9.60×10^{-3})
0.031 25	-8.11×10^{-4} (3.40×10^{-2})	1.16×10^{-5} (1.29×10^{-3})
0.015 625	-3.76×10^{-4} (1.58×10^{-2})	1.50×10^{-6} (1.67×10^{-4})

It is not immediately obvious how to form the weighted average approximation $\bar{f}$ in the eqns (4.6.7) for the SS*pj* algorithm where we are concentrating on a particular time step and in progressing from time $t = n\Delta t$ to time $t = (n+1)\Delta t$ we want only to use values of $f(t)$, $n\Delta t \le t \le (n+1)\Delta t$. (This difficulty does not arise with the GN*pj* algorithms described in §4.5 because these use collocation.)

In studying the consistency, stability, and accuracy of the SS*pj* algorithms in §4.8 we transferred attention to the corresponding p-step methods. But in order to decide how to approximate the forcing function actually in the single-step SS*pj*, i.e. how to express the weighted average $\bar{f}$, we look at the consistency condition in the approximation of the differential equation given in eqn (4.6.8). Taking α_n^p as the approximation to the pth derivative, we obtain the local truncation error here in the usual way by substituting the exact solution of the differential equation

Table 4.9.12.

Scheme: SS32B; parameters: $\theta_1 = 2.0$, $\theta_2 = 11/3$, $\theta_3 = 6.0$ (single step equivalent to Gear three-step); forcing term: $f = \sin(\pi t/20)$

Damping: $\nu = 0.5$

Δt	Error at $t = 5.0$	Error (%)
0.5	5.53×10^{-3}	9.30×10^{-1}
0.25	1.01×10^{-3}	1.70×10^{-1}
0.125	2.18×10^{-4}	3.66×10^{-2}
0.062 5	5.09×10^{-5}	8.55×10^{-3}
0.031 25	1.23×10^{-5}	2.07×10^{-3}
0.015 625	3.03×10^{-6}	5.10×10^{-4}

Damping: $\nu = 0.1$

Δt	Error at $t = 5.0$	Error (%)
0.5	1.69×10^{-2}	2.12
0.25	2.81×10^{-3}	3.52×10^{-1}
0.125	4.76×10^{-4}	5.96×10^{-2}
0.062 5	9.11×10^{-5}	1.14×10^{-2}
0.031 25	1.94×10^{-5}	2.42×10^{-3}
0.015 625	4.42×10^{-6}	5.54×10^{-4}

into the left hand side of eqn (4.6.8). Applied to the scalar eqn (4.9.7) and dropping subscripts with the understanding that all terms are evaluated at time $t = n\Delta t$, we have the local truncation error

$$\tau = m\ddot{x} + \mu\dot{x} + kx + \theta_1 \Delta t \frac{\mathrm{d}}{\mathrm{d}t}(m\ddot{x} + \mu\dot{x} + kx) + \theta_2 \frac{\Delta t^2}{2}\frac{\mathrm{d}^2}{\mathrm{d}t^2}(m\ddot{x} + \mu\dot{x} + kx) + \cdots - \bar{f}. \tag{4.9.18}$$

We know from §4.7 that for this second order equation we have in general an error in the numerical solution of $O(\Delta t^{p-1})$ and this is of the same order as the local truncation error τ. But for SS22 we need to allow for τ to be $O(\Delta t^2)$ because of the frequently used $\theta_1 = \theta_2 = 0.5$ case. Hence for SS22 and SS32 we find that

$$\bar{f}_n = f(\overline{n + \theta_1 \Delta t}), \quad \text{or } \bar{f} = f + \theta_1 \Delta t \dot{f} \tag{4.9.19}$$

is adequate. We also have

$$\theta_1 f_{n+1} + (1-\theta_1)f_n = f_n + \theta_1 \Delta t \dot{f}_n \text{ plus error } O(\Delta t^2).$$

This explains the correct order error shown in Tables 4.9.1 to 4.9.10 using the approximation (4.9.15) but for SS42, say, the local truncation error τ must be $O(\Delta t^3)$. For SS32B which corresponds to SS31, the algorithm now used on a first order equation, the error is again $O(\Delta t^3)$. Hence we need $\bar{f}$ to be represented by

$$\bar{f} = f_n + \theta_1 \Delta t \dot{f}_n + \theta_2 \Delta t^2 \ddot{f}_n \tag{4.9.20}$$

or its equivalent in terms of $f_n, f_{n+\frac{1}{2}}, f_{n+1}$, for example, whichever is convenient.

It is shown in Penry and Wood (1985) that the SS32B algorithm is $A(\alpha)$ stable with $\alpha \cong 88°$ which corresponds to unconditional stability if the natural damping ratio $\nu \geq 0.7$. Alternatively we can say that the algorithm will be stable if $\omega \Delta t > 1/12$. Figure 2.3 shows that either of these conditions corresponds to part of the left hand half-plane outside the Gear three-step stability boundary locus. The Gear four-, five- and six-step methods (Gear 1969) have smaller values of α and hence will not be any better than the three-step for the numerical solution of the vibration equation.

4.10 Generalization of SS*p*2

Thomas and Gladwell (1988) give a generalized version of the SS*p*2 algorithm for the numerical integration of the dynamic vibration equation with extra parameters included. For example, the single-step two-stage algorithm is represented by the formula

$$\boldsymbol{M\alpha}_n + \boldsymbol{C}(\dot{\boldsymbol{x}}_n + \Phi_2 \Delta t \boldsymbol{\alpha}_n) + \boldsymbol{K}(\boldsymbol{x}_n + \Phi_1 \Delta t \dot{\boldsymbol{x}}_n + \Phi_3 \Delta t^2 \boldsymbol{\alpha}_n) = \boldsymbol{f}(t_n + \Phi_1 \Delta t) \tag{4.10.1}$$

from which $\boldsymbol{\alpha}_n$ is obtained, together with

$$\boldsymbol{x}_{n+1} = \boldsymbol{x}_n + \Delta t \dot{\boldsymbol{x}}_n + \frac{\Delta t^2}{2} \boldsymbol{\alpha}_n \tag{4.10.2}$$

and

$$\dot{\boldsymbol{x}}_{n+1} = \dot{\boldsymbol{x}}_n + \Delta t \boldsymbol{\alpha}_n. \tag{4.10.3}$$

Thus there are three parameters at choice instead of two. The SS22 algorithm is given by putting $\Phi_1 = \Phi_2 = \theta_1$, and $\Phi_3 = \theta_2/2$. In the Thomas and Gladwell single-step three-stage and four-stage algorithms, there are six and nine parameters respectively. Tables are given in Thomas and Gladwell (1988) showing how these relate to the SS*pj* algorithms and to Wilson-θ (1968), Houbolt (1950) and Nørsett spline collocation (Nørsett 1981). Of particular interest in this paper are the proposals for local error estimators which can be used for time step control. These are based on the idea of using:

1. A stable two-stage first order algorithm and a stable three-stage second order algorithm on the same time step with the same matrix to be inverted for each. The difference in the two solutions is then used as an error estimator.
2. An extension of (1) to include a conditionally stable third order method to give a further estimate, still using the same matrix to be inverted. (By the Dahlquist theorem (1963) a third order method cannot be unconditionally stable.) The SS*pj* algorithms can be used for (1) but not (2).

We note that the forcing function used by Thomas and Gladwell (1988), in eqn (4.10.1), is approximated in a way which limits the possibility of identifying this method with other standard methods when the forcing function is present. Approximating the forcing function in the same pattern as the $\boldsymbol{Kx}$ term would give the method a wider applicability. In eqn (4.10.1) this would mean replacing the right hand side by

$$\boldsymbol{f}_n + \Phi_1 \Delta t \dot{\boldsymbol{f}}_n + \Phi_3 \Delta t^2 \ddot{\boldsymbol{f}}_n$$

or an expression of equivalent accuracy in terms of discrete values of $\boldsymbol{f}(t)$. It is particularly noted in Penry and Wood (1985) that the accuracy of the numerical solution depends essentially on the adequate approximation of the forcing function.

4.11. First order systems

We now consider p-step and p-stage schemes as typically coming from the finite element space discretization of a parabolic problem which as shown in Chapter 1 produces a system of equations of the form

$$\boldsymbol{C}\dot{\boldsymbol{x}} + \boldsymbol{K}\boldsymbol{x} = \boldsymbol{f}(t), \tag{4.11.1}$$

where $\boldsymbol{x}$, $\dot{\boldsymbol{x}}$ are some physical variable and its time derivative, $\boldsymbol{C}$ is now the mass matrix and $\boldsymbol{K}$ is the stiffness matrix: $\boldsymbol{f}(t)$ is the forcing function representing some external influence. We consider the numerical solution of the system (4.11.1) by the SSp1 family of algorithms obtained from eqns (4.6.4) and eqn (4.6.8) with the matrix $\boldsymbol{M} \equiv \boldsymbol{0}$. We could equally well consider the GNp1 form from eqns (4.5.16) to (4.5.19) with $\boldsymbol{M} \equiv \boldsymbol{0}$ but the simple forms which practical users prefer are so similar that it seems sufficient to restrict discussion to SSp1, $p = 1, 2, 3$.

SS11

SS11, with $\theta_1 = \theta$ is of course the θ-method which is the most popular of methods for the numerical solution of eqn (4.11.1):

$$\begin{aligned}(\boldsymbol{C} + \theta\Delta t\boldsymbol{K})\boldsymbol{x}_{n+1} = \{&\boldsymbol{C} - (1-\theta)\Delta t\boldsymbol{K}\}\boldsymbol{x}_n \\ &+ \Delta t\{(1-\theta)\boldsymbol{f}_n + \theta\boldsymbol{f}_{n+1}\}.\end{aligned} \tag{4.11.2}$$

The θ-method has been discussed in Chapter 3 as applied to the first order form of the dynamic vibration equation. For the parabolic equation (4.11.1), the eigenvalues of the matrix $\boldsymbol{C}^{-1}\boldsymbol{K}$ are real and positive (as shown for $\boldsymbol{M}^{-1}\boldsymbol{K}$ in Chapter 1) and for unconditional stability we want the algorithm to be A_0-stable.

If the system of equations (4.11.1) has come from the finite element space discretization of a physical problem we are again liable to have a system of stiff equations as explained in Chapter 1. Thus if the modal discretization of the homogeneous form of eqn (4.11.1) produces scalar equations of the form

$$\dot{x} = -\tilde{\lambda}x, \qquad \tilde{\lambda} > 0, \tag{4.11.3}$$

we have values of $\tilde{\lambda}$ which can be very large (proportional to h^{-2} where h is a measure of the size of the finite elements, as shown in Chapter 1). In the corresponding component of the numerical solution from the mode (4.11.3) where we have

$$(1 + \theta\Delta t\tilde{\lambda})x_{n+1} = \{1 - (1-\theta)\Delta t\tilde{\lambda}\}x_n, \tag{4.10.4}$$

the amplification factor is given by

$$A = \{1 - (1-\theta)\Delta t\tilde{\lambda}\}/(1 + \theta\Delta t\tilde{\lambda}). \tag{4.11.5}$$

For the Crank–Nicolson or trapezium rule method given by $\theta = 0.5$, we have

$$A = (1 - 0.5\Delta t\tilde{\lambda})/(1 + 0.5\Delta t\tilde{\lambda}) \qquad (4.11.6)$$

and for very large values of $\Delta t\tilde{\lambda}$ this is approximately equal to -1. Hence the persistent noise effects, i.e. troublesome oscillations, associated with the Crank–Nicolson method which can off-set the advantage of the higher order error obtained by taking $\theta = 0.5$.

Proposals for dealing with the Crank–Nicolson noise without losing the advantage of the higher order error are:

1. Smoothing the numerical results by averaging. This is illustrated in Wood and Lewis (1975) and Bettencourt *et al.* (1981).
2. Reducing the jerky reaction at the start by bringing in the forcing function or change in boundary conditions over several time steps. This is illustrated in Wood (1974) and (1977).
3. Allowing a 'running in' period for the noise effects to disperse. This can work if the $\tilde{\lambda}\Delta t$ in eqn (4.11.6) is not too large but experience with 'running in' varies widely and this idea is further discussed in Chapter 7 on non-linear problems.

From eqn (4.11.5) we can see that if $\theta \neq 1$ then any value of $\Delta t\tilde{\lambda} > 1/(1 - \theta)$ will make the amplification factor negative and give oscillations. The only value of θ which guarantees no oscillations is $\theta = 1$ which gives the backward difference algorithm.

SS21

For the single-step two-stage algorithm SS21 applied to the numerical solution of eqn (4.11.1), the corresponding two-step method from which we can get the information about stability and accuracy is given by using the coefficients from Table 4.8.1 in the γ_j and β_j columns, i.e.

$$(\theta_1 \boldsymbol{C} + 0.5\theta_2 \Delta t \boldsymbol{K})\boldsymbol{x}_{n+1} + \{(1 - 2\theta_1)\boldsymbol{C} + (\theta_1 - \theta_2 + 0.5)\Delta t \boldsymbol{K}\}\boldsymbol{x}_n$$
$$\{(\theta_1 - 1)\boldsymbol{C} + 0.5(\theta_2 - 2\theta_1 + 1)\Delta t \boldsymbol{K}\}\boldsymbol{x}_{n-1} = 0. \qquad (4.11.7)$$

Thus the characteristic polynomial for the SS21 algorithm is

$$(\theta_1 + 0.5\theta_2\Delta t\tilde{\lambda})r^2 + \{1 - 2\theta_1 + (\theta_1 - \theta_2 + 0.5)\Delta t\tilde{\lambda}\}r$$
$$+ \theta_1 - 1 + 0.5(\theta_2 - 2\theta_1 + 1)\Delta t\tilde{\lambda} = 0, \quad (4.11.8)$$

where $\tilde{\lambda}$ is an eigenvalue of $\boldsymbol{C}^{-1}\boldsymbol{K}$.

Equation (4.11.8) has two roots, the principal root, r_1, approximating $\exp(-\Delta t\tilde{\lambda})$ and the spurious root r_2. From eqns (4.2.9) we see that as for SS22, SS21 is consistent and the parameter $A = 1$. Also from eqn (4.8.7) we have $B_1 = B_2 = \theta_1 + 0.5$, hence the SS21 algorithm is automatically second order accurate. From the inequalities (4.8.10), omitting the m term the stability conditions for SS21 are given by

$$(2\theta_1 - 1) + (\theta_2 - \theta_1)\Delta t\tilde{\lambda} > 0, \quad (4.11.9)$$

$$\Delta t\tilde{\lambda} \geq 0, \quad \text{and} \quad 1 + (\theta_1 - 0.5)\Delta t\tilde{\lambda} \geq 0. \quad (4.11.10)$$

These conditions are satisfied for all $\Delta t\tilde{\lambda} > 0$, i.e. the algorithm is A_0-stable, if $\theta_2 > \theta_1 \geq 0.5$. We cannot have $\theta_2 = \theta_1 = 0.5$ because then the characteristic polynomial in eqn (4.11.8) is given by

$$(r + 1)\{(1 + 0.5\Delta t\tilde{\lambda})r - (1 - 0.5\Delta t\tilde{\lambda})\} = 0. \quad (4.11.11)$$

Thus the spurious root $r_2 = -1$ for all $\Delta t\tilde{\lambda}$ and the algorithm is of no practical use. We also observe that the condition for real roots of the quadratic (4.11.8) is

$$\Delta t^2\tilde{\lambda}^2\{(1 + 2\theta_1)^2 - 8\theta_2\} + 4\Delta t\tilde{\lambda}(1 - 2\theta_1) + 4 \geq 0. \quad (4.11.12)$$

For the inequality (4.11.12) to be true for all $\Delta t\tilde{\lambda}$ we must have $\theta_1 \geq \theta_2$. Hence from the results above it is not possible to have real roots and unconditional stability.

From eqns (4.8.8), the error is $O(\Delta t^3)$ when $C_1 = C_2$, i.e. when $6(\theta_1 - \theta_2) = 1$. Then the stability conditions (4.11.9) and (4.11.10) give $\theta_1 > 0.5$, $\Delta t\tilde{\lambda} < 6(2\theta_1 - 1)$.

We now consider various algorithms which have been put forward as practical propositions which are particular cases of SS21 or the corresponding two-step in eqn (4.11.7).

Inexperienced users of numerical methods who have gathered that centralized finite difference expressions are a good thing in boundary value problems frequently find the obvious numerical method for the heat conduction equation

$$\frac{\partial u}{\partial t}=\frac{\partial^2 u}{\partial x^2} \tag{4.11.13}$$

is

$$\frac{u_j^{n+1}-u_j^{n-1}}{2\Delta t}=\frac{1}{\Delta x^2}(u_{j+1}^n-2u_j^n+u_{j-1}^n). \tag{4.11.14}$$

This corresponds to the algorithm given by $\theta_1=0.5,\ \theta_2=0$ in eqn (4.11.7) giving the stability polynomial

$$r^2+2\tilde{\lambda}\Delta tr-1=0. \tag{4.11.15}$$

The roots of this quadratic are real and different and their product is -1, hence one root is greater than unity in modulus for all values of $\Delta t\tilde{\lambda}$ and the algorithm is unconditionally unstable. (In contrast the policy of centralizing works well for the second order equation such as $\ddot{x}+\omega^2x=0$ where it leads to the useful central difference method discussed in §2.3.)

Another algorithm which has been popular for use with linear and non-linear equations is Lees' algorithm (Lees 1966). This is the two-step form given by substituting $\theta_1=0.5,\ \theta_2=2/3$ in eqn (4.11.7):

$$(\tfrac{1}{2}\boldsymbol{C}+\tfrac{1}{3}\Delta t\boldsymbol{K})\boldsymbol{x}_{n+1}+\tfrac{1}{3}\Delta t\boldsymbol{K}\boldsymbol{x}_n+(-\tfrac{1}{2}\boldsymbol{C}+\tfrac{1}{3}\Delta t\boldsymbol{K})\boldsymbol{x}_{n-1}=\boldsymbol{0}. \tag{4.11.16}$$

Lees' algorithm again has a centralized format and appears to be very convenient for use with non-linear problems when it is written as

$$\boldsymbol{C}\frac{(\boldsymbol{x}_{n+1}-\boldsymbol{x}_{n-1})}{2\Delta t}+\tfrac{1}{3}\boldsymbol{K}_n(\boldsymbol{x}_{n+1}+\boldsymbol{x}_n+\boldsymbol{x}_{n-1})+\boldsymbol{F}_n=\boldsymbol{0}, \tag{4.11.17}$$

with a forcing function $\boldsymbol{F}$, where $\boldsymbol{K}_n$ is a non-linear stiffness matrix evaluated as $\boldsymbol{K}_n=\boldsymbol{K}(\boldsymbol{x}_n)$. This is the way Lees' algorithm is used for instance in Comini *et al.* (1974). It is attractive because no iterations are needed in stepping forward from $n\Delta t$ to $(n+1)\Delta t$. Substituting $\theta_1=0.5,\ \theta_2=2/3$ in eqn (4.11.8), the stability polynomial for Lees' algorithm is given by

$$(1+\tfrac{2}{3}\Delta t\tilde{\lambda})r^2+\tfrac{2}{3}\Delta t\tilde{\lambda}r+(-1+\tfrac{2}{3}\Delta t\tilde{\lambda})=0. \tag{4.11.18}$$

The principal root is

$$r_1=\{-\tfrac{1}{3}\Delta t\tilde{\lambda}+\surd(1-\tfrac{1}{3}\Delta t^2r^2)\}/(1+\tfrac{2}{3}\Delta t\tilde{\lambda}), \tag{4.11.19}$$

which expands into

$$r_1 = 1 - \Delta t\tilde{\lambda} + \tfrac{1}{2}\Delta t^2\tilde{\lambda}^2 + O(\Delta t^3).$$

This looks fine but the spurious root

$$r_2 = \{-\tfrac{1}{3}\Delta t\tilde{\lambda} - \sqrt{(1 - \tfrac{1}{3}\Delta t^2\tilde{\lambda}^2)}\}/(1 + \tfrac{2}{3}\Delta t\tilde{\lambda})$$

is always negative and if we write the roots as $r_1 = -a + b$, $r_2 = -a - b$, then a and b are positive and

$$r_1^2 - r_2^2 = -4ab < 0.$$

Hence the method is relatively unstable, the spurious root is always greater in modulus and there will always be trouble with oscillations. By 1981 in Bettencourt *et al.* (1981) Lees' algorithm is being referred to as 'notoriously oscillatory', but it was still in use in 1987 (Lewis and Schrefler).

Liniger (1969) suggests making an exponential fit by making the principal root r_1 equal to $\exp(-\Delta t\lambda_0)$ for some particular λ_0 and the spurious root r_2 equal to zero. From eqn (4.11.8) this means that θ_1 and θ_2 have to be chosen to satisfy the equations

$$\theta_1 - 1 + 0.5(\theta_2 - 2\theta_1 + 1)\Delta t\lambda_0 = 0$$

and

$$(\theta_1 + 0.5\theta_2\Delta t\lambda_0)\exp(-\Delta t\lambda_0) = -\{1 - 2\theta_1 + (\theta_1 - \theta_2 + 0.5)\Delta t\lambda_0\}. \quad (4.11.20)$$

For smaller values of $\Delta t\lambda_0$ the corresponding algorithm is only conditionally stable (see Exercise 4.4), but for $\Delta t\lambda_0 = 3$, for example, we have unconditional stability; this is given by $\theta_1 = 1.218$, $\theta_2 = 1.292$.

The backward difference method makes both roots zero when $\Delta t\tilde{\lambda}$ is infinite; this algorithm has parameters given by $\theta_1 - \theta_2 + 0.5 = 0$ and $\theta_2 - 2\theta_1 + 1 = 0$, i.e. $\theta_1 = 1.5$ and $\theta_2 = 2$. The stability polynomial is then

$$(1.5 + \Delta t\tilde{\lambda})r^2 - 2r + 0.5 = 0, \quad (4.11.21)$$

hence the roots are again complex when $\Delta t\tilde{\lambda} > 0.5$. The two-step version of this backward difference method is recommended in Richtmyer and Morton (1967) for damping high frequency components. This method is used in Wood and Lewis (1975) in experiments on noise control with the heat conduction equation. The conclusion there is that averaged Crank–Nicolson is better.

Zlamal (1975) recommends a two-step algorithm corresponding to SS21 with $\theta_1 = 5/6$, $\theta_2 = 8/9$ as giving better results than Crank–Nicolson with his numerical experiments. The characteristic polynomial is then

$$(15 + 8\Delta t\tilde{\lambda})r^2 + 4(-3 + 2\Delta t\tilde{\lambda})r + (-3 + 2\Delta t\tilde{\lambda}) = 0. \quad (4.11.21a)$$

These θ values satisfy the stability conditions: $\theta_2 > \theta_1 > 0.5$ but the roots are complex for $\Delta t\tilde{\lambda} > 1.5$ and for $\Delta t\tilde{\lambda} = 1.5$ the roots are both zero.

Table 4.11.1 summarizes these results.

SS31

For the single-step three-stage algorithm given by SS31 applied to the numerical solution of eqn (4.11.1), the corresponding three-step algorithm from which we can get information about accuracy and stability has coefficients in the γ_j and β_j columns of Table 4.8.4. From eqns (4.8.21) we have $C_1 = C_2$ and hence the

Table 4.11.1. Algorithms included in SSp1.

Algorithm	Parameters	Accuracy	Stability
SS11	$\theta_1 = \theta$	$O(\Delta t)$	$\theta \geq 0.5$
θ-method	$\theta = 0.5$	$O(\Delta t^2)$	Stable
SS21	θ_1, θ_2	$O(\Delta t^2)$	$\theta_2 > \theta_1 \geq 0.5$
	$\theta_1 - \theta_2 = 1/6$ $\theta_1 > 0.5$	$O(\Delta t^3)$	$\Delta t\tilde{\lambda} < 6(2\theta_1 - 1)$
'Central difference'	$\theta_1 = 0.5$, $\theta_2 = 0$	Unconditionally unstable	
Lees' algorithm	$\theta_1 = 0.5$, $\theta_2 = 2/3$	$O(\Delta t^2)$	Relatively unstable
Backward difference	$\theta_1 = 1.5$, $\theta_2 = 2$	$O(\Delta t^2)$	Stable
Liniger exp. fit for $\Delta t\tilde{\lambda}_0 = 3$	$\theta_1 = 1.218$, $\theta_2 = 1.292$	$O(\Delta t^2)$	Stable
Zlamal (1975)	$\theta_1 = 5/6$, $\theta_2 = 8/9$	$O(\Delta t^2)$	Stable
SS31	$\theta_1, \theta_2, \theta_3$	$O(\Delta t^3)$	See text
Backward difference	$\theta_1 = 2$, $\theta_2 = 11/3$, $\theta_3 = 6$	$O(\Delta t^3)$	Stable

SS31 algorithm is automatically third order accurate. Using the substitutions $a = \theta_1 - 0.5$, $b = \theta_2 - \theta_1 - 1/6$, $c = \theta_3 - 1.5\theta_2 + 0.25$ as in the discussion of SS32, the stability conditions given by inequalities (4.8.33) and (4.8.34) without the m terms are

$$6b + 1 + 2c\Delta t\tilde{\lambda} > 0, \tag{4.11.22}$$

$$2a + b\Delta t\tilde{\lambda} \geq 0, \tag{4.11.23}$$

$$1 + a\Delta t\tilde{\lambda} \geq 0, \tag{4.11.24}$$

and

$$2(3ab - c)(\Delta t\tilde{\lambda})^2 + (12a^2 - 1)\Delta t\tilde{\lambda} + 12a \geq 0. \tag{4.11.25}$$

These inequalities require that $a > 0$, $b > 0$, $c > 0$, $(3ab - c) > 0$ and either $12a^2 > 1$ or $(12a^2 - 1)^2 < 96a(3ab - c)$. Wood (1978) gives examples of SS31 algorithms. The main one of interest here is the backward difference algorithm obtained by making three roots equal to zero for infinite $\Delta t\tilde{\lambda}$. This is given by $\theta_1 = 2$, $\theta_2 = 11/3$, $\theta_3 = 6$, the same values as for SS32 Houbolt. These values of the parameters correspond to $a = b = 1.5$, $c = 0.75$. Thus we have $3ab - c > 0$ and $12a^2 > 1$ and as expected the backward difference algorithm is unconditionally stable but as shown in Wood (1978) the maximum modulus root is complex for larger values of $\Delta t\tilde{\lambda}$.

In SS31 we have the complication of two spurious roots and the algorithm is even less amenable to control than SS21. Gear's backward difference methods (Gear 1969) with variable time step and order are one answer but it is not surprising that many practical users prefer the simplicity of the SS11 θ-method. Table 4.11.1 summarizes these results for the SSp1 and corresponding p-step algorithms.

4.12. Park's method

Park (1975) takes the dynamic vibration equation expressed in its first order form

$$\dot{\boldsymbol{z}} + \boldsymbol{B}\boldsymbol{z} = \boldsymbol{F}, \tag{4.12.1}$$

where $\boldsymbol{z}^{\mathrm{T}} = [\boldsymbol{x}, \boldsymbol{v}]$, $\boldsymbol{F}^{\mathrm{T}} = [\boldsymbol{0}, \boldsymbol{M}^{-1}\boldsymbol{f}]$, and

$$\boldsymbol{B} = \begin{bmatrix} \boldsymbol{0} & -\boldsymbol{I} \\ \boldsymbol{M}^{-1}\boldsymbol{K} & \boldsymbol{M}^{-1}\boldsymbol{C} \end{bmatrix}. \tag{4.12.2}$$

Park shows that using the trapezium rule on the first order form given in eqn (4.12.1) (which is equivalent to Newmark with $2\beta = \gamma = 0.5$ on the second order form as shown in Chapter 2) gives unsatisfactory results for large time steps with non-linear equations. Park uses as examples the softening spring represented by the equation

$$\ddot{x} + k \tanh(x) = 0 \tag{4.12.3}$$

and the hardening spring represented by the equation

$$\ddot{x} + kx(1 + k_1x^2) = 0. \tag{4.12.4}$$

Park's results show that the trapezium rule which is unconditionally stable for linear problems, when used on the softening spring gives unstable results for a time step greater than one-sixth of the natural period, and the hardening spring, although stable, gives distorted results for a time step greater than a quarter of the natural period. (We recall the often quoted engineer's rule of thumb which says that the time step should be less than one-tenth of the natural period.) Park ascribes the unsatisfactory performance of the trapezium rule in these examples to the presence of past derivatives in the algorithm. With eqn (4.12.1), this means past values of the stiffness term. The method given by Houbolt (1950) and the Gear two-step method (Gear 1969) which are both A-stable, do not use past derivatives but they give high values of algorithmic damping. The Gear (1969) three-step method also does not use past derivatives but is only $A(\alpha)$-stable. Used for the numerical solution of

$$\dot{y} = f(t, y), \tag{4.12.5}$$

the Gear two-step is given by the formula

$$y_{n+1} = (4y_n - y_{n-1} + 2\Delta t f_{n+1})/3 \tag{4.12.6}$$

and the Gear three-step is given by

$$y_{n+1} = (18y_n - 9y_{n-1} + 2y_{n-2} + 6\Delta t f_{n+1})/11. \tag{4.12.7}$$

Taking p (Gear two-step) plus $(p - 1)$ (Gear three-step) gives an algorithm which is A-stable for $p \geq 9/20$. Park's algorithm takes $p = 9/20$. Hence Park's method is A-stable but it is only of the same order of accuracy as the Gear two-step (that is, the error is $O(\Delta t^2)$) while necessitating the storage of the extra vector.

The three-step method of Park does not correspond to a member of the single-step SS31 family of algorithms because SS31 automatically has an error $O(\Delta t^3)$ or better. But it is interesting to note that the generalization of the SSp2 family of algorithms given by Thomas and Gladwell (1988) for use with the second order equation

$$\boldsymbol{M}\ddot{\boldsymbol{x}} + \boldsymbol{C}\dot{\boldsymbol{x}} + \boldsymbol{K}\boldsymbol{x} = \boldsymbol{f} \tag{4.12.8}$$

can also be used for the first order equation

$$\boldsymbol{C}\dot{\boldsymbol{x}} + \boldsymbol{K}\boldsymbol{x} = \boldsymbol{f} \tag{4.12.9}$$

simply by putting the matrix $\boldsymbol{M} \equiv \boldsymbol{0}$ in their formulae. We can then identify a single-step three-stage method corresponding to the three-step single-stage method of Park. For application to eqn (4.12.9) this is given by:

$$\boldsymbol{x}_{n+1} = \boldsymbol{x}_n + \Delta t\dot{\boldsymbol{x}}_n + \tfrac{1}{2}\Delta t^2\ddot{\boldsymbol{x}}_n + \tfrac{1}{6}\Delta t^3\boldsymbol{\beta}_n, \tag{4.12.10}$$

$$\dot{\boldsymbol{x}}_{n+1} = \dot{\boldsymbol{x}}_n + \Delta t\ddot{\boldsymbol{x}}_n + \tfrac{1}{2}\Delta t^2\boldsymbol{\beta}_n, \tag{4.12.11}$$

$$\ddot{\boldsymbol{x}}_{n+1} = \ddot{\boldsymbol{x}}_n + \Delta t\boldsymbol{\beta}_n, \tag{4.12.12}$$

where for the homogeneous equation (4.12.9) $\boldsymbol{\beta}_n$ is given by

$$\begin{aligned}\Delta t^2(\Psi_5\boldsymbol{C} + \Psi_6\Delta t\boldsymbol{K})\boldsymbol{\beta}_n = &-\boldsymbol{C}(\dot{\boldsymbol{x}}_n + \Psi_3\Delta t\ddot{\boldsymbol{x}}_n)\\ &-\boldsymbol{K}(\boldsymbol{x}_n + \Psi_1\Delta t\dot{\boldsymbol{x}}_n + \Psi_2\Delta t^2\ddot{\boldsymbol{x}}_n)\end{aligned} \tag{4.12.13}$$

(the parameter Ψ_4 has disappeared along with the $\boldsymbol{M}$ term). The single-step method corresponding to Park's three-step applied to the homogeneous form of eqn (4.12.9) then consists of eqns (4.12.10) to (4.12.12) together with

$$\begin{aligned}\Delta t^2(\tfrac{5}{3}\boldsymbol{C} + \Delta t\boldsymbol{K})\boldsymbol{\beta}_n = &-\boldsymbol{C}(\dot{\boldsymbol{x}}_n + 2\Delta t\ddot{\boldsymbol{x}}_n)\\ &-\boldsymbol{K}(\boldsymbol{x}_n + 2\Delta t\dot{\boldsymbol{x}}_n + \tfrac{11}{6}\Delta t^2\ddot{\boldsymbol{x}}_n).\end{aligned} \tag{4.12.14}$$

Of course it is difficult to avoid using past derivatives with a single-step method. Taking the first order equation (4.12.1) and treating the forcing function in with the $-\boldsymbol{B}\boldsymbol{z}$ term as in the usual mathematical approach to equations of the form $\dot{y} = f(t, y)$ gives the single-step corresponding to Park's method as

$$\begin{aligned}\boldsymbol{z}_{n+1} &= \boldsymbol{z}_n + \Delta t\dot{\boldsymbol{z}}_n + \tfrac{1}{2}\Delta t^2\ddot{\boldsymbol{z}}_n + \tfrac{1}{6}\Delta t^3\hat{\boldsymbol{\beta}}_n\\ \dot{\boldsymbol{z}}_{n+1} &= \dot{\boldsymbol{z}}_n + \Delta t\ddot{\boldsymbol{z}}_n + \tfrac{1}{2}\Delta t^2\hat{\boldsymbol{\beta}}_n\\ \ddot{\boldsymbol{z}}_{n+1} &= \ddot{\boldsymbol{z}}_n + \Delta t\hat{\boldsymbol{\beta}}_n\end{aligned}$$

together with

$$\Delta t^2(\tfrac{5}{3}\boldsymbol{I}+\Delta t\boldsymbol{B})\hat{\boldsymbol{\beta}}_n = -(\dot{\boldsymbol{z}}_n + 2\Delta t\ddot{\boldsymbol{z}}_n) - \boldsymbol{B}(\boldsymbol{z}_n + 2\Delta t\dot{\boldsymbol{z}}_n + \tfrac{11}{6}\Delta t^2\ddot{\boldsymbol{z}}_n) + \boldsymbol{F}_n + 2\Delta t\dot{\boldsymbol{F}}_n + \tfrac{11}{6}\Delta t^2\ddot{\boldsymbol{F}}_n - \Delta t^3\dddot{\boldsymbol{F}}_n. \quad (4.12.15)$$

The representation of the forcing function could alternatively be an expression of equal accuracy in terms of its values at particular times.

Park's method in the context of methods for non-linear equations is discussed further in Chapter 7.

Exercises

4.1. Apply the SS22 algorithm to the equation

$$m\ddot{x} + kx = 0 \quad (1)$$

and obtain the amplification matrix $\boldsymbol{A}$ in $\boldsymbol{Y}_{n+1} = \boldsymbol{A}\boldsymbol{Y}_n$, where

$$\boldsymbol{Y}_n^{\mathrm{T}} = [x_n, \Delta t\dot{x}_n].$$

Find the characteristic polynomial of $\boldsymbol{A}$. Compare these results with the amplification matrix and characteristic polynomial of the θ-method applied to the first order form of eqn (1) given in §2.3. Find the conditions

(a) for the equivalent two-step recurrences in the displacements to give the same result;

(b) for the two one-step algorithms to give the same result.

4.2. Show that if r_1 and r_2 are the roots of the stability polynomial of the SS22 algorithm in Exercise 4.1, then $r_1 - r_2$ is $\mathrm{O}(\Delta t)$ and the effect on the convergence proof is similar to that with Newmark in Exercise 2.4.

4.3. Find the local truncation error (eqn (2.2.4)) for the Aboudi (1976) algorithm given by eqn (4.4.43) for the solution of $\ddot{x} + \omega^2 x = 0$.

4.4. Show that making a Liniger (1968) exponential fit on the SS21 algorithm in eqn (4.11.7) by choosing θ_1, θ_2, such that the roots of eqn (4.11.8) are exp(−0.5) and zero, gives an algorithm which is only conditionally stable. Find the stability limit.

4.5. Check the accuracy of SS22 with $\theta_1 = 1/2$, $\theta_2 = 1/6$, for the numerical solution of $\ddot{x} + 2v\omega\dot{x} + \omega^2 x = 0$.

4.6. Consider the numerical solution of the equation $\ddot{x} + \omega^2 x = 0$ with initial conditions $x(0) = a$, $\dot{x}(0) = 0$. Find the condition on $\Omega = \omega\Delta t$ for overshoot on the first time step in the sense that $x_1 < -a$ for the SS32 equivalents of (i) Wilson $\theta = 1.4$, (ii) Houbolt. (The θ values are given in Table 4.8.6.)

References

Aboudi, J. (1976). Two-dimensional wave propagation in non-linear elastic half-space. *Computer Methods in Applied Mechanics and Engineering,* **9,** 25–46.

Adams, D. D. and Wood, W. L. (1983). Comparison of Hilber–Hughes–Taylor and Bossak 'α-methods' for the numerical integration of vibration equations. *International Journal of Numerical Methods in Engineering,* **19,** 765–771.

Bathe, K. J. and Wilson, E. L. (1973). Stability and accuracy analysis of direct integration methods. *Earthquake Engineering and Structural Dynamics,* **1,** 283–291.

Bettencourt, J. M., Zienkiewicz, O. C., and Cantin, G. (1981). Consistent use of finite elements in time and the performance of various recurrence schemes for the heat diffusion equation. *International Journal of Numerical Methods in Engineering,* **17,** 931–938.

Comini, G., DelGuidice, S., Lewis, R. W., and Zienkiewicz, O. C. (1974). Finite element solution of non-linear heat conduction problems with special reference to phase change. *International Journal of Numerical Methods in Engineering,* **8,** 613–624.

Dahlquist, G. G. (1963). A special stability problem for linear multistep methods. *BIT,* **3,** 27–43.

Enright, W. H. (1986). A new error-control for initial value solvers. *Numerical Analysis Report no. 122.* Department of Mathematics, University of Manchester.

Farhoomand, I. (1970). Nonlinear dynamic stress analysis of two-dimensional solids. *Ph.D. Dissertation.* University of California, Berkeley.

Fox, L. and Goodwin, E. T. (1949). Some new methods for the numerical integration of ordinary differential equations. *Proceedings of the Cambridge Philosophical Society,* **45,** 373–388.

Gear, C. W. (1969). The automatic integration of stiff ordinary differential equations. in *Information processing 68* (ed, A. J. H. Morrell), pp. 187–193 North-Holland, Amsterdam.

Goudreau, G. L. and Taylor, R. L. (1972). Evaluation of numerical integration methods in elastodynamics. *Computer Methods in Applied Mechanics and Engineering,* **2,** 69–97.

Hilber, H. M., Hughes, T. J. R., and Taylor, R. L. (1977). Improved numerical dissipation for time integration algorithms in structural dynamics. *Earthquake Engineering and Structural Dynamics,* **5,** 283–292.

Hilber, H. M. and Hughes, T. J. R. (1978). Collocation, dissipation and 'overshoot' for time integration schemes in structural dynamics. *Earthquake Engineering and Structural Dynamics,* **6,** 99–117.

Houbolt, J. C. (1950). A recurrence matrix solution for the dynamic response of elastic aircraft. *Journal of the Aeronautical Sciences,* **17,** 540–550.

Hughes, T. J. R. (1983). Transient algorithms and stability. In *Computational methods for transient analysis* (eds. T. Belytschko and T. J. R. Hughes). Elsevier Science Publishers B.V.

Katona, M. G. (1985). A general family of single-step methods for numerical time integration of structural dynamic equations. *Proceedings of the International Conference on Numerical Methods in Engineering: Theory and Applications (NUMETA 85), Swansea,* Vol. 1, pp. 213–225.

Katona, M. G. and Zienkiewicz, O. C. (1985). A unified set of single step algorithms. Part 3: The beta-*m* method, a generalization of the Newmark scheme. *International Journal of Numerical Methods in Engineering,* **21,** 1345–1359.

Lees, M. (1966). A linear three-level difference scheme for quasi-linear parabolic equations. *Mathematics of Computation,* **20,** 516–622.

Lewis, R. W. and Schrefler, B. A. (1987). *The finite element method in the deformation and consolidation of porous media.* Wiley.

Liniger, W. (1969). Global accuracy and A-stability of one and two step integration formulae for stiff ordinary differential equations. *Proceedings of Conference on Numerical Solution of Differential Equation, Dundee,* (ed. J. Ll. Morris). Springer.

Newmark, N. M. (1959). A method of computation for structural dynamics. *Proceedings of ASCE 85*, **EM3**, 67–94.

Nørsett, S. P. (1981). The Wilson-θ method as a spline collocation method for solving dynamic problems. *Mathematics and Computation Report No. 1/81.* Division of Numerical Mathematics, University of Trondheim.

Park, K. C. (1975). An improved stiffly stable method for direct integration of non-linear structural dynamic equations. *ASME Journal of Applied Mechanics,* **42,** 464–470.

Penry, S. N. and Wood, W. L. (1985). Comparison of some single-step methods for the numerical solution of the structural dynamic equation. *International Journal of Numerical Methods in Engineering,* **21,** 1951–1955.

Richtmyer, R. D. and Morton, K. W. (1967). *Difference methods for initial value problems.* Wiley.

Strang, G. and Fix, G. J. (1973). *An analysis of the finite element method.* Prentice-Hall.

Symon, K. R. (1977). *Mechanics* (3rd edn). Addison-Wesley, Reading, Mass.

Thomas, R. M. and Gladwell, I. (1988). Variable-order variable-step algorithms for second order systems. *International Journal of Numerical Methods in Engineering,* **26,** 39–53 and 55–80.

Wilson, E. L. (1968). A computer program for the dynamic stress analysis of underground structures. *SEL Report 68-1.* University of California, Berkeley.

Wood, W. L. (1974). Marginal stability: experiments with the diffusion equation. *Finite Elements Symposium.* Atlas Laboratory.

Wood, W. L. (1977). Control of Crank–Nicolson noise in the numerical solution of the heat conduction equation. *International Journal of Numerical Methods in Engineering,* **11,** 1059–1065.

Wood, W. L. (1978). On the Zienkiewicz three- and four-time level schemes applied to the numerical integration of parabolic equations. *International Journal of Numerical Methods in Engineering,* **12,** 1717–1726.

Wood, W. L. (1981). Numerical integration of the structural dynamics equations including natural damping and periodic forcing terms. *International Journal of Numerical Methods in Engineering,* **17,** 281–289.

Wood, W. L. (1984a). A further look at Newmark, Houbolt, etc., time-stepping schemes. *International Journal of Numerical Methods in Engineering,* **20,** 1009–1017.

Wood, W. L. (1984b). A unified set of single-step algorithms. Part 2: Theory. *International Journal of Numerical Methods in Engineering,* **20,** 2302–2309.

Wood, W. L. (1985). Addendum to 'A unified set of single-step algorithms, Part 2: Theory. *International Journal of Numerical Methods in Engineering,* **21.**

Wood, W. L. (1986). On matching the SS*pj* and GN*pj* algorithms. *Communications in Applied Numerical Methods,* **2,** 471–475.

Wood, W. L. (1987a). On the effect of natural damping on the stability of a time-stepping scheme. *International Journal of Numerical Methods in Engineering,* **3,** 141–144.

Wood, W. L. (1987b). Further results on the accuracy of the SS*pj* algorithms. *Communications in Applied Numerical Methods,* **3,** 215–221.

Wood, W. L. and Lewis, R. W. (1975). A comparison of time-marching schemes for the transient heat conduction equation. *International Journal of Numerical Methods in Engineering,* **9,** 679–689.

Wood, W. L., Bossak, M. and Zienkiewicz, O. C. (1981). An alpha modification of Newmark's method. *International Journal of Numerical Methods in Engineering,* **15,** 1562–1566.

Zienkiewicz, O. C. (1976). A new look at Newmark, Houbolt and other time stepping formulae. A weighted residual approach. *Report C/R/273/76.* University of Wales, Department of Civil Engineering.

Zienkiewicz, O. C. (1977). *The finite element method.* McGraw-Hill.

Zienkiewicz, O. C., Wood, W. L., Hine, N. W. and Taylor, R. L., (1984). A unified set of single step algorithms. *International Journal of Numerical Methods in Engineering,* **20,** 1529–1552.

Zlamal, M. (1975). Finite element methods in heat conduction problems. *Proceedings of Conference on Finite Element Methods.* Brunel.

5 Other single-step methods. Global crunch or divide and conquer?

5.1. Introduction

In the previous three chapters we have discussed in detail p-step and p-stage algorithms for the numerical integration of the dynamic vibration equation

$$\boldsymbol{M}\ddot{\boldsymbol{x}} + \boldsymbol{C}\dot{\boldsymbol{x}} + \boldsymbol{K}\boldsymbol{x} = \boldsymbol{f}(t) \tag{5.1.1}$$

and the first order equation

$$\boldsymbol{C}\dot{\boldsymbol{x}} + \boldsymbol{K}\boldsymbol{x} = \boldsymbol{f}(t). \tag{5.1.2}$$

In this chapter we consider some other single-step algorithms proposed by engineers and applied scientists for the numerical solution of eqns (5.1.1) and (5.1.2). The emphasis here is still on the situation where these equations have constant coefficients, but the methods particularly relevant to non-linear equations are discussed further in Chapter 7.

The idea of Hermitian interpolation described in §5.2 seems to be a logical extension of the Lagrange interpolation used in the SSpj-type algorithms described in previous chapters, but the results are not so easily controllable. There is a close connection between these and the methods using the Padé and Nørsett approximations to the exponential for application to first order equations described in §5.3. The Nørsett method reduces the work involved by using repeated factors in the denominator. The Runge–Kutta methods described in §5.4 are also shown to be allied to those in §5.3. A disadvantage of these methods is that they do seem to be further removed from the physics of the problem than the methods described in earlier chapters, and with the dynamic vibration equation it is more difficult to see what is happening to the amplitude and phase errors.

Rather than use the global crunch as in the methods described in §§5.2, 5.3, and 5.4, many practical users prefer methods which employ the simple algorithms from previous chapters used in more sophisticated ways. With the advent of parallel processors these appear to be the methods with potential for future development. We quote S. Hammarling (NAG Ltd) (seminar at Reading University, November 1988) giving paradigms for algorithms for high-performance computers as

'think in blocks—divide and conquer'.

This advice seems to lead directly to the approaches described in §5.5 (partitioning by blocks of nodes or elements), §5.6 (element by element methods), and §5.7 (operator splitting methods).

5.2. Methods using Hermitian interpolation

We start with these methods because the idea of Hermitian interpolation is very close to the interpolation implicit in the p-step and p-stage algorithms discussed in Chapter 4.

Argyris *et al.* (1973a) introduce a Hermitian cubic interpolation of the inertia forces in the dynamic vibration equation (5.1.1) where the inertia force is denoted by $\boldsymbol{R}$ and the equation in our notation is written as

$$\boldsymbol{M}\ddot{\boldsymbol{x}} = \boldsymbol{R} = -\boldsymbol{K}\boldsymbol{x} - \boldsymbol{C}\dot{\boldsymbol{x}} + \boldsymbol{f}. \tag{5.2.1}$$

This method is explained in a complicated way but it reduces to a Hermitian cubic approximation for the acceleration. We illustrate the method briefly here by applying it to the homogeneous equation

$$\boldsymbol{M}\ddot{\boldsymbol{x}} + \boldsymbol{C}\dot{\boldsymbol{x}} + \boldsymbol{K}\boldsymbol{x} = \boldsymbol{0}. \tag{5.2.2}$$

It is convenient to use the symbols $\boldsymbol{v}$ and $\boldsymbol{a}$ for the approximate velocity and acceleration to fit in with the notation used elsewhere in this text rather than the symbols used in the original. The acceleration within a time step $n\Delta t \le t \le (n+1)\Delta t$ is expressed as

$$\boldsymbol{a} = \omega_1(\zeta)\boldsymbol{a}_n + \omega_2(\zeta)\Delta t\dot{\boldsymbol{a}}_n + \omega_3(\zeta)\boldsymbol{a}_{n+1} + \omega_4(\zeta)\Delta t\dot{\boldsymbol{a}}_{n+1}, \tag{5.2.3}$$

where $t = (n+\zeta)\Delta t$ and the interpolation functions are given by

$$\omega_1(\zeta) = 1 - 3\zeta^2 + 2\zeta^3, \qquad \omega_2(\zeta) = \zeta - 2\zeta^2 + \zeta^3,$$
$$\omega_3(\zeta) = 3\zeta^2 - 2\zeta^3, \qquad \omega_4(\zeta) = -\zeta^2 + \zeta^3. \tag{5.2.4}$$

Equation (5.2.4) is integrated over the time step from $t = n\Delta t$ to $t = (n+1)\Delta t$ to obtain the approximations $\boldsymbol{v}$ and $\boldsymbol{x}$ to the velocity and displacement in terms of $\boldsymbol{a}_n$, $\dot{\boldsymbol{a}}_n$, $\boldsymbol{a}_{n+1}$, $\dot{\boldsymbol{a}}_{n+1}$, and the values $\boldsymbol{v}_n$ and $\boldsymbol{x}_n$ at the beginning of the time step. The values $\boldsymbol{v}_{n+1}$ and $\boldsymbol{x}_{n+1}$ obtained by putting $\zeta = 1$ in the integrated forms are then

$$\boldsymbol{v}_{n+1} = \boldsymbol{v}_n + \tfrac{1}{12}\Delta t(6\boldsymbol{a}_n + \Delta t\dot{\boldsymbol{a}}_n + 6\boldsymbol{a}_{n+1} - \Delta t\dot{\boldsymbol{a}}_{n+1}) \quad (5.2.5)$$

and

$$\boldsymbol{x}_{n+1} = \boldsymbol{x}_n + \Delta t\boldsymbol{v}_n + \tfrac{1}{60}\Delta t^2 \times (21\boldsymbol{a}_n + 3\Delta t\dot{\boldsymbol{a}}_n + 9\boldsymbol{a}_{n+1} - 2\Delta t\dot{\boldsymbol{a}}_{n+1}) \quad (5.2.6)$$

These expressions for $\boldsymbol{x}_{n+1}$ and $\boldsymbol{v}_{n+1}$ are substituted into the equations

$$\boldsymbol{M}\boldsymbol{a}_{n+1} + \boldsymbol{C}\boldsymbol{v}_{n+1} + \boldsymbol{K}\boldsymbol{x}_{n+1} = \boldsymbol{0}$$

and

$$\boldsymbol{M}\dot{\boldsymbol{a}}_{n+1} + \boldsymbol{C}\boldsymbol{a}_{n+1} + \boldsymbol{K}\boldsymbol{v}_{n+1} = \boldsymbol{0} \quad (5.2.7)$$

which represent eqn (5.2.2) and its derivative at $t = (n+1)\Delta t$. The odd $\boldsymbol{x}_n$ and $\boldsymbol{v}_n$ terms from eqns (5.2.5) and (5.2.6) are removed by using eqns (5.2.7) at $t = n\Delta t$ also. The result is a one-step two-stage recurrence in the vector $[\boldsymbol{a}_n, \dot{\boldsymbol{a}}_n]^{\mathrm{T}}$. With the same equations switched round, this is then changed to a one-step two-stage recurrence in the vector $[\boldsymbol{x}_n, \boldsymbol{v}_n]^{\mathrm{T}}$. Argyris *et al.* (1973a) then find that this algorithm is not *A*-stable so they retain the same equation (5.2.5) for $\boldsymbol{v}_{n+1}$ but modify eqn (5.2.6) for $\boldsymbol{x}_{n+1}$ to

$$\boldsymbol{x}_{n+1} = \boldsymbol{x}_n + \Delta t\boldsymbol{v}_n + \frac{\Delta t^2}{60} \times (20\boldsymbol{a}_n + 2.5\Delta t\dot{\boldsymbol{a}}_n + 10\boldsymbol{a}_{n+1} - 2.5\dot{\boldsymbol{a}}_{n+1}). \quad (5.2.8)$$

The algorithm when applied to the numerical solution of $\ddot{x} + \omega^2 x = 0$ now has an amplification matrix with complex eigenvalues of modulus unity for all values of the time step Δt, that is, with no algorithmic damping. This amplification matrix may be expressed as

$$\boldsymbol{A} = \begin{bmatrix} \cos\Phi & \omega^{-1}\sin\Phi \\ -\omega\sin\Phi & \cos\Phi \end{bmatrix},$$

where

$$\Phi = 2\tan^{-1}\{0.5\omega\Delta t/(1 - \omega^2\Delta t^2/12)\}.$$

By the method using the Cayley–Hamilton theorem described in

§4.4 this method is also equivalent to a two-step recurrence in the displacement:

$$x_{n+1} - 2\cos \Phi x_n + x_{n-1} = 0$$

(but this is not a linear two-step because of the higher powers of $\omega\Delta t$ in the coefficients). The authors state that the addition of natural damping will reduce the modulus of the eigenvalues of the amplification matrix to less than unity. They do not discuss the problem of damping out the inaccurate higher frequency modes while retaining those with the lower frequencies. When written out for application to the system of equations (5.1.1) with natural damping present the algorithm as given in Argyris *et al.* (1973b) has awkward terms in $\boldsymbol{CM}^{-1}\boldsymbol{K}$ and $\boldsymbol{CM}^{-1}\boldsymbol{C}$. This is not so with the algorithm introduced by Gellert (1978) which also uses Hermitian interpolation.

We illustrate the Gellert (1978) method on the equation

$$\boldsymbol{M}\ddot{\boldsymbol{x}} + \boldsymbol{C}\dot{\boldsymbol{x}} + \boldsymbol{K}\boldsymbol{x} = \boldsymbol{f}(t). \tag{5.2.9}$$

Gellert uses a Hermitian interpolation formula for the displacement, that is, with the same notation as above,

$$\boldsymbol{x} = \boldsymbol{x}_n\omega_1 + \Delta t\boldsymbol{v}_n\omega_2 + \boldsymbol{x}_{n+1}\omega_3 + \Delta t\boldsymbol{v}_{n+1}\omega_4, \tag{5.2.10}$$

where ω_1, ω_2, ω_3, ω_4 are again given by eqns (5.2.4). Gellert also specifies that the forcing function $\boldsymbol{f}(t)$ is represented by a quadratic Lagrange interpolation using $\boldsymbol{f}(n\Delta t)$, $\boldsymbol{f}((n+\frac{1}{2})\Delta t)$, and $\boldsymbol{f}((n+1)\Delta t)$:

$$\boldsymbol{f} = (1 - 3\zeta + 2\zeta^2)\boldsymbol{f}_n + 4(\zeta - \zeta^2)\boldsymbol{f}_{n+\frac{1}{2}} + (-\zeta + 2\zeta^2)\boldsymbol{f}_{n+1}. \tag{5.2.11}$$

The algorithm uses eqn (5.2.9), its first derivative and the results of integrating eqn (5.2.9) once and then twice over the time step. Into these four equations as at time $t = (n+1)\Delta t$ are substituted the expressions for $\boldsymbol{x}_{n+1}$, $\boldsymbol{v}_{n+1}$, $\boldsymbol{a}_{n+1}$, $\dot{\boldsymbol{a}}_{n+1}$ obtained from the derivatives of eqn (5.2.10). The results give the four equations

$$\left(\boldsymbol{M} + \frac{\Delta t}{2}\boldsymbol{C} + \frac{3\Delta t^2}{20}\boldsymbol{K}\right)\boldsymbol{x}_{n+1} - \left(\frac{\Delta t}{12}\boldsymbol{C} + \frac{\Delta t^2}{30}\boldsymbol{K}\right)\Delta t\boldsymbol{v}_{n+1}$$
$$- \left(\boldsymbol{M} + \frac{\Delta t}{2}\boldsymbol{C} - \frac{7\Delta t^2}{20}\boldsymbol{K}\right)\boldsymbol{x}_n - \left(\boldsymbol{M} - \frac{\Delta t}{12}\boldsymbol{C} - \frac{\Delta t^2}{20}\boldsymbol{K}\right)\Delta t\boldsymbol{v}_n$$
$$= \frac{\Delta t^2}{6}(\boldsymbol{f}_n + 2\boldsymbol{f}_{n+\frac{1}{2}}) \tag{5.2.12}$$

$$\left(C + \frac{\Delta t}{2} K\right) x_{n+1} + \left(\frac{1}{\Delta t} M - \frac{\Delta t}{12} K\right) \Delta t v_{n+1} - \left(C - \frac{\Delta t}{2} K\right) x_n - \left(\frac{1}{\Delta t} M - \frac{\Delta t}{12} K\right) \Delta t v_n = \frac{\Delta t}{6} (f_n + 4 f_{n+\frac{1}{2}} + f_{n+1}) \quad (5.2.13)$$

$$-\left(\frac{6}{\Delta t^2} M - K\right) x_{n+1} + \left(\frac{4}{\Delta t^2} M + \frac{1}{\Delta t} C\right) \Delta t v_{n+1} + \frac{6}{\Delta t^2} M x_n + \frac{2}{\Delta t} M v_n = f_{n+1} \quad (5.2.14)$$

$$-\left(\frac{12}{\Delta t^3} M + \frac{6}{\Delta t^2} C\right) x_{n+1} + \left(\frac{6}{\Delta t^3} M + \frac{4}{\Delta t^2} C + \frac{1}{\Delta t} K\right) \Delta t v_{n+1} + \left(\frac{12}{\Delta t^3} M + \frac{6}{\Delta t^2} C\right) x_n + \left(\frac{6}{\Delta t^3} M + \frac{2}{\Delta t^2} C\right) \Delta t v_n - \frac{1}{\Delta t} (f_n - 4 f_{n+\frac{1}{2}} + 3 f_{n+1}). \quad (5.2.15)$$

There are four equations here and only two equations are needed to make an algorithm of the form

$$A_1 Y_{n+1} = A_0 Y_n + F_n, \quad (5.2.16)$$

where the vector $Y_n = [x_n, v_n]^{\mathrm{T}}$. Gellert uses eqn (5.2.13) which came from the original differential equation (5.2.9) integrated once and a linear combination of the other three equations (5.2.12), (5.2.14), and (5.2.15). Gellert presumably chose this particular combination because he judged it to give the best result (see Exercise 5.1 for the result of taking simply (5.2.14) and (5.2.15) in the case where C and f are zero). Gellert's first equation is formed by taking

$$\tfrac{5}{3}\{\text{eqn (5.2.12)}\} + \frac{\Delta t^2}{6}\{\text{eqn (5.2.14)}\} - \frac{\Delta t^3}{36}\{\text{eqn (5.2.15)}\}.$$

The result is an algorithm of the form (5.2.16) with

$$A_1 = \begin{pmatrix} M + \Delta t C + 5\Delta t^2 K/12 & \Delta t M/2 - \Delta t^2 C/12 - \Delta t^3 K/12 \\ C + \Delta t K/2 & M - \Delta t^2 K/12 \end{pmatrix} \quad (5.2.17)$$

$$A_0 = \begin{pmatrix} M + \Delta t C - 7\Delta t^2 K/12 & 3\Delta t M/2 - \Delta t^2 C/12 - \Delta t^3 K/12 \\ C - \Delta t K/2 & M - \Delta t^2 K/12 \end{pmatrix} \quad (5.2.18)$$

and

$$\boldsymbol{F} = \begin{bmatrix} \Delta t^2(3f_n + 8f_{n+\frac{1}{2}} + f_{n+1})/12 \\ \Delta t(f_n + 4f_{n+\frac{1}{2}} + f_{n+1})/6 \end{bmatrix}, \tag{5.2.19}$$

Gellert remarks that a major advantage of his method over that of Argyris *et al.* (1973a,b) is the sparse and usually banded form of the matrices $\boldsymbol{A}_1$ and $\boldsymbol{A}_0$ when the algorithm is applied to the system of equations (5.1.1). The matrix $\boldsymbol{A}_1$ can be decomposed into two triangular matrices $\boldsymbol{X}_1$, $\boldsymbol{X}_2$, such that

$$\boldsymbol{A}_1 = \boldsymbol{X}_1^{\mathrm{T}}\mathbf{X}_2.$$

Hence the recurrence can be solved easily with no need of the storage required by a full matrix inversion. Gellert also shows that the eigenvalues of the amplification matrix $\boldsymbol{A} = \boldsymbol{A}_1^{-1}\boldsymbol{A}_0$ are of unit modulus if the natural damping matrix $\boldsymbol{C} = \mathbf{0}$. They are also complex conjugates with modulus less than unity if the damping is less than critical. Hence the eigenvalues of the amplification matrix model the exact solution with no algorithmic damping and Gellert shows that the error in the solution is $\mathrm{O}(\Delta t^4)$. When the natural damping is zero, the equations for the eigenvalues of the amplification matrices are the same for the algorithms of Gellert (with Hermitian cubic approximation of the displacement) and of Argyris *et al.* (1973a,b) (with a modified Hermitian cubic approximation of the acceleration).

Gellert seems surprised to find that the displacement and velocity have the same order error. We can see that this is another situation where we can use the Cayley–Hamilton theorem on the scalar form of eqn (5.2.9). If we have

$$\boldsymbol{Y}_{n+1} = \boldsymbol{A}\boldsymbol{Y}_n$$

where $\boldsymbol{Y}_n = [x_n, v_n]^{\mathrm{T}}$ and $\boldsymbol{A}$ is the 2×2 amplification matrix, then as shown in §4.4 we have

$$p\boldsymbol{Y}_{n+1} + q\boldsymbol{Y}_n + r\boldsymbol{Y}_{n-1} = \mathbf{0},$$

where $p\lambda^2 + q\lambda + r = 0$ is the characteristic polynomial of the matrix $\boldsymbol{A}$. Hence both x_n and v_n satisfy the same recurrence of the form

$$pu_{n+1} + qu_n + ru_{n-1} = 0$$

and given sufficiently accurate starting conditions must be obtained to the same order of accuracy. In the next section we show the relationship of Gellert's method to the method using Padé (2.2) approximant for the exponential.

5.3. Rational approximations to the exponential

We consider the linear first order differential equation

$$\dot{\boldsymbol{z}} + \boldsymbol{B}\boldsymbol{z} = \boldsymbol{g}, \tag{5.3.1.}$$

which may be derived from a spatial discretization of a parabolic problem with the matrix $\boldsymbol{B} = \boldsymbol{C}^{-1}\boldsymbol{K}$ or it may represent the first order form of the dynamic vibration equation with

$$\boldsymbol{B} = \begin{bmatrix} \boldsymbol{0} & -\boldsymbol{I} \\ \boldsymbol{M}^{-1}\boldsymbol{K} & \boldsymbol{M}^{-1}\boldsymbol{C} \end{bmatrix}, \tag{5.3.2}$$

$\boldsymbol{z} = [\boldsymbol{x}, \boldsymbol{v}]^{\mathrm{T}}$ and $\boldsymbol{g}^{\mathrm{T}} = [\boldsymbol{0}, \boldsymbol{M}^{-1}\boldsymbol{f}]$ in the usual notation. The exact solution of eqn (5.3.1) may be written as

$$\boldsymbol{z}((n+1)\Delta t) = \exp(-\boldsymbol{B}\Delta t)\boldsymbol{z}(n\Delta t) + \exp\{-\boldsymbol{B}(n+1)\Delta t\} \times \int_{n\Delta t}^{(n+1)\Delta t} \boldsymbol{g}(\tau)\exp \boldsymbol{B}\tau \,\mathrm{d}\tau \tag{5.3.3}$$

obtained by integrating over the interval $t = n\Delta t$ to $t = (n+1)\Delta t$. The linear p-step and p-stage methods discussed previously depend on approximating $\exp(-\boldsymbol{B}\Delta t)$ by the principal roots of the characteristic polynomial of the amplification matrix. The methods discussed in this section depend instead on making rational approximations to the exponential in eqn (5.3.3). If r is a complex scalar we consider approximations to $\exp(r)$ of the form

$$R_m^n(r) = \left(\sum_{j=0}^{n} a_j r^j\right) \bigg/ \left(\sum_{j=0}^{m} b_j r^j\right), \tag{5.3.4}$$

where all the coefficients a_j and b_j are real. If we put

$$\left(\sum_{j=0}^{n} a_j r^j\right) = \left(\sum_{j=0}^{m} b_j r^j\right)\left(\sum_{j=0}^{\infty} r^j/j!\right), \tag{5.3.5}$$

then we can select for each pair of non-negative integers n and m the polynomials in the numerator and denominator in eqn (5.3.4) which agree with as many leading terms in the expansion of the exponential as possible. Since the ratio of polynomials in eqn (5.3.4) has $(n+m+1)$ unknown coefficients, we can write down $(n+m+1)$ equations to determine these coefficients such that

$$R_m^n = \exp(r) + O(r^{n+m+1}) \tag{5.3.6}$$

and then the rational approximation R_m^n is said to be of order $(n+m)$. This rational approximation of the highest possible order for given n, m is called the (n, m) Padé approximation after Padé (1892). Varga (1962) discusses the use of Padé approximations in the numerical solution of parabolic differential equations with the space discretization done by finite differences. Lambert (1973) discusses their use in the solution of ordinary differential equations.

We start by looking at ideas of stability in connection with rational approximations to the exponential in the solution of the first order equation. For this purpose we suppose we have a modal discretization of eqn (5.3.1) into scalar equations of the form

$$\dot{z} = -\lambda z + g, \tag{5.3.7}$$

where λ is an eigenvalue of the matrix $\boldsymbol{B}$. Then we are approximating $\exp(-\lambda\Delta t)$ in the numerical solution of eqn (5.3.7) by the rational function $R_m^n(-\lambda\Delta t)$. We want to categorize these rational approximations according to the stability of the solutions produced by them. We recall from §2.8 that there is some difference of opinion in the literature about the labelling of the different kinds of stability. Dahlquist (1963), when he introduced the term *A*-stability first in connection with linear multistep methods for the numerical solution of a first order equation

$$\dot{y} = \mu y \tag{5.3.8}$$

defined *A*-stability as 'y_n tends to zero as n tends to infinity with fixed Δt whenever $y(t)$ tends to zero as t tends to infinity', i.e. for μ anywhere in the left-hand half-plane in the Argand diagram: $\mathrm{Re}(\mu) < 0$. This definition excludes the imaginary axis because if $\mu = \mathrm{i}\omega$ the solution does not decay. However, Dahlquist added a note at the end of this (1963) paper to the effect that an *A*-stable method used on the eqn (5.3.8) when the eigenvalue μ is on the imaginary axis will give a bounded solution. This means that an A-stable algorithm produces a solution which tends to zero whenever the exact solution tends to zero and remains bounded when the exact solution remains bounded, i.e. to this extent it reproduces the behaviour of the exact solution. This concept of A-stability introduced originally in Dahlquist (1963) in connection with linear multistep methods is now extended to cover other methods such as those which depend on rational approximations of the exponential which are being considered in this section. We incorporate A_0-stability and L-stability in the same way. Hence we say that the rational approximation $R^n_m(-\lambda\Delta t)$ to the exponential $\exp(-\lambda\Delta t)$ gives an algorithm which is

(1) A-stable if

$$|R^n_m(-\lambda\Delta t)| \leq 1 \quad \text{whenever} \quad \text{Re}(-\lambda) \geq 0;$$

(2) A_0-stable if

$|R^n_m(-\lambda\Delta t)| < 1$ whenever λ is real and positive (i.e. $-\lambda$ is on the negative real axis in the Argand diagram);

(3) L-stable if it is A-stable and

$|R^n_m(-\lambda\Delta t)|$ tends to zero as λ tends to infinity.

The results for the (n, m) Padé approximations (Wanner 1987 and references therein) show that they are

(1) not A- or A_0-stable for $n > m$;

(2) A-stable for $m = n$, $n + 1$, or $n + 2$;

(3) A_0-stable for $n \leq m$;

(4) L-stable for $m = n + 1$ or $n + 2$.

The Padé approximations with $m \geq n + 3$ are not A-stable. For example, the $(0, 3)$ Padé approximates $\exp(-\lambda)$ by

$$R = (1 + \lambda + \tfrac{1}{2}\lambda^2 + \tfrac{1}{6}\lambda^3)^{-1}, \tag{5.3.9}$$

which certainly has modulus less than unity for positive λ and hence is A_0-stable. But if $\lambda = \mathrm{i}\omega$ then $|R| < 1$ requires $(1 - \tfrac{1}{2}\omega^2)^2 + \omega^2(\tfrac{1}{6}\omega^2 - 1)^2 > 1$, i.e. $\omega^2 > 3$. This means that the

Table 5.3.1. Padé table for $\exp(-z)$.

	n			
m	0	1	2	3
0	1	$1-z$ Euler	$1-z+z^2/2$ explicit RK2	$1-z+z^2/2-z^3/6$ explicit RK3
1	$\dfrac{1}{1+z}$ Backward difference	$\dfrac{1-z/2}{1+z/2}$ C–N-Trap	$\dfrac{1-2z/3+z^2/6}{1+z/3}$	...
2	$\dfrac{1}{1+z+z^2/6}$	$\dfrac{1-z/3}{1+2z/3+z^2/6}$	$\dfrac{1-z/2+z^2/12}{1+z/2+z^2/12}$	...
3	$\dfrac{1}{1+z+z^2/2+z^3/6}$	...	...	

stability boundary cuts the imaginary axis at $\pm i\sqrt{3}$ and hence it obtrudes into the left-hand half-plane while leaving the negative real axis clear like the Gear three-step in Fig. 2.4. Thus the $(0, 3)$ Padé is not A-stable. Table 5.3.1 gives the Padé approximations of most interest to practical users.

Trujillo (1976) in a paper entitled 'The direct numerical integration of linear matrix differential equations using Padé approximations' seems to have been the first to use the method on the dynamic vibration equation. He illustrates the method on the modally discretized scalar form corresponding in our notation to

$$\ddot{x} + 2\nu\omega\dot{x} + \omega^2 x = f \tag{5.3.10}$$

and puts it into the first order form (5.3.1) with

$$\boldsymbol{B} = \begin{bmatrix} 0 & -1 \\ \omega^2 & 2\nu\omega \end{bmatrix}. \tag{5.3.11}$$

Trujillo gives a table of Padé approximations for the exponential function in the formula (5.3.3) which he is using for his survey. These include

(1, 0), Euler forward difference, (2, 0),
(0, 1) backward difference, (0, 2),
(2, 1)

and what he calls the evenly balanced formulae (1, 1) Crank–Nicolson (1947), also equivalent to Newmark (1959) with $2\beta = \gamma = 0.5$ and (2, 2), (3, 3), and (4, 4).

With all these, the forcing function $f(t)$ is represented as piecewise linear. Trujillo recognizes the difficulty with the powers of matrices that appear in the higher order methods. For instance, the (2, 2) Padé used on the homogeneous form of eqn (5.3.1) produces the recurrence

$$\left(\boldsymbol{I} + \frac{\Delta t}{2}\boldsymbol{B} + \frac{\Delta t^2}{12}\boldsymbol{B}^2\right)\boldsymbol{Y}_{n+1} = \left(\boldsymbol{I} - \frac{\Delta t}{2}\boldsymbol{B} + \frac{\Delta t^2}{12}\boldsymbol{B}^2\right)\boldsymbol{Y}_n \tag{5.3.12}$$

and Trujillo suggests putting the squared matrices on the right hand side and solving iteratively.

Thomas and Gladwell (1984) point out that if the recurrence (5.3.12) is multiplied by the matrix

$$\begin{bmatrix} \boldsymbol{M} & \boldsymbol{0} \\ \boldsymbol{C} & \boldsymbol{M} \end{bmatrix}$$

and the matrix $\boldsymbol{B}$ substituted from eqn (5.3.2), the result is

$$\begin{bmatrix} \boldsymbol{M} - \Delta t^2\boldsymbol{K}/12 & -\Delta t\boldsymbol{M}/2 - \Delta t^2\boldsymbol{C}/12 \\ \boldsymbol{C} + \Delta t\boldsymbol{K}/2 & \boldsymbol{M} - \Delta t^2\boldsymbol{K}/12 \end{bmatrix}\boldsymbol{Y}_{n+1}$$

$$= \begin{bmatrix} \boldsymbol{M} - \Delta t^2\boldsymbol{K}/12 & \Delta t\boldsymbol{M}/2 - \Delta t^2\boldsymbol{C}/12 \\ \boldsymbol{C} - \Delta t\boldsymbol{K}/2 & \boldsymbol{M} - \Delta t^2\boldsymbol{K}/12 \end{bmatrix}\boldsymbol{Y}_n.$$

Then adding Δt times row 2 on to row 1 gives the matrices $\boldsymbol{A}_1$ and $\boldsymbol{A}_0$ in the Gellert method as in eqns (5.2.17) and (5.2.18). Thus Gellert's method is equivalent to using the (2, 2) Padé approximation for the exponential term in eqn (5.3.3) whch is to be expected as he manipulated his equations to obtain the highest possible order for the accuracy of an R_2^2 type expression. Thomas and Gladwell (1984) also give the corresponding form of Gellert's method including the forcing function as

$$\begin{aligned}\left(\boldsymbol{I} + \frac{\Delta t}{2}\boldsymbol{B} + \frac{\Delta t^2}{12}\boldsymbol{B}^2\right)\boldsymbol{Y}_{n+1} = \left(\boldsymbol{I} - \frac{\Delta t}{2}\boldsymbol{B} + \frac{\Delta t^2}{12}\boldsymbol{B}^2\right)\boldsymbol{Y}_n \\ + \frac{\Delta t}{6}(\boldsymbol{g}_n + 4\boldsymbol{g}_{n+\frac{1}{2}} + \boldsymbol{g}_{n+1}) \\ - \frac{\Delta t^2}{12}\boldsymbol{B}(\boldsymbol{g}_n - \boldsymbol{g}_{n+1}).\end{aligned} \tag{5.3.13}$$

We can compare this representation of the forcing function with that obtained from a Runge–Kutta method which also produces the (2, 2) Padé approximation to the exponential; this is discussed in §5.4.

Another way of avoiding the difficulty with the powers of matrices is to use the method of Nørsett (1974) in which the matrix polynomials corresponding to the denominators of the rational approximations to the exponential are replaced by products of linear factors $(\boldsymbol{I} + \alpha_p \Delta t \boldsymbol{B})^p$. This is discussed further later.

Argyris *et al.* (1977) in a very comprehensive paper entitled 'Higher order methods for transient diffusion analysis', relate Hermitian expansions of increasing order to methods using rational approximations of the exponential including Padé and Nørsett. The implication of the title is that the first order equation, of the form of (5.3.1), dealt with there is parabolic, but much of the analysis also applies to the first order form of the dynamic vibration equation.

Argyris *et al.* (1977) propose a series of approximations for the unknown function in terms of the values of the function and its derivatives at the ends of the time interval $[n\Delta t, (n+1)\Delta t]$. The first member of the series is of course the simple form given by

$$\boldsymbol{z} = (1 - \zeta)\boldsymbol{z}_n + \zeta \boldsymbol{z}_{n+1}. \tag{5.3.14}$$

Differentiating this expression for $\boldsymbol{z}$ with respect to time t $(=(n + \zeta)\Delta t)$, gives

$$\dot{\boldsymbol{z}} = \frac{1}{\Delta t}(\boldsymbol{z}_{n+1} - \boldsymbol{z}_n). \tag{5.3.15}$$

These expressions for $\boldsymbol{z}$ and $\dot{\boldsymbol{z}}$ are substituted into the homogeneous differential equation (5.3.1) with $\boldsymbol{g} \equiv 0$ which gives

$$(\boldsymbol{I} + \zeta \Delta t \boldsymbol{B})\boldsymbol{z}_{n+1} = (\boldsymbol{I} - (1 - \zeta)\Delta t \boldsymbol{B})\boldsymbol{z}_n, \tag{5.3.16}$$

which effectively means collocation at $t = \zeta \Delta t$. Equation (5.3.16) is the same as that given by the θ-method $(\zeta = \theta)$ or SS11 $(\zeta = \theta_1)$ already covered in previous chapters. Comparing eqns (5.3.16) and (5.3.3) it also means that the matrix $\exp(-\Delta t \boldsymbol{B})$ is approximated by the rational function

$$(\boldsymbol{I} + \zeta \Delta t \boldsymbol{B})^{-1}(\boldsymbol{I} - (1 - \zeta)\Delta t \boldsymbol{B}). \tag{5.3.17}$$

Then $\zeta = 0$, 0.5, 1 give the (1, 0), (1, 1), (0, 1) Padé approximants respectively. The result of putting $\zeta = 0.5$ also corresponds to the Crank–Nicolson or trapezium rule or to Newmark with $2\beta = \gamma = 0.5$ or SS21 with $\theta_1 = \theta_2 = 0.5$.

The next member of the series is obtained by using the Hermitian cubic expansion like that of Gellert (1978):

$$z = \omega_1 z_n + \omega_2 \Delta t \dot{z}_n + \omega_3 z_{n+1} + \omega_4 \Delta t \dot{z}_{n+1}, \qquad (5.3.18)$$

where the ω_j, $j = 1$ to 4, are as in eqns (5.2.4). The expression for z given by eqn (5.3.18) is then differentiated with respect to time and substituted into the homogeneous equation (5.3.1) collocated at $t = \zeta \Delta t$. The forcing function is dealt with separately so that the emphasis here is on the representation of the exponential function. Writing $\dot{\omega}_j$ for the derivative of ω_j with respect to time the substitution gives:

$$(\dot{\omega}_3 \boldsymbol{I} + \omega_3 \boldsymbol{B}) z_{n+1} + (\dot{\omega}_4 \boldsymbol{I} + \omega_4 \boldsymbol{B}) \Delta t \dot{z}_{n+1}$$
$$= -(\dot{\omega}_1 \boldsymbol{I} + \omega_1 \boldsymbol{B}) z_n - (\dot{\omega}_2 \boldsymbol{I} + \omega_2 \boldsymbol{B}) \Delta t \dot{z}_n. \qquad (5.3.19)$$

In order to remove the nodal derivatives the homogeneous differential equation is used to give $\dot{z}_n = -\boldsymbol{B} z_n$, and $\dot{z}_{n+1} = -\boldsymbol{B} z_{n+1}$, i.e. for $0 < \zeta < 1$ the homogeneous equation is effectively collocated at three points. The result is a one-step recurrence:

$$(\boldsymbol{I} + b_1 \Delta t \boldsymbol{B} + b_2 \Delta t^2 \boldsymbol{B}^2) z_{n+1} = (\boldsymbol{I} + a_1 \Delta t \boldsymbol{B} + a_2 \Delta t^2 \boldsymbol{B}^2) z_n \qquad (5.3.20)$$

where

$$a_1 = (-\omega_1 + \Delta t \dot{\omega}_2)/\Delta t \dot{\omega}_3, \qquad a_2 = \omega_2/\Delta t \dot{\omega}_3,$$
$$b_1 = (\omega_3 - \Delta t \dot{\omega}_4)/\Delta t \dot{\omega}_3, \qquad b_2 = -\omega_4/\Delta t \dot{\omega}_3. \qquad (5.3.21)$$

Then putting in different values of ζ gives various R_2^2 rational approximations to the exponential function $\exp(-\Delta t \boldsymbol{B})$. For example, putting $\zeta = 0.5$ gives the highest order accuracy (2, 2) Padé approximation which for the scalar equation (5.3.7) approximates $\exp(-\lambda \Delta t)$ by:

$$(1 - \tfrac{1}{2}\lambda \Delta t + \tfrac{1}{12}\lambda^2 \Delta t^2)/(1 + \tfrac{1}{2}\lambda \Delta t + \tfrac{1}{12}\lambda^2 \Delta t^2). \qquad (5.3.22)$$

The expression (5.3.22) has fourth order accuracy as an approximation to $\exp(-\lambda \Delta t)$, i.e. the truncation error when it is expanded in powers of $\lambda \Delta t$ is $O(\lambda^5 \Delta t^5)$. Unlike the Crank–Nicolson-trapezium rule higher order approximation obtained by using $\zeta = 0.5$ in the linear expansion the (2.2) Padé is non-oscillatory. The expression in the numerator:

$$1 - \tfrac{1}{2}\lambda\Delta t + \tfrac{1}{12}\lambda^2\Delta t^2 = \tfrac{1}{12}\{(\lambda\Delta t - 3)^2 + 3\}$$

is always positive for real values of λ and hence the expression (5.3.22) is always positive also as is the exponential. But the limit when $\lambda\Delta t \rightarrow$ infinity is unity instead of zero.

Taking the limit when $\zeta \rightarrow 1$ gives the L-stable (1, 2) Padé approximation to $\exp(-\lambda\Delta t)$:

$$(1 - \tfrac{1}{3}\lambda\Delta t)/(1 + \tfrac{2}{3}\lambda\Delta t + \tfrac{1}{6}\lambda^2\Delta t^2). \qquad (5.3.23)$$

This (1, 2) Padé approximation is negative for $\lambda\Delta t > 3$. Lambert (1973) gives an expression for R_2^2 in terms of two parameters α and β which is equivalent to writing the approximation for $\exp(-\lambda\Delta t)$ as

$$R_2^2(\alpha, \beta) = \{1 - \tfrac{1}{2}(1 - \alpha)\lambda\Delta t + \tfrac{1}{4}(\beta - \alpha)\lambda^2\Delta t^2\}/$$
$$\{1 + \tfrac{1}{2}(1 + \alpha)\lambda\Delta t + \tfrac{1}{4}(\beta + \alpha)\lambda^2\Delta t^2\}. \qquad (5.3.24)$$

This approximation has order two in general, order three if $\beta = \frac{1}{3}$, $\alpha \neq 0$ and order four if $\alpha = 0$ and $\beta = \frac{1}{3}$ which gives the (2, 2) Padé. Since the L-stable (1, 2) Padé is given by $\alpha = \beta = 1/3$, this is of order three. Figure 5.3.1 shows some Padé approximations for $\exp(-z)$.

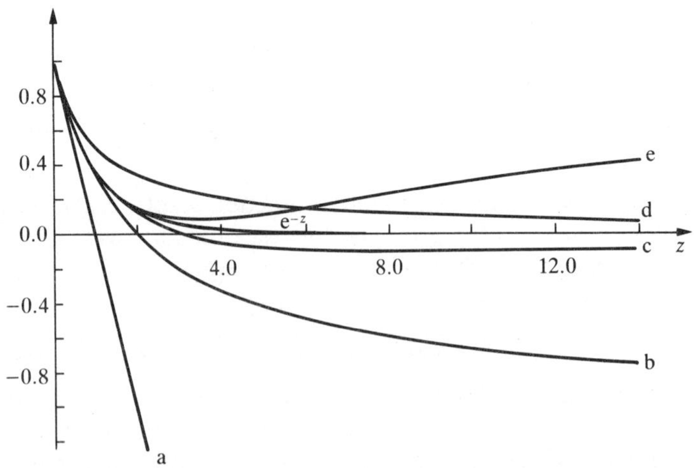

Fig. 5.3.1. Padé approximations for $\exp(-z)$: (a) R_0^1, (b) R_1^1, (Crank–Nicolson), (c) R_2^1, (d) R_1^0, (e) R_2^2.

Argyris *et al.* (1977) also consider a quintic Hermitian approximation for the function z in eqn (5.3.1) expressing it in terms of the six quantities $z_n, \dot{z}_n, \ddot{z}_n$, and $z_{n+1}, \dot{z}_{n+1}, \ddot{z}_{n+1}$ and a parameter ζ. The expression is differentiated and substituted into the homogeneous differential equation. The derivatives are removed by using the homogeneous differential equation and its derivative at $t = n\Delta t$ and $t = (n+1)\Delta t$. The result is an R_3^3 approximation for $\exp(-\Delta t\boldsymbol{B})$ which for $\zeta = 0.5$ gives the (3, 3) Padé and in the limit as $\zeta \to 1$ gives an L-stable method with an approximation which now stays positive. The interested reader is referred to Argyris *et al.* (1977) for further details.

An approximation to the exponential of another type is implicitly used by Zienkiewicz and Lewis (1973) in a frequently quoted paper. They take the equation

$$\boldsymbol{C}\dot{\boldsymbol{\Phi}} + \boldsymbol{K}\boldsymbol{\Phi} + \boldsymbol{f}(t) = \boldsymbol{0} \tag{5.3.25}$$

and approximate it with a 'least squares weighting' in the time step to give the formula

$$\begin{aligned}\boldsymbol{\Phi}_{n+1} = {} & \{\boldsymbol{C}^{\mathrm{T}}\boldsymbol{C} + \Delta t(\boldsymbol{K}^{\mathrm{T}}\boldsymbol{C} + \boldsymbol{C}^{\mathrm{T}}\boldsymbol{K})/2 + \Delta t^2\boldsymbol{K}^{\mathrm{T}}\boldsymbol{K}/3\}^{-1} \\ & \times \{\boldsymbol{C}^{\mathrm{T}}\boldsymbol{C} + \Delta t(\boldsymbol{K}^{\mathrm{T}}\boldsymbol{C} - \boldsymbol{C}^{\mathrm{T}}\boldsymbol{K})/2 - \Delta t^2\boldsymbol{K}^{\mathrm{T}}\boldsymbol{K}/6\}\boldsymbol{\Phi}_n \\ & + \text{forcing function terms.}\end{aligned} \tag{5.3.26}$$

Comparing this with eqn (5.3.3) we see that the expression multiplying $\boldsymbol{\Phi}_n$ on the right-hand side is effectively an approximation to $\exp(-\Delta t\boldsymbol{C}^{-1}\boldsymbol{K})$. If we call the numerator of this expression N and the denominator M, we can look at the accuracy of this approximation by considering the difference

$$\begin{aligned}& \boldsymbol{N} - \boldsymbol{M}\{\boldsymbol{I} - \boldsymbol{C}^{-1}\boldsymbol{K}\Delta t + (\boldsymbol{C}^{-1}\boldsymbol{K})^2\Delta t^2/2 - (\boldsymbol{C}^{-1}\boldsymbol{K})^3\Delta t^3/6 + \cdots\} \\ & \qquad = \frac{\Delta t^3}{12}\{-\boldsymbol{K}^{\mathrm{T}} + \boldsymbol{C}^{\mathrm{T}}\boldsymbol{K}\boldsymbol{C}^{-1}\}\boldsymbol{K}\boldsymbol{C}^{-1}\boldsymbol{K} + \mathrm{O}(\Delta t^4).\end{aligned} \tag{5.3.27}$$

If the matrices C and K are the standard mass and stiffness matrices respectively they are symmetric but the Δt^3 will not disappear unless they also commute. Hence the error will be of the same order as with Crank–Nicolson. The numerical example given by Zienkiewicz and Lewis takes a one-dimensional problem with $\boldsymbol{K} = \boldsymbol{C} = 1$. With these values the error given by eqn (5.3.27) is $\mathrm{O}(\Delta t^4)$ and this is why their numerical results look so good compared with Crank–Nicolson but these results do not justify a general recommendation of the least squares method.

A major disadvantage of the Padé approximations of the exponential in eqn (5.3.3) is the computational difficulty with the powers of the matrix (except in the (1, 1) Padé of course). Trujillo (1975) suggests putting the powers of the matrix on the right-hand side of the equation and iterating. (The convergence of the iteration is likely to limit the size of the time step Δt.) Another approach is to use Nørsett's method (1974a,b) which is also used by Argyris *et al.* (1977). The basic idea of Nørsett's method is to form rational approximations for the exponential such that the denominator is expressed as a power of $(\boldsymbol{I} + \alpha\Delta t\boldsymbol{B})$ where the parameter α is real. The algorithm is then arranged in a series of steps such that each uses the inverse of the same matrix $(\boldsymbol{I} + \alpha\Delta t\boldsymbol{B})$. Illustrating the method on the scalar equation, we can say that we want to choose the polynomial of degree p, $P_p(-\lambda\Delta t)$, such that

$$\exp(-\lambda\Delta t) = P_p(-\lambda\Delta t)/(1 + \alpha\lambda\Delta t)^p \tag{5.3.28}$$

to as high an accuracy as possible. The coefficients of the polynomial depend on the parameter α, and we look for the vaue of α which gives the most accurate A-stable method in this class. Alternatively we can form an L-stable method by reducing the degree of the polynomial in the numerator to $(p-1)$. Nørsett is following up a method put forward by Makinson (1968) which uses Hermite polynomials (the same kind of idea is an Argyris *et al.* 1973a,b, and 1977 and Gellert 1978). The Nørsett A-stable algorithm which gives the highest accuracy for a particular p in the rational approximation R_p^p for $\exp(-\lambda\Delta t)$ is given by

$$N_p^p(\alpha_p) = 1 + \sum_{j=1}^{p} \frac{L_j'(1/\alpha_p)}{j\alpha_p}\left(\frac{\alpha_p\lambda\Delta t}{1 + \alpha_p\lambda\Delta t}\right)^j \tag{5.3.29}$$

where $L_j(x)$ is the Laguerre polynomial of degree j and $L_j'(x)$ is its derivative. The Laguerre polynomials are given by the recurrence

$$L_{k+1}(x) = A_k(x - B_k)L_k(x) - C_kL_{k-1}(x), \tag{5.3.30}$$

where $A_k = -1/(k+1)$, $B_k = 2k+1$, $C_k = k/(k+1)$, and the starting values are $L_{-1} \equiv 0$ and $L_0 = 1$. We thus have $L_1(x) = 1 - x$, $L_2(x) = 1 - 2x + x^2/2$, $L_3(x) = 1 - 3x + 3x^2/2 - x^3/6$, etc. (For further information on Laguerre polynomials see, for example, Conte and de Boor 1980.)

The parameter α has to be chosen so that $L'_{p+1}(1/\alpha) = 0$. The polynomial $L'_{p+1}(x)$ is of degree p and has p distinct real positive roots, hence for each value of p there are p values of α_p and thus p possible approximations to the exponential of the form given in (5.3.13). With any of these values of α_p we have

$$|N_p^p(\alpha_p) - \exp(-\lambda\Delta t)| = O(|\lambda\Delta t|^{p+2}). \tag{5.3.31}$$

Nørsett (1974a,b) shows that

(1) There are A_0-stable methods of this type for all values of p;

(2) A-stable methods exist for $p = 1, 2, 3, 5$;

(3) when $p = 4$ there are no A-stable methods;

(4) for $6 \leq p \leq 15$ the methods are $A(\alpha)$-stable with the angle α about 84°.

The algorithm from the Nørsett method for the solution of the equation $\dot{\boldsymbol{z}} + \boldsymbol{B}\boldsymbol{z} = \boldsymbol{0}$ is as follows:

$$\begin{aligned} (\boldsymbol{I} + \alpha\Delta t\boldsymbol{B}\boldsymbol{u}_1 &= \alpha\Delta t\boldsymbol{B}\boldsymbol{z}_n \\ (\boldsymbol{I} + \alpha\Delta t\boldsymbol{B})\boldsymbol{u}_{j+1} &= \alpha\Delta t\boldsymbol{B}\boldsymbol{u}_j, \qquad j = 1, 2, \ldots, p-1, \\ \boldsymbol{z}_{n+1} &= \boldsymbol{z}_n + \frac{1}{\alpha}\sum_{j=1}^{p} L'_j(1/\alpha)\boldsymbol{u}_j/j. \end{aligned} \tag{5.3.32}$$

Except for $p = 1$, the Nørsett approximation of the exponential is less accurate than the (p, p) Padé approximation but the factors in the denominator make the Nørsett approximation more convenient. When $p = 1$, we have $L'_{p+1}(x) = L'_2(x) = x - 2$, hence in this case there is only the one root $\alpha = \frac{1}{2}$ and the approximation for the $\exp(-\lambda\Delta t)$ is

$$N_1^1(\alpha) = (1 - 0.5\lambda\Delta t)/(1 + 0.5\lambda\Delta t).$$

This is the Crank–Nicolson-trapezium rule which is A-stable as shown in Chapter 2.

When $p = 2$ the Nørsett approximation for $\exp(-\lambda\Delta t)$ is given by

$$N_2^2(\alpha) = \{1 + (2\alpha - 1)\lambda\Delta t - (\alpha - \tfrac{1}{3})\lambda^2\Delta t^2\}(1 + \alpha\lambda\Delta t)^{-2}, \tag{5.3.33}$$

where $1/\alpha$ is a root of

$$L'_3(x) = -3 + 3x - x^2/2 = 0. \tag{5.3.34}$$

The two values of α are $(3 \pm \sqrt{3})/6$; we order these as $\alpha_1 < \alpha_2$. The larger value, α_2 gives an approximation for the exponential which is the same as that in a method proposed by Calahan (1968). Calahan sets out to construct a 'half-plane stable' (A-stable) method with local truncation error $O(\Delta t^4)$ for the non-linear equation

$$\dot{z} = f(z, t). \tag{5.3.35}$$

With the Jacobian $\partial f(z, t)/\partial z = A(z, t)$, the solution Calahan gives corresponds to approximating $\exp(\Delta t A(z, t))$ by the $N_2^2(\alpha)$ expression with $\alpha = (3 + \sqrt{3})/6$. Calahan is frequently quoted as if this were all there is to his method but it is worth noting that this is a method which takes account of the forcing function as well in the non-linear equation (5.3.35). For the solution of eqn (5.3.35), the Calahan method gives:

$$z_{n+1} = z_n + R_1 k_{n+1} + R_1 l_{n+1}, \tag{5.3.36}$$

where

$$\begin{aligned} k_{n+1} &= \Delta t \{I - \Delta t a A(z_n, t_n)\}^{-1} \\ &\quad \times \left\{ f(z_n, t_n) + a \Delta t \frac{\partial}{\partial t} f(z_n, t_n) \right\}, \\ l_{n+1} &= \Delta t \{I - \Delta t a A(z_n, t_n)\}^{-1} \\ &\quad \times \left\{ f(z_n + b k_{n+1}, t_n + b \Delta t) + a \Delta t \frac{\partial}{\partial t} f(z_n, t_n) \right\}, \end{aligned}$$

and $a = \alpha_2 = (3 + \sqrt{3})/6$, $b = 2(1 - 2\alpha_2) = -2\sqrt{3}/3$, $R_1 = 0.75$, $R_2 = 0.25$.

Returning to the Nørsett approximation for $\exp(-z)$, if we expand in power series in order to look at the error, we have

$$\begin{aligned} N_2^2(\alpha) - \exp(-z) &= \{1 + (2\alpha - 1)z - (\alpha - \tfrac{1}{3})z^2\} \\ &\quad \times (1 - 2\alpha z + 3\alpha^2 z^2 - 4\alpha^3 z^3 + \cdots) - \exp(-z) \\ &= \tfrac{1}{24}(1 - 4\alpha)z^4 + \text{higher order terms} \end{aligned}$$

where we have removed second and higher powers of α by using the equation for α from (5.3.34), i.e. $\alpha^2 = \alpha - \frac{1}{6}$. Thus for the Calahan $\alpha = \alpha_2 = 0.7887$, the coefficient of the leading error term is -8.98×10^{-2}, approximately, and for $\alpha = \alpha_1 = 0.2113$, the coefficient is $+6.45 \times 10^{-3}$, approximately.

For real z the $N_2^2(\alpha_1)$ expression is always positive because the numerator:

$$1+(2\alpha-1)z+(\tfrac{1}{3}-\alpha)z^2=(\tfrac{1}{3}-\alpha)\{z+(\alpha-\tfrac{1}{2})/(\tfrac{1}{3}-\alpha)\}^2 + (\tfrac{1}{4}-\alpha)/(\tfrac{1}{3}-\alpha)$$

which is positive for $\alpha=\alpha_1=0.2113$. Hence the approximation $N_2^2(\alpha_1)$ for $\exp(-z)$ with real z stays above the axis but for large values of z we have $N_2^2(\alpha_1)>1$. The Calahan approximation crosses the axis where $(\tfrac{1}{3}-\alpha_1)z^2+(2\alpha_2-1)z+1=0$, i.e. where $z=2.246$ approximately. Hence the Calahan approximation becomes negative and is liable to give oscillations. By the theory in Nørsett (1974a,b), the Calahan value gives an A-stable approximation. We can easily see that happens when we use the $N_2^2(\alpha)$ approximation for $\exp(i\omega)$. We have stability if

$$|1-(2\alpha-1)i\omega-(\tfrac{1}{3}-\alpha)\omega^2|<|(1-i\omega\alpha)^2|.$$

This condition simplifies into $\omega^2(4\alpha-1)>0$, which is true for all real ω with $\alpha=\alpha_2$.

In order to generate L-stable methods with $(1+\alpha\lambda\Delta t)^p$ in the denominator of the expression for $\exp(-\lambda\Delta t)$, we obtain a numerator of degree $(p-1)$ by forming

$$N_p^{p-1}(\alpha)=\sum_{j=0}^{p-1} L_j(1/\alpha)(\alpha\lambda\Delta t)^j/(1+\alpha\lambda\Delta t)^{j+1}, \quad (5.3.37)$$

where α satisfies $L_p(1/\alpha)=0$. The error now is given by

$$|N_p^{p-1}(\alpha)-\exp(-\lambda\Delta t)|=O(|\lambda\Delta t|^{p+1}).$$

The algorithm is formed by taking

$$\begin{aligned}(\boldsymbol{I}+\alpha\Delta t\boldsymbol{B})\boldsymbol{u}_1&=\alpha\Delta t\boldsymbol{B}\boldsymbol{z}_n\\(\boldsymbol{I}+\alpha\Delta t\boldsymbol{B})\boldsymbol{u}_{j+1}&=\alpha\Delta t\boldsymbol{B}\boldsymbol{u}_j, \qquad j=1,2,\ldots,p-2,\\(\boldsymbol{I}+\alpha\Delta t\boldsymbol{B})\boldsymbol{z}_{n+1}&=\boldsymbol{B}\Big\{\boldsymbol{z}_n+\sum_{j=1}^{p-1}L_j(1/\alpha)\boldsymbol{u}_j\Big\}.\end{aligned} \quad (5.3.38)$$

For $p=1$, $L_1(x)=1-x$, hence $\alpha=1$, and the corresponding approximation for $\exp(-\lambda\Delta t)$ is just $(1+\lambda\Delta t)^{-1}$, the same as the θ-method with $\theta=1$ or SS11 with $\theta_1=1$. For $p=2$ we have $L_2(x)=0$ when $x=2\pm\sqrt{2}$ and we take $\alpha=1+\sqrt{2}/2$. The approximation for $\exp(-\lambda\Delta t)$ is then

$$N_2^1(\alpha)=\{1+(2\alpha-1)\lambda\Delta t\}/(1+\alpha\lambda\Delta t)^2. \quad (5.3.39)$$

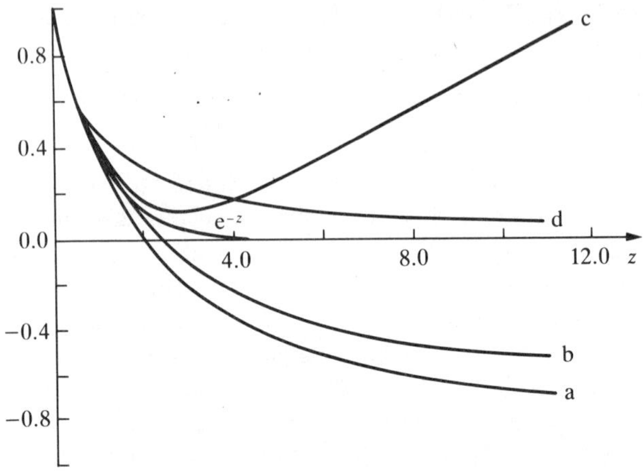

Fig. 5.3.2. Nørsett approximations for exp$(-z)$: (a) $N_1^1(\alpha)$, $\alpha = 1$, (b) $N_2^2(\alpha_2)$, $\alpha_2 = (3+\sqrt{3})/6$, Calahan. (c) $N_2^2(\alpha_1)$, $\alpha_1 = (3-\sqrt{3})/6$, (d) $N_2^1(\alpha_2)$, $\alpha_2 = 1 + \sqrt{2}/2$.

This expression is always positive because $(2\alpha - 1) = 1 + \sqrt{2} > 0$. The approximations $N_p^{p-1}(\alpha)$ with $p = 3, 4, 5$, change sign.

Figue 5.3.2 shows some examples of the Nørsett approximations for exp$(-z)$.

Nørsett (1974a,b) gives a general discussion of A-stable Nørsett methods. Siemieniuch (1975) covers L-stable Nørsett methods in the solution of civil engineering problems. The papers Smith (1975, 1977, 1978) give examples of their use in the solution of the diffusion and diffusion–convection equations of water resources problems, the coupled equations representing the generation and dissipation of excess pore–water pressures beneath gravity structures and driven piles and dynamic vibration problems.

The emphasis in the preceding discussion has been on the approximation for the exponential in eqn (5.3.3) for the exact solution of eqn (5.3.1) over a time step. It is also important to treat the forcing function term such that the accuracy of the solution is not reduced. Trujillo (1975) simply uses a piecewise linear representation for $f(t)$. Gellert (1978) with his Hermitian cubic approximation for the unknown function (which is actually equivalent to using a (2, 2) Padé approximation for the exponential), specifies a Lagrange quadratic interpolation for the forcing function. He then obtains $O(\Delta t^4)$ for the cumulative error which is consistent with that expected from using the Padé approximation on a time step. Smith *et al.* (1977) discuss the use of Nørsett methods for the numerical solution of the convection–diffusion equation

$$\frac{\partial u}{\partial t}=\frac{\partial^2 u}{\partial x^2}-\mu\frac{\partial u}{\partial x},\ \mu>0 \tag{5.3.40}$$

with constant boundary conditions. This problem, when space-discretized, produces a system of equations of the form (5.3.1) with a matrix $\boldsymbol{B}$ which can have complex eigenvalues for large values of the Reynolds' number type term μ. Smith *et al.*, compare the performance of various Nørsett algorithms with that of Crank–Nicolson for the numerical solution of this problem. They assume that an nth order Nørsett method takes at most n times more arithmetic operations per time step than Crank–Nicolson. Hence if a Nørsett method with a time step Δt produces an error smaller than that given by Crank–Nicolson with time step $\Delta t/n$ it can be considered an improvement. On this basis they conclude that for larger values of Δt the third order L-stable Nørsett method is more cost-effective than Crank–Nicolson.

Brusa and Nigro (1980) take the dynamic vibration equation (5.1.1) and reduce it to the first order form in the usual way, i.e.

$$\dot{\boldsymbol{z}}=-\boldsymbol{B}\boldsymbol{z}+\boldsymbol{F}$$

where

$$\boldsymbol{B}=\begin{bmatrix}\boldsymbol{0} & -\boldsymbol{I}\\ \boldsymbol{M}^{-1}\boldsymbol{K} & \boldsymbol{M}^{-1}\boldsymbol{C}\end{bmatrix}\quad\text{and}\quad \boldsymbol{F}=\begin{bmatrix}0\\ \boldsymbol{M}^{-1}\boldsymbol{f}(t)\end{bmatrix},\qquad \boldsymbol{z}=\begin{bmatrix}\boldsymbol{x}\\ \dot{\boldsymbol{x}}\end{bmatrix}.$$

Brusa and Nigro then look for a one-step scheme of the form

$$\boldsymbol{A}_0\boldsymbol{z}_{n+1}=\boldsymbol{A}_1\boldsymbol{z}_n+\boldsymbol{S}_n, \tag{5.3.41}$$

where they assume from analogy with the corresponding trapezium rule formula the pattern required is given by

$$\boldsymbol{A}_0=\begin{bmatrix}\boldsymbol{M}+a_1\Delta t\boldsymbol{C}+a_2\Delta t^2\boldsymbol{K} & \Delta t(a_3\boldsymbol{M}+a_4\Delta t\boldsymbol{C}+a_5\Delta t^2\boldsymbol{K})\\ a_6\boldsymbol{C}+a_7\Delta t\boldsymbol{K} & \boldsymbol{M}+a_8\Delta t\boldsymbol{C}+a_9\Delta t^2\boldsymbol{K}\end{bmatrix}$$

(5.3.42)

$$\boldsymbol{A}_1=\begin{bmatrix}b_1\boldsymbol{M}+b_2\Delta t\boldsymbol{C}+b_3\Delta t^2\boldsymbol{K} & \Delta t(b_4\boldsymbol{M}+b_5\Delta t\boldsymbol{C}+b_6\Delta t^2\boldsymbol{K})\\ b_7\boldsymbol{C}+b_8\Delta t\boldsymbol{K} & b_9\boldsymbol{M}+b_{10}\Delta t\boldsymbol{C}+b_{11}\Delta t^2\boldsymbol{K}\end{bmatrix}.$$

The parameters a_j, b_j and the representation of the forcing function $\boldsymbol{S}_n$ are then determined such that (i) the local truncation error is at least as good as $O(\Delta t^4)$, (ii) only one factorization of a matrix of the same order and sparsity as the matrices $\boldsymbol{M}$, $\boldsymbol{C}$, and $\boldsymbol{K}$ is required per time step, and (iii) the method is unconditionally A-stable. The result has been shown by Thomas and Gladwell (1984) to produce the equivalent of using the Nørsett $N_2^2(\alpha_2)$ approximation for the $\exp(-\Delta t\boldsymbol{B})$ in eqn (5.3.3). Brusa and Nigro make a consistent approximation to the forcing function by representing it by a cubic using the values $\boldsymbol{f}_n, \boldsymbol{f}_{n+\frac{1}{3}}, \boldsymbol{f}_{n+\frac{2}{3}}, \boldsymbol{f}_{n+1}$. Thus their representation of the exponential is the same as that of Calahan (1968) but the representation of the forcing function is different. Addison and Gladwell (1984) discuss the implementation of the algorithm of Brusa and Nigro in the context of a discussion of algorithms of second derivative type. These are algorithms for the solution of $\dot{z} = f(t, z)$ of the form

$$z_{n+1} = z_n + \Delta t \sum_{j=0}^{k} b_j \dot{z}_{n+1-j} + \Delta t^2 (c_0 \ddot{z}_{n+1} + c_1 \ddot{z}_n) \quad (5.3.43)$$

derived by Enright (1974). For the linear equation

$$\dot{\boldsymbol{z}} = -\boldsymbol{B}\boldsymbol{z} + \boldsymbol{F}, \quad (5.3.44)$$

we have, $\dot{\boldsymbol{z}}_n = -\boldsymbol{B}\boldsymbol{z}_n + \boldsymbol{F}_n$, and $\ddot{\boldsymbol{z}}_n = \boldsymbol{B}^2\boldsymbol{z}_n - \boldsymbol{B}\boldsymbol{F}_n + \dot{\boldsymbol{F}}_n$. Addison and Gladwell comment that the need for the derivative of the forcing function may be a major disadvantage of these formulae. This must depend on how the information about the forcing function is supplied. For $k = 1$, the second derivative formula (5.3.43) applied to the linear equation (5.3.44) gives the result

$$\begin{aligned}(\boldsymbol{I} + b_0\Delta t\boldsymbol{B} - c_0\Delta t^2\boldsymbol{B}^2)\boldsymbol{z}_{n+1} = {} & (\boldsymbol{I} - b_1\Delta t\boldsymbol{B} + c_1\Delta t^2\boldsymbol{B}^2)\boldsymbol{z}_n \\ & + \Delta t(b_1\boldsymbol{I} - c_1\Delta t\boldsymbol{B})\boldsymbol{F}_n \\ & + \Delta t(b_0\boldsymbol{I} - c_0\Delta t\boldsymbol{B})\boldsymbol{F}_{n+1} \\ & + c_0\Delta t^2\dot{\boldsymbol{F}}_{n+1} + c_1\Delta t^2\dot{\boldsymbol{F}}_n. \quad (5.3.45)\end{aligned}$$

We can immediately see the link between the second derivative methods and the rational approximation to the exponential in eqn (5.3.3). The (2, 2) Padé is given by

$$b_0 = b_1 = \tfrac{1}{2}, \qquad c_1 = -c_0 = \tfrac{1}{12},$$

the (1, 2) Padé is given by

$$b_0 = \tfrac{2}{3}, \qquad b_1 = \tfrac{1}{3}, \qquad c_0 = -\tfrac{1}{6},\ c_1 = 0,$$

and the Nørsett $N_2^2(\alpha_2)$, Calahan approximation is given by

$$b_0 = (3 + \sqrt{3})/3, \quad b_1 = -\sqrt{3}/3, \quad c_0 = -(2 + \sqrt{3})/6,$$
$$c_1 = -(1 + \sqrt{3})/6.$$

Addison and Gladwell (1984) also discuss the links between the second derivative formulae (Enright 1974, 1980) and the Runge–Kutta formulae which are introduced in the next section.

5.4. Runge–Kutta methods

Runge–Kutta methods are one-step methods which obtain a high order of accuracy by making extra function evaluations per time step. Since they are one-step methods it is easy to change the time-step size and there are no spurious solutions as in the higher order linear multistep and the equivalent SS*pj* methods but the error analysis is much more difficult. Dekker and Verwer (1984) give a comprehensive discussion of Runge–Kutta methods. Here we give a brief description in order to introduce some of the suggestions made in the engineering literature for the use of these methods.

The Runge–Kutta methods were first proposed by Runge (1895) and later developed by Heun (1900) and Kutta (1901). They are defined here for application to the numerical solution of the equation

$$\dot{z} = g(t, z) \tag{5.4.1}$$

as this fits in with the notation previously used here. Equation (5.4.1) could represent a space-discretized parabolic partial differential equation or the first order form of the dynamic vibration equation. We note that the function on the right-hand side of eqn (5.4.1) includes the forcing function as well as the non-differentiated term in z. This means that there is no question of the separate consideration of the forcing function here with the implied 'complementary function and particular integral' approach; the forcing function is immediately included. The Runge–Kutta method is an example of methods of the form:

$$z_{n+1} - z_n = \Delta t \Phi(t_n, z_n, \Delta t). \tag{5.4.2}$$

The method (5.4.2) is said to have order p if p is the largest integer for which

$$\frac{1}{\Delta t}\{z(t+\Delta t)-z(t)\}-\Phi(t, z(t), \Delta t)=\mathrm{O}(\Delta t^p), \quad (5.4.3)$$

where $z(t)$ is the exact solution of the differential equation (5.4.1). This corresponds to the definition previously used for the local truncation error in the approximation of the differential equation, that is according to the convention in Richtmyer and Morton (1967) rather than Lambert (1973). The method (5.4.2) is consistent with the differential equation (5.4.1) if

$$\Phi(t, z, 0) \equiv g(t, z). \quad (5.4.4)$$

Thus if the method (5.4.2) is consistent then we can expand the left-hand side of eqn (5.4.3) as

$$\dot{z}(t)-\Phi(t, z(t), 0)+\mathrm{O}(\Delta t)=\dot{z}(t)-g(t, z)+\mathrm{O}(\Delta t)=\mathrm{O}(\Delta t)$$

by eqn (5.4.4). The definition of consistency (5.4.4) corresponds to the idea we have used in previous chapters that consistency means that the differential equation is approximated with an error $\mathrm{O}(\Delta t^p)$ where $p \geq 1$.

A theorem proved in Henrici (1962), p. 71, gives necessary and sufficient conditions for the method (5.4.2) to be convergent in the sense that

$$\lim_{\substack{\Delta t \to 0 \\ n\Delta t = T, \text{fixed}}} z_n = z(T)$$

$z_0 = z(0)$. The theorem is stated as follows:

Theorem 5.4.1

(1) Let $\Phi(t, z, \Delta t)$ be a function which is continuous jointly as a function of its three arguments in the region D defined by $t \in [0, T]$, $z \in (-\infty, \infty)$, and $\Delta t \in [0, \Delta t_0]$, where $\Delta t_0 > 0$,

(2) Let $\Phi(t, z, \Delta t)$ satisfy the Lipschitz condition such that

$$|\Phi(t, z^*, \Delta t)-\Phi(t, z, \Delta t)| \leq L\,|z^*-z|$$

for all points $(t, z^*, \Delta t)$, $(t, z, \Delta t)$ in the region D, then the method (5.4.2) is convergent if and only if it is consistent.

The general s-stage Runge–Kutta algorithm is defined by

$$z_{n+1} - z_n = \Delta t \Phi(t_n, z_n, \Delta t)$$

where

$$\Phi(t_n, z_n, \Delta t) = \sum_{i=1}^{s} b_i k_i \tag{5.4.5}$$

$$k_i = g\left(t_n + c_i \Delta t,\ z_n + \Delta t \sum_{j=1}^{s} a_{ij} k_j\right), \quad i = 1, 2, \ldots, s \tag{5.4.6}$$

and

$$c_i = \sum_{j=1}^{s} a_{ij}, \qquad i = 1, 2, 3, \ldots, s. \tag{5.4.7}$$

The algorithm is called s-stage because it requires s evaluations of the function $g(t, z)$. The coefficients for the algorithm can be represented in the Butcher-array (Butcher 1964) as

$$\begin{array}{c|c} c & A \\ \hline & b \end{array} = \begin{array}{c|c} c_i & a_{ij} \\ \hline & b_j^{\mathrm{T}} \end{array}. \tag{5.4.8}$$

If $a_{ij} = 0$, for $j \geq i$ and all values of i, then each vector k_i can be computed in an explicit way from the preceding $k_1, k_2, \ldots, k_{i-1}$ and the algorithm is called explicit Runge–Kutta. If $a_{ij} = 0$ for $j > i$ and some $a_{ii} \neq 0$ the algorithm is called semi-explicit.

This means that for some i, k_i is defined by

$$k_i = g\left(t_n + c_i \Delta t,\ z_n + \Delta t \sum_{j=1}^{i} a_{ij} k_j\right). \tag{5.4.9}$$

Then if $g(t, z)$ is not a linear function of z, eqn (5.4.9) for k_i has to be solved by an iterative method. If $a_{ij} = 0$ for $j > i$ and all $a_{ii} \neq 0$ the algorithm is called diagonally implicit Runge–Kutta. If the algorithm is not explicit, nor semi-explicit, nor diagonally implicit it is just called implicit Runge–Kutta. In the pre-computer era the only Runge–Kutta methods used were explicit. Some examples of explicit Runge–Kutta methods are:

$s = 1$. Here we have $\Phi(t, z, \Delta t) \equiv g(t, z)$, that is, the forward difference Euler rule, i.e. the SS11 θ-method with $\theta = 0$ applied to eqn (5.4.1), giving:

$$z_{n+1} - z_n = \Delta t g(t, z_n),$$

which is the only SSpj method which coincides with a Runge–Kutta method.

$s = 2$. Two particular examples are

$$(1)\quad \begin{array}{c|cc} 0 & 0 & 0 \\ 0.5 & 0.5 & 0 \\ \hline & 0 & 1 \end{array},$$

that is, $k_1 = g(t_n, z_n)$, $k_2 = g(t_n + 0.5\Delta t, z_n + 0.5\Delta t k_1)$, $z_{n+1} - z_n = \Delta t k_2$, which gives the 'modified Euler' method

$$z_{n+1} - z_n = \Delta t g(t_n + 0.5\Delta t, z_n + 0.5\Delta t g(t_n, z_n)) \quad (5.4.10)$$

$$(2)\quad \begin{array}{c|cc} 0 & 0 & 0 \\ 1 & 1 & 0 \\ \hline & 0.5 & 0.5 \end{array},$$

that is $k_1 = g(t_n, z_n)$, $k_2 = g(t_n + \Delta t, z_n + \Delta t k_1)$ which gives the 'improved' Euler

$$z_{n+1} - z_n = 0.5\Delta t\{g(t_n, z_n) + g(t_n + \Delta t, z_n + \Delta t g(t_n, z_n))\}. \quad (5.4.11)$$

Both the modified and the improved Euler when applied to the equation $\dot{z} = -\lambda z$, give

$$z_{n+1} = (1 - \lambda\Delta t + \lambda^2\Delta t^2/2)z_n, \quad (5.4.12)$$

which we can see from Table 5.3.1 corresponds to the (2.0) Padé approximation to the $\exp(-\lambda\Delta t)$.

$s = 4$. Probably the most popular of all the Runge–Kutta methods is given by

$$\begin{array}{c|cccc} 0 & 0 & 0 & 0 & 0 \\ 1/2 & 1/2 & 0 & 0 & 0 \\ 1/2 & 0 & 1/2 & 0 & 0 \\ 1 & 0 & 0 & 1 & 0 \\ \hline & 1/6 & 1/3 & 1/3 & 1/6 \end{array},$$

which corresponds to

$$
\begin{aligned}
k_1 &= g(t_n, z_n)\\
k_2 &= g(t_n + 0.5\Delta t,\ z_n + 0.5\Delta t k_1)\\
k_3 &= g(t_n + 0.5\Delta t,\ z_n + 0.5\Delta t k_2)\\
k_4 &= g(t_n + \Delta t,\ z_n + \Delta t k_3)\\
z_{n+1} &= z_n + \Delta t(k_1 + 2k_2 + 2k_3 + k_4)/6. \qquad (5.4.13)
\end{aligned}
$$

This is often called 'the fourth order Runge–Kutta method'. The cumulative error in the solution is $O(\Delta t^4)$ (Lambert 1973) and it is frequently used for generating starting values for a multistep scheme. For example, in Ashford and Wood (1978) the Zienkiewicz 'four-time-level scheme' (Zienkiewicz 1977, Wood 1977), is used for the numerical integration of the motion of a tethered buoyant platform and the Runge–Kutta fourth order scheme is used to produce the required extra starting values. The 'four-time-level', i.e. three time step scheme used here is equivalent to the three-step scheme analysed in connection with the SS32 algorithm in Chapter 4. The local truncation error is $O(\Delta t^2)$ and the fourth order Runge–Kutta is perfectly safe as a generator of starting values.

Used on the equation $\dot{z} = -\lambda z$, the Runge–Kutta fourth order scheme gives a result corresponding to the use of the $(4, 0)$ Padé approximation to the $\exp(-\lambda\Delta t)$.

$$z_{n+1} = (1 - \lambda\Delta t + \lambda^2\Delta t^2/2 - \lambda^3\Delta t^3/6 + \lambda^4\Delta t^4/24)z_n. \qquad (5.4.14)$$

Dekker and Verwer (1984) give a stability analysis of the Runge–Kutta methods by applying them to the scalar test equation of the form

$$\dot{z}(t) = -\lambda z(t), \quad t \geq 0, \quad z(0) = z_0, \qquad (5.4.15)$$

λ complex. If the Runge–Kutta method given by eqns (5.4.5) to (5.4.7) applied to eqn (5.4.15) gives

$$z_{n+1} = R(-\lambda\Delta t)z_n, \qquad (5.4.16)$$

then $R(-\lambda\Delta t)$ is called the stability function of the method. Using the Butcher-array notation given in eqn (5.4.8) we have

$$R(\zeta) = 1 + \boldsymbol{b}^{\mathrm{T}}\zeta(\boldsymbol{I} - \boldsymbol{A}\zeta)^{-1}\boldsymbol{e}, \qquad (5.4.17)$$

where $\zeta = -\lambda\Delta t$ and the vector $\boldsymbol{e}$ is the vector of length s with all entries equal to unity. For an explicit Runge–Kutta method $R(\zeta)$ is a polynomial as in the examples above; otherwise it is a rational function. The Runge–Kutta method is defined to be

(1) A-stable if $|R(\zeta)| \leq 1$ for all $\text{Re}(\zeta) \leq 0$;

(2) A_0-stable if $|R(\zeta)| \leq 1$ for all real $\zeta < 0$;

(3) $A(\alpha)$-stable if $|R(\zeta)| \leq 1$ for all ζ lying in the wedge of angle 2α shown in the Argand diagram Fig. 2.4;

(4) L-stable if the method is A-stable and $|R(\zeta)| \to 0$ as $|\zeta| \to \infty$.

These definitions conform with those in Chapter 2. As expected from their affinity with the $(p, 0)$ Padé approximants illustrated in eqns (5.4.12) and (5.4.14) the explicit Runge–Kutta methods are conditionally stable.

In order to relate Runge–Kutta methods with those proposed by Brusa and Nigro (1980) and Calahan (1968) we look at the diagonally implicit method given by the array

$$\begin{array}{c|cc} \alpha & \alpha & 0 \\ 1-\alpha & 1-2\alpha & \alpha \\ \hline & 0.5 & 0.5 \end{array} . \tag{5.4.18}$$

This method is given by Nørsett (1974a,b) who refers to it as semi-explicit or semi-implicit. When applied to the equation

$$\dot{\boldsymbol{z}} = -\boldsymbol{B}\boldsymbol{z} + \boldsymbol{F}(t) \tag{5.4.19}$$

the array (5.4.18) gives

$$\begin{aligned} (\boldsymbol{I} + \alpha\Delta t\boldsymbol{B})\boldsymbol{k}_1 &= \boldsymbol{B}\boldsymbol{z}_n + \boldsymbol{F}(t_n + \alpha\Delta t) \\ (\boldsymbol{I} + \alpha\Delta t\boldsymbol{B})\boldsymbol{k}_2 &= -\boldsymbol{B}\boldsymbol{z}_n - \Delta t\boldsymbol{B}(1-2\alpha)\boldsymbol{k}_1 + \boldsymbol{F}(t_n + (1-\alpha)\Delta t) \\ \boldsymbol{z}_{n+1} &= \boldsymbol{z}_n + 0.5\Delta t(\boldsymbol{k}_1 + \boldsymbol{k}_2). \end{aligned} \tag{5.4.20}$$

Because we have $a_{11} = a_{22}$ here there is only one matrix to invert in this two-stage method. Dekker and Verwer remark that several authors restrict the term 'diagonally implicit' to the subclass of methods where the diagonal entries of the matrix $\boldsymbol{A}$ are equal. The stability function is given by

$$R(\zeta) = \{1 + (1-2\alpha)\zeta + (0.5 - 2\alpha + \alpha^2)\zeta^2\}/(1-\alpha\zeta)^2. \tag{5.4.21}$$

For general values of α the order of this method is two. For $\alpha^2 = \alpha - 1/6$, i.e. for $\alpha = (3 \pm \sqrt{3})/6$, we have the Nørsett $N_2^2(\alpha)$ approximation for the exponential term in the corresponding eqn (5.3.3) given by

$$R(\zeta) = \{1 + (1 - 2\alpha)\zeta + (1/3 - \alpha)\zeta^2\}/(1 - \alpha\zeta)^2 \quad (5.4.22)$$

(compare with eqn (5.3.33) with $\lambda\Delta t = -\zeta$), and the order of the method is now three. We know from the results in §5.3 that this Runge–Kutta method is A_0-stable but only A-stable for the larger vaue of $\alpha = (3 + \sqrt{3})/6$ which corresponds to the approximation for the exponential used by both Calahan (1968) and Brusa and Nigro (1980). From eqn (5.4.22) and also from the expressions for the Nørsett approximation in §5.3 we can see from the coefficients of the ζ^2 term in the numerator and denominator that this A-stable Runge–Kutta method is not L-stable.

Writing the three methods, Calahan, Brusa and Nigro, and Nørsett–Runge–Kutta as applied to eqn (5.4.19), in the form

$$\boldsymbol{A}_0\boldsymbol{z}_{n+1} = \boldsymbol{A}_1\boldsymbol{z}_n + \bar{\boldsymbol{F}} \quad (5.4.23)$$

they have the same matrices $\boldsymbol{A}_0$ and $\boldsymbol{A}_1$ and forcing function representations:

$$\boldsymbol{A}_0 = \left(\boldsymbol{I} + \frac{(3+\sqrt{3})}{6}\Delta t\boldsymbol{B}\right)^2$$

$$\boldsymbol{A}_1 = \boldsymbol{I} + \frac{\sqrt{3}}{3}\Delta t\boldsymbol{B} - \frac{(1+\sqrt{3})}{6}\Delta t^2\boldsymbol{B}^2$$

Calahan:

$$\begin{aligned}\bar{\boldsymbol{F}} &= 0.75\Delta t\left\{\boldsymbol{I} + \frac{(9+7\sqrt{3})}{18}\Delta t\boldsymbol{B}\right\}\boldsymbol{F}(t_n)\\ &+ 0.25\Delta t\left\{\boldsymbol{I} + \frac{(3+\sqrt{3})}{6}\Delta t\boldsymbol{B}\right\}\boldsymbol{F}\left(t_n - \frac{2\sqrt{3}}{3}\Delta t\right)\\ &+ \frac{(3+\sqrt{3})}{6}\Delta t^2\left\{\boldsymbol{I} + \frac{(3+2\sqrt{3})}{6}\Delta t\boldsymbol{B}\right\}\dot{\boldsymbol{F}}(t_n). \quad (5.4.24)\end{aligned}$$

Brusa and Nigro:

$$\begin{aligned}\bar{\boldsymbol{F}} &= \{2(3+\sqrt{3})\boldsymbol{I} + (2+\sqrt{3})\Delta t\boldsymbol{B}\}\Delta t\boldsymbol{F}(t_n + \Delta t)/6\\ &+ \{-2\sqrt{3}\boldsymbol{I} + (1+\sqrt{3})\Delta t\boldsymbol{B}\}\Delta t\boldsymbol{F}(t_n)/6\\ &- \{(2+\sqrt{3})\Delta t^2\dot{\boldsymbol{F}}(t_n+\Delta t) + (1+\sqrt{3})\Delta t^2\dot{\boldsymbol{F}}(t_n)\}/6 \quad (5.4.25)\end{aligned}$$

Nørsett–Runge–Kutta:

$$\bar{\boldsymbol{F}} = 0.5\Delta t\left\{\boldsymbol{I} + \frac{(1+\sqrt{3})}{2}\Delta t\boldsymbol{B}\right\}\boldsymbol{F}\left(t_n + \frac{(3+\sqrt{3})}{6}\Delta t\right)$$
$$+ 0.5\Delta t\left\{\boldsymbol{I} + \frac{(3+\sqrt{3})}{6}\Delta t\boldsymbol{B}\right\}\boldsymbol{F}\left(t_n + \frac{(3-\sqrt{3})}{6}\Delta t\right). \quad (5.4.26)$$

These three ways of dealing with the forcing function all give the same accuracy, i.e. the local truncation error in the approximation of the differential equation as defined in §2.2 is $O(\Delta t^3)$ and hence by the theory in §2.9 the global error also is $O(\Delta t^3)$ for the numerical solution of the linear eqn (5.4.19)

Dekker and Verwer (1984) give Runge–Kutta methods derived from various quadrature formulae for the integration of

$$\dot{z} = \boldsymbol{F}(t).$$

In particular two of these which appear in the engineering literature come from the Gauss–Legendre one-point and two-point quadrature formulae:

1. From the Butcher array

$$\begin{array}{c|c} 0.5 & 0.5 \\ \hline & 1 \end{array},$$

we have for eqn (5.4.1)

$$z_{n+1} = z_n + \Delta t\boldsymbol{g}(t_n + 0.5\Delta t, 0.5(z_n + z_{n+1})), \quad (5.4.27)$$

which is the implicit mid-point rule. Applied to the linear equation (5.4.19) it gives

$$(\boldsymbol{I} + 0.5\Delta t\boldsymbol{B})z_{n+1} = (\boldsymbol{I} - 0.5\Delta t\boldsymbol{B})z_n + \Delta t\boldsymbol{F}(t_n + 0.5\Delta t), \quad (5.4.28)$$

which differs from the θ-method with $\theta = 0.5$ discussed in Chapter 3 in the treatment of the forcing function.

2. From the Butcher array

$$\begin{array}{c|cc} (3-\sqrt{3})/6 & 1/4 & (3-2\sqrt{3})/12 \\ (3+\sqrt{3})/6 & (3+2\sqrt{3})/12 & 1/4 \\ \hline & 1/2 & 1/2 \end{array},$$

the stability function for this two stage Runge–Kutta can be calculated from a formula given by Dekker and Verwer:

$$R(\zeta) = \det(I - \zeta A + \zeta e b^{\mathrm{T}})/\det(I - \zeta A), \tag{5.4.29}$$

and this works out to be

$$R(\zeta) = (1 + \tfrac{1}{2}\zeta + \tfrac{1}{12}\zeta^2)/(1 - \tfrac{1}{2}\zeta + \tfrac{1}{12}\zeta^2),$$

which is the (2,2) Padé approximation to $\exp(\zeta)$ discussed in §5.3. Thus we can compare the Gellert (1978) algorithm for the numerical integration of the dynamic vibration equation, which as shown in §5.3 uses the equivalent of the (2, 2) Padé approximation to the exponential in the first order form of this equation, with the Runge–Kutta Gauss–Legendre two-stage method. The Runge–Kutta method given by the Butcher array (b) as applied to eqn (5.4.19) gives

$$\begin{aligned} \boldsymbol{k}_1 &= -\boldsymbol{B}\left\{\boldsymbol{z}_n + \frac{\Delta t}{4}\boldsymbol{k}_1 + \frac{(3-2\sqrt{3})}{12}\Delta t\boldsymbol{k}_2\right\} + \boldsymbol{F}(t_n + c_1\Delta t) \\ \boldsymbol{k}_2 &= -\boldsymbol{B}\left\{\boldsymbol{z}_n + \frac{(3+2\sqrt{3})}{12}\Delta t\boldsymbol{k}_1 + \frac{\Delta t}{4}\boldsymbol{k}_2\right\} + \boldsymbol{F}(t_n + c_2\Delta t) \end{aligned} \tag{5.4.30}$$

$$\boldsymbol{z}_{n+1} = \boldsymbol{z}_n + 0.5(\boldsymbol{k}_1 + \boldsymbol{k}_2).$$

Eliminating $\boldsymbol{k}_1$ and $\boldsymbol{k}_2$ this becomes

$$\begin{aligned} &\left(\boldsymbol{I} + \frac{\Delta t}{2}\boldsymbol{B} + \frac{\Delta t^2}{12}\boldsymbol{B}^2\right)\boldsymbol{z}_{n+1} \\ &\quad = \left(\boldsymbol{I} - \frac{\Delta t}{2}\boldsymbol{B} + \frac{\Delta t^2}{12}\boldsymbol{B}^2\right)\boldsymbol{z}_n + \frac{\Delta t}{2}\{\boldsymbol{F}(t_n + c_1\Delta t) + \boldsymbol{F}(t_n + c_2\Delta t)\} \\ &\qquad + \frac{\sqrt{3}\Delta t^2}{12}\boldsymbol{B}\{\boldsymbol{F}(t_n + c_2\Delta t) - \boldsymbol{F}(t_n + c_1\Delta t)\}, \end{aligned} \tag{5.4.31}$$

where $c_1 = (3 - \sqrt{3})/6$, $c_2 = (3 + \sqrt{3})/6$.

We can compare eqn (5.4.31) with eqn (5.3.13) from the Gellert method which put into the same notation gives

$$\begin{aligned} \left(\boldsymbol{I} + \frac{\Delta t}{2}\boldsymbol{B} + \frac{\Delta t^2}{12}\boldsymbol{B}^2\right)\boldsymbol{z}_{n+1} &= \left(\boldsymbol{I} - \frac{\Delta t}{2}\boldsymbol{B} + \frac{\Delta t^2}{12}\boldsymbol{B}^2\right)\boldsymbol{z}_n \\ &\quad + \frac{\Delta t}{6}(\boldsymbol{F}_n + 4\boldsymbol{F}_{n+\frac{1}{2}} + \boldsymbol{F}_{n+1}) \\ &\quad + \frac{\Delta t^2}{12}\boldsymbol{B}(\boldsymbol{F}_{n+1} - \boldsymbol{F}_n). \end{aligned} \tag{5.4.32}$$

These two formulae both give the same accuracy as that found by Gellert, that is, $O(\Delta t^4)$ global error, but the Runge–Kutta formula (5.4.31) only requires the forcing function evaluation at two points rather than three. Whether it is convenient to evaluate the forcing function at the Gauss points depends on the problem.

We also consider in this section a different type of Runge–Kutta method called a 'rational' Runge–Kutta method as proposed in a paper by Hairer (1980) on 'unconditionally stable explicit methods for parabolic equations'. The reason for including this method here is that an adaptation of it has been used in a paper by Liu *et al.* (1984) as part of a family of coupled time integration operators. The general idea of coupled time integration operators is discussed in §5.5. Here we look briefly at the method of Hairer (1980) and the adaptation of it by Liu *et al.* The great attraction of the method of Hairer is that it is unconditionally stable and explicit, which is not possible with the linear multistep type schemes.

The Hairer (1980) methods are for the system of differential equations of the form

$$\dot{y} = f(y), \qquad y(0) = y_0. \tag{5.4.33}$$

The method relevant here is a consistent two-stage rational Runge–Kutta method of the form

$$\begin{aligned} g_1 &= \Delta t f(y_n) \\ g_2 &= \Delta t f(y_n + c g_1) \\ y_{n+1} &= y_n + \frac{g_1 g_1}{2g_1 - g_2}, \end{aligned} \tag{5.4.34}$$

where an expression of the form aa/b is defined for real vectors as

$$\frac{aa}{b} = \{2a(a^{\mathrm{T}}b) - b(a^{\mathrm{T}}a)\}/(b^{\mathrm{T}}b). \tag{5.4.35}$$

Hairer shows that this method is A_0-stable and of second order if $c = 0.5$ (see Exercise 5.5). Hairer only states his algorithm for an equation of the form (5.4.33) and Liu *et al.* want to use it for an equation of the form $\dot{y} = f(t, y)$. Their equation is

$$\boldsymbol{M}\dot{\boldsymbol{\theta}} + \boldsymbol{K}\boldsymbol{\theta} = \boldsymbol{f}(t) \tag{5.4.36}$$

and their modified form of the Hairer algorithm is given by

$$\boldsymbol{v}_1 = -\boldsymbol{M}^{-1}\boldsymbol{K}\boldsymbol{\theta}_n + \boldsymbol{M}^{-1}\boldsymbol{f}_{n+1}$$

and

$$\boldsymbol{v}_2 = -\boldsymbol{M}^{-1}\boldsymbol{K}(\boldsymbol{\theta}_n + 0.5\Delta t\boldsymbol{v}_1) + \boldsymbol{M}^{-1}\boldsymbol{f}_{n+1} \tag{5.4.37}$$

$$\boldsymbol{\theta}_{n+1} = \boldsymbol{\theta}_n + \Delta t\boldsymbol{e}/\boldsymbol{b}^{\mathrm{T}}\boldsymbol{b},$$

where $\boldsymbol{b} = 2\boldsymbol{v}_1 - \boldsymbol{v}_2$, and $\boldsymbol{e} = 2(\boldsymbol{v}_1^{\mathrm{T}}\boldsymbol{b})\boldsymbol{v}_1 - (\boldsymbol{v}_1^{\mathrm{T}}\boldsymbol{v}_1)\boldsymbol{b}$. If $\boldsymbol{f}(t) \equiv \boldsymbol{0}$ this is the same as the Hairer algorithm. We can look at the effect of the presence of the forcing term when the algorithm given by eqns (5.4.37) is applied to the scalar test equation $\dot{\theta} = -\lambda\theta + f$. The result is

$$\begin{aligned} v_1 &= -\lambda\theta_n + f_{n+1} \\ v_2 &= -\lambda(\theta_n + 0.5\Delta t v_1) + f_{n+1} \\ &= (1 - 0.5\lambda\Delta t)v_1. \end{aligned}$$

Hence $b = (1 + 0.5\lambda\Delta t)v_1$, and $e = (1 + 0.5\lambda\Delta t)v_1^3$. These values substituted into the last of the eqns (5.4.37) give

$$\theta_{n+1} = \theta_n + \Delta t v_1/(1 + 0.5\lambda\Delta t)$$

that is,

$$(1 + 0.5\lambda\Delta t)\theta_{n+1} = (1 - 0.5\lambda\Delta t)\theta_n + \Delta t f_{n+1}. \tag{5.4.38}$$

Comparing eqn (5.4.38) with the second order Crank–Nicolson-trapezium rule algorithm applied to the same equation:

$$(1 + 0.5\lambda\Delta t)\theta_{n+1} = (1 - 0.5\lambda\Delta t)\theta_n + 0.5\Delta t(f_n + f_{n+1}) \tag{5.4.39}$$

we can see that (5.4.38) will only give a first order approximation to the scalar test equation. This may explain why Liu *et al.* find that the acuracy of their algorithm deteriorates at a time step only five to ten times the critical time-step for the explicit Euler method. It does not appear to be easy to form the algorithm so that the average value of the forcing function in eqn (5.4.39) appears but a compromise is to replace f_{n+1} in the algorithm by $f_{n+\frac{1}{2}}$. Equation (5.4.38) is then the second order implicit midpoint rule as given in eqn (5.4.28). A further very relevant paper in this context is that of Sottas (1984) which shows that rational Runge–Kutta methods are not suitable for stiff systems of equations of the autonomous form (5.4.33). Sottas shows that, although the rational Runge–Kutta methods have A- or A_0-stability, the numerical solution can possess completely wrong components because of the way the rational function involves all the components at the beginning of the time step. The stability analysis cannot be done on a scalar test equation here but Hairer (1980) shows that the stability behaviour of the rational Runge–Kutta method applied to the equation

$$\dot{y} = \boldsymbol{A}y, \qquad y(0) = y_0, \tag{5.4.40}$$

where $\boldsymbol{A}$ is a constant $N \times N$ matrix reducible to diagonal form by a unitary transformation, is equivalent to its behaviour for the system

$$\dot{\boldsymbol{y}} = \boldsymbol{\Lambda}\boldsymbol{y}, \quad \boldsymbol{y}(0) = \boldsymbol{y}_0, \quad \boldsymbol{\Lambda} = \mathrm{diag}(\lambda_1, \lambda_2, \ldots, \lambda_N). \tag{5.4.41}$$

Suppose that applying the rational Runge–Kutta method for one time-step to the diagonal system of equations (5.4.41) gives

$$\boldsymbol{y}_1 = \boldsymbol{R}\boldsymbol{y}_0, \tag{5.4.42}$$

where $\boldsymbol{R} = \mathrm{diag}(r_1, r_2, \ldots, r_N)$, then the r_j depend on all the $\Delta t\lambda_j$ as well as on $\boldsymbol{y}_0$. Hence we have $\boldsymbol{R} = \boldsymbol{R}(z; \boldsymbol{y}_0)$ where $z = (\Delta t\lambda_1, \Delta t\lambda_2, \ldots, \Delta t\lambda_N)^{\mathrm{T}}$. Scottas gives numerical examples to show that, with the rational Runge–Kutta method given by eqns (5.4.34) with $c = 1$ that is, with the second equation replaced by

$$g_2 = \Delta t f(y_n + g_1) \tag{5.4.43}$$

(which is also A_0-stable), the property

$$\|R(z; y_0)y_0\| \leq \|y_0\| \tag{5.4.44}$$

is not sufficient to ensure that $\|R(z; y_0\| \leq 1$. This means that, for a problem where all the components should decrease, although the A_0-stability ensures that the norm of the solution is decreasing it is possible to have a component which is actually increasing. This property is not only bad for the numerical solution but also for any automatic error estimate and dependent step size adjustment. Sottas introduces the concept of componentwise contractivity. A rational Runge–Kutta method is called CA_0-stable if and only if in every dimension N its application to the system (5.4.41) with the $\Delta t\lambda_j$ all real and non-positive and an arbitrary initial vector y_0 (which may be complex), gives rational functions r_j which satisfy $|r_j| \leq 1$, $j = 1, 2, \ldots, N$. Sottas then proves that there are no CA_0-stable rational Runge–Kutta methods of the general type given in Hairer (1980) and the two-stage rational

Runge–Kutta method of the type (5.4.34) can only be CA_0-stable when applied to the diagonal system (5.4.41) if this system is not stiff. For the particular case when $N=2$ it is required that $0.5 \leq \lambda_2/\lambda_1 \leq 2$. The consequence of using the rational Runge–Kutta method (5.4.34) on a system which does not satisfy this condition is that an automatic time step size control makes the time step so small that it becomes comparable with that which would be chosen by a program using a conditionally stable explicit Runge–Kutta method.

Seward *et al.* (1984) give a useful survey of higher order methods for the numerical integration of semi-discretized parabolic problems with many references.

5.5. Partitioning methods

Many engineers and practical scientists shy away from the higher order methods described in §§5.1 to 5.4. They prefer to use the simplest time-stepping schemes in combinations which exploit their most attractive features. The idea is to obtain the required accuracy without complicated programming or excessive computer time or core storage. For the first order system of equations:

$$\boldsymbol{C}\dot{\boldsymbol{x}} + \boldsymbol{K}\boldsymbol{x} = \boldsymbol{f}, \tag{5.5.1}$$

the simplest time-stepping scheme is the Euler explicit method (i.e. the θ-method described in §§2.6, 3.3 with $\theta = 0$) which gives

$$\boldsymbol{C}\boldsymbol{x}_{n+1} = (\boldsymbol{C} + \Delta t\boldsymbol{K})\boldsymbol{x}_n + \Delta t\boldsymbol{f}_n \tag{5.5.2}$$

but we know this is only conditionally stable with the time step Δt limited by

$$\lambda \Delta t \leq 2 \tag{5.5.3}$$

where λ is the maximum eigenvalue of $\boldsymbol{C}^{-1}\boldsymbol{K}$. We know from the Irons theorem bounding the eigenvalues (Irons 1970, Chapter 1) that we can get an estimate of λ from the eigenvalues of the element matrices:

$$\lambda \leq \max\{\lambda^{(e)}\}. \tag{5.5.4}$$

From a study of the element eigenvalues it may be apparent that in part of the region the elements are such that an explicit scheme with a reasonable time step could be used but in the rest of the region the material properties and element sizes are such that an implicit scheme is necessary. Similarly for the second order system

$$M\ddot{x} + C\dot{x} + Kx = f \tag{5.5.5}$$

the simplest time-stepping scheme is the central difference method (Chapter 2) but we know this is only conditionally stable with the time step again limited by (5.5.3) with $\lambda = \omega$, where ω^2 is the maximum eigenvalue of the matrix $M^{-1}K$. We have the estimate corresponding to (5.5.4) and again the possible situation where an explicit scheme could be used in part of the region but not the whole. Two ways of dealing with this are by nodal or element partitioning. To quote Park and Felippa (1983), 'rather than augment and drown in complexity, divide and conquer.'

Nodal partitioning

This is the idea of partitioning the region into the two parts A and B as in Fig. 5.5.1 and using different time-stepping schemes on the two parts of the mesh. This procedure can be combined with subcycling, with different size time steps used in the different parts of the mesh, (Belytschko 1983).

1. For the first order equation (5.5.1), using as a basis the θ-method which if applied equally to the whole region would give

$$(C + \theta\Delta t K)x_{n+1} = \{C - (1-\theta)\Delta t K\}x_n + \Delta t\{\theta f_{n+1} + (1-\theta)f_n\} \tag{5.5.6}$$

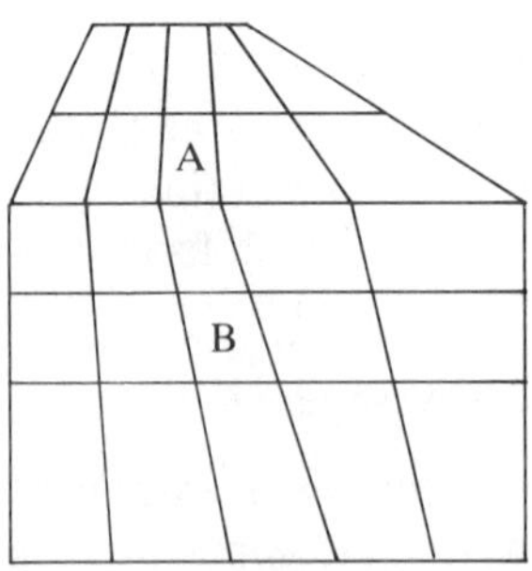

Fig. 5.5.1. Mesh partitioning.

we partition the vectors $\boldsymbol{x}_n$, $\boldsymbol{f}_n$ each into two parts:

$$\boldsymbol{x}_n = [\boldsymbol{x}_n^{\mathrm{A}}, \boldsymbol{x}_n^{\mathrm{B}}]^{\mathrm{T}}, \qquad \boldsymbol{f}_n = [\boldsymbol{f}_n^{\mathrm{A}}, \boldsymbol{f}_n^{\mathrm{B}}]$$

where the nodes in the regions A and B are to be processed with time steps $N\Delta t$ and Δt respectively and N is an integer. We partition the matrices $\boldsymbol{C}$ and $\boldsymbol{K}$ to correspond, that is with

$$\boldsymbol{C} = \begin{bmatrix} \boldsymbol{C}^{11} & \boldsymbol{C}^{12} \\ \boldsymbol{C}^{21} & \boldsymbol{C}^{22} \end{bmatrix}$$

and similarly for $\boldsymbol{K}$. Then if the nodes in regions A and B are advanced with the θ algorithm with $\theta = \theta_{\mathrm{A}}$ and θ_{B} respectively, we can write the scheme as follows.

(a) For $n \neq Np$, where p is an integer, the entries in $\boldsymbol{x}^{\mathrm{A}}$ are left unchanged and the entries in $\boldsymbol{x}^{\mathrm{B}}$ are updated with the $\boldsymbol{x}^{\mathrm{A}}$ nodes on the boundary of the region B treated as Dirichlet boundary conditions. Thus the vector $\boldsymbol{x}^{\mathrm{B}}$ is updated by the formula:

$$\begin{aligned}(\boldsymbol{C}^{22} + \theta_{\mathrm{B}}\Delta t\boldsymbol{K}^{22})\boldsymbol{x}_{n+1}^{\mathrm{B}} = {} & \{\boldsymbol{C}^{22} - (1-\theta_{\mathrm{B}})\Delta t\boldsymbol{K}^{22}\}\boldsymbol{x}_n^{\mathrm{B}} \\ & + \Delta t\{-\boldsymbol{K}^{21}\boldsymbol{x}_n^{\mathrm{A}} + \theta_{\mathrm{B}}\boldsymbol{f}_{n+1}^{\mathrm{B}} \\ & + (1-\theta_{\mathrm{B}})\boldsymbol{f}_n^{\mathrm{B}}\}. \end{aligned} \tag{5.5.7}$$

(b) If $n = Np$, we have

$$\begin{aligned} & \begin{bmatrix} \boldsymbol{C}^{11} + N\theta_{\mathrm{A}}\Delta t\boldsymbol{K}^{11} & \boldsymbol{C}^{12} + N\theta_{\mathrm{A}}\Delta t\boldsymbol{K}^{12} \\ \boldsymbol{C}^{21} + \theta_{\mathrm{B}}\Delta t\boldsymbol{K}^{21} & \boldsymbol{C}^{22} + \theta_{\mathrm{B}}\Delta t\boldsymbol{K}^{22} \end{bmatrix} \begin{bmatrix} \boldsymbol{x}_{n+1}^{\mathrm{A}} \\ \boldsymbol{x}_{n+1}^{\mathrm{B}} \end{bmatrix} \\ & = \begin{bmatrix} \boldsymbol{C}^{11} - N\Delta t(1-\theta_{\mathrm{A}})\boldsymbol{K}^{11} & \boldsymbol{C}^{12} - N\Delta t(1-\theta_{\mathrm{A}})\boldsymbol{K}^{12} \\ \boldsymbol{C}^{21} - \Delta t(1-\theta_{\mathrm{B}})\boldsymbol{K}^{21} & \boldsymbol{C}^{22} - (1-\theta_{\mathrm{B}})\Delta t\boldsymbol{K}^{22} \end{bmatrix} \begin{bmatrix} \boldsymbol{x}_n^{\mathrm{A}} \\ \boldsymbol{x}_n^{\mathrm{B}} \end{bmatrix} \\ & \quad + \Delta t \begin{bmatrix} N[\theta_{\mathrm{A}}\boldsymbol{f}_{n+1}^{\mathrm{A}} + (1-\theta_{\mathrm{A}})\boldsymbol{f}_n^{\mathrm{A}}] \\ \theta_{\mathrm{B}}\boldsymbol{f}_{n+1}^{\mathrm{B}} + (1-\theta_{\mathrm{B}})\boldsymbol{f}_n^{\mathrm{B}} \end{bmatrix}. \end{aligned} \tag{5.5.8}$$

Of course this procedure is much simpler if the mass matrix $\boldsymbol{C}$ is lumped so that it is diagonal; if $\theta_{\mathrm{B}} = 0$ so that the subcycling is done with an explicit scheme, the procedure is even simpler. If $N = 1$ so that the same time step Δt is used in both parts of the mesh and $\theta_{\mathrm{B}} = 0$ we have to treat the explicit part of the mesh first with

$$\boldsymbol{C}^{22}\boldsymbol{x}_{n+1}^{\mathrm{B}} = (\boldsymbol{C}^{22} - \Delta t\boldsymbol{K}^{22})\boldsymbol{x}_n^{\mathrm{B}} - \Delta t\boldsymbol{K}^{21}\boldsymbol{x}_n^{\mathrm{A}} + \Delta t\boldsymbol{f}_n^{\mathrm{B}}. \tag{5.5.9}$$

In eqn (5.5.9) the right-hand side uses only values at the $n\Delta t$ time which are already known. Then the rest of the nodes are updated with

$$(\boldsymbol{C}^{11}+\theta_{\mathrm{A}}\Delta t\boldsymbol{K}^{11})\boldsymbol{x}_{n+1}^{\mathrm{A}}=\{\boldsymbol{C}^{11}-(1-\theta_{\mathrm{A}})\Delta t\boldsymbol{K}^{11}\}\boldsymbol{x}_{n}^{\mathrm{A}}$$
$$+\Delta t\{\theta_{\mathrm{A}}(-\boldsymbol{K}^{12}\boldsymbol{x}_{n+1}^{\mathrm{B}}+\boldsymbol{f}_{n+1}^{\mathrm{A}})+(1-\theta_{\mathrm{A}})(-\boldsymbol{K}^{12}\boldsymbol{x}_{n}^{\mathrm{B}}+\boldsymbol{f}_{\mathrm{n}}^{\mathrm{A}})\} \quad (5.5.10)$$

where the matrix to be inverted is symmetric. In eqn (5.5.10) the right-hand side includes the value of $\boldsymbol{x}_{n+1}^{\mathrm{B}}$ which has been calculated in eqn (5.5.9). The stability limit on the time step Δt is given by

$$\lambda_{\mathrm{B}}\Delta t \le 2, \quad (5.5.11)$$

where

$$\lambda_{\mathrm{B}} = \max_{\text{all } e\in S_{\mathrm{B}}} \lambda^{(e)},$$

and S_{B} is the set of all elements with at least one node in the B nodal partition. This means that the nodes on the boundary are identified as belonging to the nodal partition A.

There are various possible combinations of explicit–explicit, explicit–implicit or implicit–implicit with or without subcycling. Belytschko (1983) says that it is common to use anything from two to twenty partitions.

Neuman *et al.* (1982) use explicit–implicit partitioning with the first order time-dependent equations from a quasi-three-dimensional finite element model of flow and subsidence in multi-aquifer systems. There is a large ratio between the hydraulic diffusivities of the aquifer and the aquitard (the aquitard transmits water at a much lower rate than the aquifer—see Bear (1979) for details). Thus the aquitard nodes can be solved explicitly with a convenient value of the time step and the rest of the equations are solved implicitly.

2. Belytschko and Mullen (1978) use the same nodal partitioning idea in connection with the solution of the second order system of equations

$$\boldsymbol{M}\ddot{\boldsymbol{x}}+\boldsymbol{K}\boldsymbol{x}=\boldsymbol{f}. \quad (5.5.12)$$

They cite the example of a stiff structure in a fluid where it makes sense to integrate the structure implicitly and the fluid explicitly hence the nodal partition. The mass matrix $\boldsymbol{M}$ is lumped so that it is diagonal and there is no natural damping. The explicit scheme used is the central difference, that is, Newmark with $\beta=0$ and $\gamma=0.5$; the implicit scheme is the Newmark trapezium rule equivalent with $2\beta=\gamma=0.5$. These Newmark algorithms are discussed in Chapters 2 and 3.

Using the usual notation $\boldsymbol{a}_n$ to denote the approximate acceleration at time $t=n\Delta t$ and partitioning the matrices and vectors with superscripts E and I to label the entries from the parts of the mesh to be treated respectively explicitly and implicitly we can represent eqn (5.5.12) at $t=n\Delta t$ as

$$\begin{bmatrix} \boldsymbol{M}^{\mathrm{E}} & \boldsymbol{0} \\ \boldsymbol{0} & \boldsymbol{M}^{\mathrm{I}} \end{bmatrix} \begin{bmatrix} \boldsymbol{a}_n^{\mathrm{E}} \\ \boldsymbol{a}_n^{\mathrm{I}} \end{bmatrix} + \begin{bmatrix} \boldsymbol{K}^{\mathrm{E}} & \boldsymbol{K}^{\mathrm{EI}} \\ \boldsymbol{K}^{\mathrm{IE}} & \boldsymbol{K}^{\mathrm{II}} \end{bmatrix} \begin{bmatrix} \boldsymbol{x}_n^{\mathrm{E}} \\ \boldsymbol{x}_n^{\mathrm{I}} \end{bmatrix} = \begin{bmatrix} \boldsymbol{f}_n^{\mathrm{E}} \\ \boldsymbol{f}_n^{\mathrm{I}} \end{bmatrix}. \tag{5.5.13}$$

For the nodes in the explicit subset the central difference expression for the acceleration gives

$$\boldsymbol{a}_n^{\mathrm{E}} = (\boldsymbol{x}_{n+1}^{\mathrm{E}} - 2\boldsymbol{x}_n^{\mathrm{E}} + \boldsymbol{x}_{n-1}^{\mathrm{E}})/\Delta t^2. \tag{5.5.14}$$

This expression for $\boldsymbol{a}_n^{\mathrm{E}}$ when substituted into the explicit part of eqn (5.5.13) gives the formula for $\boldsymbol{x}_{n+1}^{\mathrm{E}}$ as

$$\boldsymbol{x}_{n+1}^{\mathrm{E}} = 2\boldsymbol{x}_n^{\mathrm{E}} - \boldsymbol{x}_{n-1}^{\mathrm{E}} - \Delta t^2(\boldsymbol{M}^{\mathrm{E}})^{-1}(\boldsymbol{f}_n^{\mathrm{E}} - \boldsymbol{K}^{\mathrm{E}}\boldsymbol{x}_n^{\mathrm{E}} - \boldsymbol{K}^{\mathrm{EI}}\boldsymbol{x}_n^{\mathrm{I}}). \tag{5.5.15}$$

Thus $\boldsymbol{x}_{n+1}^{\mathrm{E}}$ is given in terms of values known at the beginning of the time step. Of course, a special procedure must be used at the start. For the nodes in the implicit part of the mesh equation (5.5.13) is used now at $t=(n+1)\Delta t$ and the substitution for the acceleration is

$$\boldsymbol{a}_{n+1}^{\mathrm{I}} = \frac{4}{\Delta t^2}(\boldsymbol{x}_{n+1}^{\mathrm{I}} - \boldsymbol{x}_n^{\mathrm{I}}) - \frac{4}{\Delta t}\boldsymbol{v}_n^{\mathrm{I}} - \boldsymbol{a}_n^{\mathrm{I}}, \tag{5.5.16}$$

where $\boldsymbol{v}_n$ is obtained from

$$\boldsymbol{v}_n^{\mathrm{I}} = \boldsymbol{v}_{n-1}^{\mathrm{I}} + \frac{\Delta t}{2}(\boldsymbol{a}_n^{\mathrm{I}} + \boldsymbol{a}_{n-1}^{\mathrm{I}}). \tag{5.5.17}$$

(These are like the Newmark ($2\beta=\gamma=0.5$ approximations.) The result is that $\boldsymbol{x}_{n+1}^{\mathrm{I}}$ is given by

$$\left(\frac{4}{\Delta t^2}\boldsymbol{M}^{\mathrm{I}} + \boldsymbol{K}^{\mathrm{II}}\right)\boldsymbol{x}_{n+1}^{\mathrm{I}} = \boldsymbol{f}_{n+1}^{\mathrm{I}} - \boldsymbol{K}^{\mathrm{IE}}\boldsymbol{x}_{n+1}^{\mathrm{E}} + \boldsymbol{M}^{\mathrm{I}} \times (\Delta t^2\boldsymbol{a}_n^{\mathrm{I}} + 4\Delta t\boldsymbol{v}_n^{\mathrm{I}} + 4\boldsymbol{x}_n^{\mathrm{I}})/\Delta t^2, \tag{5.5.18}$$

where everything on the right-hand side is now known and a symmetric matrix has to be inverted. Belytschko and Mullen demonstrate the stability of the scheme using the energy method (Chapter 6) and again produce the result

$$\omega \Delta t \leq 2,$$

where ω^2 is specified as the maximum eigenvalue of the elements in the explicit part of the mesh with the implicit nodes on its boundary held fixed. They give some results of numerical experiments with subcycling: N explicit time steps to one implicit, and different treatment of the nodes on the interface between the two regions: weak coupling where the nodes remain distinct, and strong coupling where the interface includes N elements, with the idea that in the explicit scheme the information can travel across N elements in the time $N\Delta t$ which is one time step for the implicit scheme. It is evident from the matrix-vector operation involved in an explicit scheme that the information from node j at a particular time can only have travelled after one further time step to the nodes with which node j is coupled; with an implicit scheme this information can travel throughout the mesh. The strong coupling is using this notion but there does not seem to be a definite conclusion about the effect.

Element partitioning

The nodal partitioning idea discussed above presupposes that the matrices are assembled and then partitioned into blocks but in element partitioning the contributions from the elements in the different regions A and B can be assembled separately. Illustrating this on a one-dimensional problem with two linear elements shown in Fig. 5.5.2, the global stiffness matrix is

$$\boldsymbol{K} = \begin{bmatrix} 1/h_1 & -1/h_1 & 0 \\ -1/h_1 & 1/h_1 + 1/h_2 & -1/h_2 \\ 0 & -1/h_2 & 1/h_2 \end{bmatrix}. \tag{5.5.19}$$

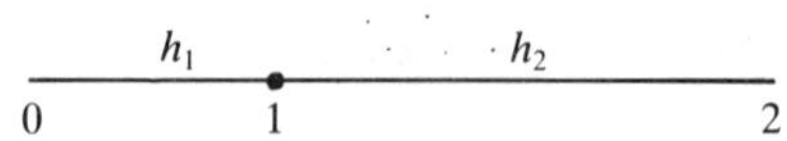

Fig. 5.5.2. Two linear elements (eqn 5.5.19).

The nodal partitioning might take

$$\boldsymbol{K}^{\mathrm{E}} = \begin{bmatrix} 1/h_1 & -1/h_1 \\ -1/h_1 & 1/h_1 + 1/h_2 \end{bmatrix} \quad \text{and } \boldsymbol{K}^{\mathrm{I}} = 1/h_2,$$

whereas the element partitioning would take

$$\boldsymbol{K}^{\mathrm{E}} = \begin{bmatrix} 1/h_1 & -1/h_1 & 0 \\ -1/h_1 & 1/h_1 & 0 \\ 0 & 0 & 0 \end{bmatrix}$$

and

$$\boldsymbol{K}^{\mathrm{I}} = \begin{bmatrix} 0 & 0 & 0 \\ 0 & 1/h_2 & -1/h_2 \\ 0 & -1/h_2 & 1/h_2 \end{bmatrix}$$

so that in the element partitioning each matrix is of the full order. Now the boundary nodes such as node 1 in Fig. 5.5.2 do not have to be assigned to one of the other region; their entries in the global matrix are split between the two regions and this simplifies the programming.

1. For the first order system of equations (5.5.1) with a lumped diagonal mass matrix, Hughes and Belytschko (1983) suggest a predictor–corrector form which in our notation gives

$$\begin{aligned} \tilde{\boldsymbol{x}}_{n+1} &= \boldsymbol{x}_n + (1-\theta)\Delta t \boldsymbol{v}_n \\ \boldsymbol{x}_{n+1} &= \tilde{\boldsymbol{x}}_{n+1} + \theta \Delta t \boldsymbol{v}_{n+1}. \end{aligned} \tag{5.5.20}$$

Equation (5.5.1) at time $t = (n+1)\Delta t$ can be written as

$$\boldsymbol{M}\boldsymbol{v}_{n+1} + \boldsymbol{K}\boldsymbol{x}_{n+1} = \boldsymbol{f}_{n+1}.$$

Then multiplying through by $\theta\Delta t$, substituting for $\boldsymbol{v}_{n+1}$, and replacing $\boldsymbol{K}\boldsymbol{x}_{n+1}$ by $\boldsymbol{K}^{\mathrm{I}}\boldsymbol{x}_{n+1} + \boldsymbol{K}^{\mathrm{E}}\tilde{\boldsymbol{x}}_{n+1}$, gives the compact form

$$(\boldsymbol{M} + \theta\Delta t \boldsymbol{K}^{\mathrm{I}})\boldsymbol{x}_{n+1} = (\boldsymbol{M} - \theta\Delta t \boldsymbol{K}^{\mathrm{E}})\tilde{\boldsymbol{x}}_{n+1} + \theta\Delta t \boldsymbol{f}_{n+1}, \tag{5.5.21}$$

where any element included in the $\boldsymbol{K}^{\mathrm{I}}$ assembly will be integrated implicitly. The rest are integrated explicitly with only the diagonal mass matrix on the left-hand side. The stability condition on the time step is the same as before, inequality (5.5.3) with the eigenvalue $\lambda \leq \max_{e \in S_{\mathrm{E}}} (\lambda^{(e)})$ where S_{E} is the set of elements being treated explicitly. Liu and Zhang (1983) give a formula for a bound on the eigenvalues of a general quadrilateral bilinear element which can be used in this context to estimate the bound on the time step from the explicit region. They demonstrate the method on the numerical solution of the equations representing the thermal protection system of a space shuttle, using a combination of implicit and explicit groups of elements and subcycling and again assuming a lumped mass matrix.

2. Hughes and Liu (1978) use implicit–explicit element partitioning for the dynamic vibration equation (5.5.5) using an explicit predictor–corrector adaptation of Newmark's method given by:

$$\boldsymbol{M}\boldsymbol{a}_{n+1}+\boldsymbol{C}\tilde{\boldsymbol{v}}_{n+1}-\boldsymbol{K}\tilde{\boldsymbol{x}}_{n+1}=\boldsymbol{f}_{n+1} \tag{5.5.22}$$

with

$$\tilde{\boldsymbol{x}}_{n+1}=\boldsymbol{x}_n+\Delta t\boldsymbol{v}_n+\Delta t^2(1-2\beta)\boldsymbol{a}_n/2 \tag{5.5.23}$$

$$\tilde{\boldsymbol{v}}_{n+1}=\boldsymbol{v}_n+\Delta t(1-\gamma)\boldsymbol{a}_n \tag{5.5.24}$$

and

$$\boldsymbol{x}_{n+1}=\tilde{\boldsymbol{x}}_{n+1}+\Delta t^2\beta\boldsymbol{a}_{n+1} \tag{5.5.25}$$

$$\boldsymbol{v}_{n+1}=\tilde{\boldsymbol{v}}_{n+1}+\Delta t\gamma\boldsymbol{a}_{n+1} \tag{5.5.26}$$

with the usual notation $\boldsymbol{x}_n$, $\boldsymbol{v}_n$, $\boldsymbol{a}_n$ for the approximate displacement, velocity and acceleration at time $t=n\Delta t$. Here $\tilde{\boldsymbol{x}}_{n+1}$ and $\tilde{\boldsymbol{v}}_{n+1}$ may be regarded as predictor values. Equation (5.5.22) is different from the standard Newmark formula because of the presence of these predictor values and this means that only the mass matrix $\boldsymbol{M}$ has to be inverted in order to obtain $\boldsymbol{a}_{n+1}$ instead of the usual matrix $\boldsymbol{M}+\gamma\Delta t\boldsymbol{C}+\beta\Delta t^2\boldsymbol{K}$.

In the Hughes and Liu (1978) implicit–explicit algorithm, the matrices $\boldsymbol{M}$, $\boldsymbol{C}$, $\boldsymbol{K}$ are split corresponding to the two groups of elements as

$\boldsymbol{M}=\boldsymbol{M}^{\mathrm{I}}+\boldsymbol{M}^{\mathrm{E}}$, with only $\boldsymbol{M}^{\mathrm{E}}$ assumed to be diagonal.

$\boldsymbol{C}=\boldsymbol{C}^{\mathrm{I}}+\boldsymbol{C}^{\mathrm{E}}$ and $\boldsymbol{K}=\boldsymbol{K}^{\mathrm{I}}+\boldsymbol{K}^{\mathrm{E}}$.

Then the algorithm is given by

$$\boldsymbol{M}\boldsymbol{a}_{n+1}+\boldsymbol{C}^{\mathrm{I}}\boldsymbol{v}_{n+1}+\boldsymbol{C}^{\mathrm{E}}\tilde{\boldsymbol{v}}_{n+1}+\boldsymbol{K}^{\mathrm{I}}\boldsymbol{x}_{n+1}+\boldsymbol{K}^{\mathrm{E}}\tilde{\boldsymbol{x}}_{n+1}=\boldsymbol{f}_{n+1} \tag{5.5.27}$$

together with eqns (5.5.23) to (5.6.26). Hughes investigates the stability of the predictor–corrector algorithm given by eqns (5.5.22) to (5.5.26) by the energy method (Chapter 6). We can also investigate the stability conditions by the method used in previous chapters of putting $x_{n+1} = rx_n$, $v_{n+1} = rv_n$, $a_{n+1} = ra_n$ in the corresponding scalar equations:

$$ma_{n+1} + c\tilde{v}_{n+1} + k\tilde{x}_{n+1} = 0, \text{ etc.} \tag{5.5.28}$$

From eqns (5.5.23) and (5.5.25) we then have

$$(r-1)x_n = \Delta t v_n + \frac{\Delta t^2}{2}(1 - 2\beta + 2\beta r)a_n. \tag{5.5.29}$$

From eqns (5.5.24) and (5.5.26) we have

$$(r-1)v_n = (1 - \gamma + \gamma r)\Delta t a_n. \tag{5.5.30}$$

Hence substituting for v_n from (5.5.30) into (5.5.29) gives

$$x_n = \{2(1-\gamma+\gamma r) + (r-1)(1-2\beta+2\beta r)\}\Delta t^2 a_n / 2(r-1)^2. \tag{5.5.31}$$

Then writing eqn (5.5.28) in the form

$$rma_n + c\{v_n + (1-\gamma)\Delta t a_n\} + k\{x_n + \Delta t v_n + (0.5-\beta)\Delta t^2 a_n\} = 0,$$

substituting for x_n and v_n from eqns (5.5.30) and (5.5.31), and cancelling the non-trivial a_n, we have the characteristic polynomial of the Hughes predictor–corrector method given by the equation

$$\begin{aligned}(r-1)^2\{rm + (1-\gamma)\Delta tc + (0.5-\beta)\Delta t^2 k\} \\ + \{1-\gamma+\gamma r + (r-1)(0.5-\beta+\beta r)\}\Delta t^2 k \\ +(r-1)(1-\gamma+\gamma r)\Delta t(c+\Delta tk) = 0\end{aligned}$$

The term independent of r is zero so this is a quadratic not a cubic and after sorting out the algebra this stability polynomial reduces to

$$\begin{aligned}r^2 m + r\{-2m + \Delta tc + (0.5+\gamma)\Delta t^2 k\} \\ + m - \Delta tc + (0.5-\gamma)\Delta t^2 k = 0.\end{aligned} \tag{5.5.32}$$

It is interesting to note that (a) the parameter β has disappeared, (b) the corresponding two-step algorithm in the displacements only which we can immediately deduce using the analysis in §2.1 is given by

$$\begin{aligned}mx_{n+1} + \{-2m + \Delta tc + (0.5+\gamma)\Delta t^2 k\}x_n \\ + \{m - \Delta tc + (0.5-\gamma)\Delta t^2 k\}x_{n-1} = 0\end{aligned} \tag{5.5.33}$$

that is,

$$m(x_{n+1} - 2x_n + x_{n-1})/\Delta t^2 + c(x_n - x_{n-1})/\Delta t + k\{(0.5+\gamma)x_n + (0.5-\gamma)x_{n-1}\} = 0. \quad (5.6.34)$$

With $\gamma = 0.5$ the two-step algorithm gives

$$m(x_{n+1} - 2x_n + x_{n-1})/\Delta t^2 + c(x_n - x_{n-1})/\Delta t + kx_n = 0, \quad (5.5.35)$$

which is representing the differential equation with a central difference for the acceleration and a backward difference for the velocity in order to make it explicit. Thus the accuracy can only be first order, whatever the value of the parameter γ, unless the natural damping is zero, that is we have $c = 0$, when we have again the central difference algorithm discussed in §2.3 (the central difference algorithm for $c \neq 0$ uses a centred difference for the velocity term).

Returning to the stability polynomial (5.5.32), we can investigate the stability by making the substitution $r = (1+z)/(1-z)$ to give

$$m(1+z)^2 + \{-2m + \Delta tc + (0.5+\gamma)\Delta t^2 k\}(1-z^2) + \{m - \Delta tc + (0.5-\gamma)\Delta t^2 k\}(1-z)^2 = 0. \quad (5.5.36)$$

Then using the Routh–Hurwitz conditions (Appendix 1) for the roots of the eqn (5.5.36) to have non-positive real parts gives the conditions

$$\left.\begin{aligned} 2m - \Delta tc - \gamma\Delta t^2 k &> 0 \\ \Delta tc + (\gamma - 0.5)\Delta t^2 k &\geq 0 \\ \text{and} \qquad \Delta t^2 k &\geq 0 \end{aligned}\right\} \quad (5.5.37)$$

The second of inequalities (5.5.37) means we must have $\gamma \geq 0.5$ and the first is satisfied if

$$\gamma\Delta t^2 k + \Delta tc - 2m < 0$$

that is, (5.5.38)

$$\Delta t < \{-c + \sqrt{(c^2 + 8mk\gamma)}\}/(2\gamma k).$$

Putting $c^2 = 4\nu^2 km$, so that ν is the fraction of critical damping and $\omega^2 = k/m$, we have

$$\omega\Delta t < \Omega_{\text{crit}} = \{-\nu + \sqrt{(\nu^2 + 2\gamma)}\}/\gamma \quad (5.5.39)$$

as the condition on the time step for stability of this explicit method. We can also see that the roots of the stability polynomial in (5.5.32) are complex (which means that the difference equation has an oscillatory solution like that of the differential equation it is approximating) if

$$\{-2m + \Delta tc + (0.5+\gamma)\Delta t^2 k\}^2 < 4m\{m - \Delta tc + (0.5-\gamma)\Delta t^2 k\}. \tag{5.5.40}$$

After some algebra this condition reduces to

$$\omega^2 \Delta t^2 (0.5+\gamma)^2 + 4\nu\omega\Delta t(0.5+\gamma) + 4(\nu^2 - 1) < 0, \tag{5.5.41}$$

substituting $\omega^2 = k/m$ and $c = 2\nu m\omega$ as before. Hence the condition on the time step for an oscillatory solution is

$$\omega\Delta t < \Omega_b = 2(1-\nu)/(\gamma+0.5). \tag{5.5.42}$$

Comparing conditions (5.5.39) and (5.5.42), we see that

$$\Omega_{crit} - \Omega_b = \{(\gamma+0.5)\sqrt{(\nu^2+2\gamma)} + \nu(\gamma-0.5) - 2\gamma\}/\{\gamma(\gamma+0.5)\}$$

and

$$(\gamma+0.5)\sqrt{(\nu^2+2\gamma)} + \nu(\gamma-0.5) - 2\gamma \geq 0$$

if

$$(\gamma+0.5)\sqrt{(\nu^2+2\gamma)} \geq 2\gamma - \nu(\gamma-0.5),$$

which is true if

$$(\gamma+0.5)^2(\nu^2+2\gamma) \geq 4\gamma^2 - 4\gamma\nu(\gamma-0.5) + \nu^2(\gamma-0.5)^2,$$

that is

$$\gamma(\nu+\gamma-0.5)^2 \geq 0.$$

Hence $\Omega_{crit} \geq \Omega_b$, as we expect, since $\omega\Delta t = \Omega_b$ is the value at which the roots of the stability polynomial change from complex conjugates to two real roots which can still have modulus less than unity. Thus to ensure C-stability (Chapter 2), it is necessary to restrict $\omega\Delta t$ by a slightly lower limit than that for stability. When the natural damping is zero, that is, $\nu = 0$, we have $\Omega_{crit} = \sqrt{(2/\gamma)}$ and $\Omega_b = 2/(\gamma+0.5)$. With $\gamma = 0.5$ these each correspond to the condition $\omega\Delta t < 2$ so that we are back to the central difference scheme situation.

Returning to the Hughes and Liu (1978) implicit–explicit

element partitioned algorithm given by eqn (5.5.27) together with eqns (5.5.23) to (5.5.26), in the implicit element group $2\beta = \gamma = 0.5$ gives the unconditionally stable second order trapezium rule equivalent and in the explicit group the time step is restricted by inequality (5.5.39) or inequality (5.5.42).

Hughes (1983) indicates a proof of convergence for the explicit scheme (5.5.22) to (5.5.26) which shows as suggested by the scalar equations (5.5.34) and (5.5.35) that with non-zero natural damping the error is $O(\Delta t)$. For the implicit–explicit scheme with $2\beta = \gamma = 0.5$ and $C^{E} = 0$ the error is $O(\Delta t^2)$. Hughes (1983) also gives a predictor–multi-corrector algorithm, iterating on the explicit stage, which has second order convergence with $\boldsymbol{C}^{E} \neq 0$.

5.6. Element by element methods

The element by element (EBE) methods are the logical extension of the element partitioning schemes. The basic idea is that the global matrices are never assembled but all the working is done with the element matrices. For example, the global stiffness matrix $\boldsymbol{K}$ is expressed as

$$\boldsymbol{K} = \sum_{e=1}^{N} \boldsymbol{K}^{(e)} \tag{5.6.1}$$

where the $\boldsymbol{K}^{(e)}$ are matrices of the same order as $\boldsymbol{K}$, that is, their order equals the number of unknowns, each with the entries from an element inserted into the correct positions and the rest of the array filled up with zeros, and N is the number of elements. The $\boldsymbol{K}^{(e)}$ are called the element stiffness matrices. Hughes *et al.* (1983a,b) give an example of the EBE method used on the parabolic heat conduction problem represented by the equation

$$\boldsymbol{M}\dot{\boldsymbol{x}} + \boldsymbol{K}\boldsymbol{x} = \boldsymbol{f}(t), \tag{5.6.2}$$

where $\boldsymbol{M}$ and $\boldsymbol{K}$ are the mass and stiffness matrices, respectively, supposed constant. The mass matrix $\boldsymbol{M}$ is as usual supposed symmetric and positive definite and the stiffness matrix $\boldsymbol{K}$ is supposed symmetric and positive semi-definite. Hughes *et al.* assume that the mass matrix is diagonal. The generalized θ-method (Chapter 3) applied to eqn (5.6.2) is written as

$$(\boldsymbol{M} + \theta\Delta t\boldsymbol{K})\boldsymbol{x}_{n+1} = \{\boldsymbol{M} - (1-\theta)\Delta t\boldsymbol{K}\}\boldsymbol{x}_n + \Delta t\boldsymbol{f}_{n+\theta}, \tag{5.6.3}$$

where either

$$f_{n+\theta} = (1-\theta)f_n + \theta f_{n+1}$$

or (5.6.4)

$$f_{n+\theta} = f(t_n + \theta\Delta t).$$

We define $M^{\frac{1}{2}}$ as the positive definite matrix square root of the diagonal mass matrix M, multiply eqn (5.6.3) through by $M^{-\frac{1}{2}}$ and rewrite it in the form:

$$(I + \theta\Delta t C)u_{n+1} = \{I - (1-\theta)\Delta t C\}u_n + \Delta t M^{-\frac{1}{2}} f_{n+\theta}, \quad (5.6.5)$$

where

$$u_n = M^{\frac{1}{2}} x_n \quad \text{and} \quad C = M^{-\frac{1}{2}} K M^{-\frac{1}{2}}. \quad (5.6.6)$$

Next we define

$$B = (I + \theta\Delta t C)^{-1} \quad (5.6.7)$$

and

$$A = B\{I - (1-\theta)\Delta t C\}, \quad (5.6.8)$$

then eqn (5.6.5) now appears as

$$u_{n+1} = Au_n + \Delta t B M^{-\frac{1}{2}} f_{n+\theta}. \quad (5.6.9)$$

The matrices A and B given by eqns (5.6.7) and (5.6.8) are functions of Δt and θ as well as of the matrix C; we can write $A = A(\theta, \Delta t)$, $B = B(\theta, \Delta t)$. The element stiffness matrices as defined in eqn (5.6.1) are symmetric and positive semi-definite. the element counterparts of the matrices C, B, and A are given by

$$C^{(e)} = M^{-\frac{1}{2}} K^{(e)} M^{-\frac{1}{2}} \quad (5.6.10)$$

$$B^{(e)} = (I + \theta\Delta t C^{(e)})^{-1} \quad (5.6.11)$$

and

$$A^{(e)} = B^{(e)}\{I - (1-\theta)\Delta t C^{(e)}\}, \quad (5.6.12)$$

where $e = 1, 2, \ldots, N$, the number of elements. The EBE method uses approximations to the matrices A and B as follows

$$\bar{A} = \prod_{e=1}^{N} A^{(e)} \approx A \quad (5.6.13)$$

and

$$\bar{B} = \prod_{e=1}^{N} B^{(e)} \approx B. \quad (5.6.14)$$

Substituting $\bar{\boldsymbol{A}}$ and $\bar{\boldsymbol{B}}$ for $\boldsymbol{A}$ and $\boldsymbol{B}$ in eqn (5.6.9) gives the one-pass EBE algorithm

$$\boldsymbol{u}_{n+1} = \left\{\prod_{e=1}^{N} \boldsymbol{A}^{(e)}\right\}\boldsymbol{u}_n + \Delta t\left\{\prod_{e=1}^{N} \boldsymbol{B}^{(e)}\right\}\boldsymbol{M}^{-\frac{1}{2}}\boldsymbol{f}_{n+\theta}. \qquad (5.6.15)$$

The element matrices $\boldsymbol{K}^{(e)}$ can be formed as they are required. The algorithm is implemented by a series of element size matrix inversions and element matrix–vector multiplications.

We know from Chapter 3 that the θ-method given in eqn (5.6.3) or (5.6.9) is unconditionally stable for $\theta \geq 0.5$, is second order accurate for $\theta = 0.5$ and first order accurate for $\theta \neq 0.5$. The effect of the one-pass EBE algorithm thus depends on the accuracy with which the matrices are approximated by the product formulae in eqns (5.6.13) and (5.6.14). Hughes *et al.* show that the one-pass EBE algorithm is only first order accurate even if $\theta = 0.5$. Their proof depends on expanding the matrices $\boldsymbol{A}$ and $\boldsymbol{B}$ and their approximations $\bar{\boldsymbol{A}}$, $\bar{\boldsymbol{B}}$ in powers of Δt. For example from eqn (5.6.7) we can expand $\boldsymbol{B}(\theta, \Delta t)$ as a power series in Δt given by

$$\boldsymbol{B}(\theta, \Delta t) = \boldsymbol{I} - \theta \Delta t \boldsymbol{C} + \mathbf{O}(\Delta t^2). \qquad (5.6.16)$$

We also expand $\boldsymbol{B}^{(e)}$ as a power series in Δt given by

$$\boldsymbol{B}^{(e)}(\theta, \Delta t) = \boldsymbol{I} - \theta \Delta t \boldsymbol{C}^{(e)} + \mathbf{O}(\Delta t^2). \qquad (5.6.17)$$

It is interesting to note that the convergence of the power series in eqn (5.6.17) depends on the condition

$$\|\theta \Delta t \boldsymbol{C}^{(e)}\| < 1 \qquad (5.6.18)$$

(Faddeeva 1959). Since from eqn (5.6.10) the matrix $\boldsymbol{C}^{(e)} = \boldsymbol{M}^{-\frac{1}{2}}\boldsymbol{K}^{(e)}\boldsymbol{M}^{-\frac{1}{2}}$, its entries are $O(h^{-2})$, where h is the measure of the element (Chapter 1), and thus we have implicitly here a bound on $\Delta t/h^2$. There is a corresponding bound implied by the global matrix expansion in eqn (5.6.16).

The limitation on the accuracy of the one-pass algorithm motivates Hughes *et al.* (1983a) to introduce the double-pass

EBE algorithm given by

$$\boldsymbol{u}_{n+1} = \left[\prod_{e=1}^{N} \boldsymbol{A}^{(e)}(\theta, \Delta t/2)\right]\left[\prod_{e=N}^{1} \boldsymbol{A}^{(e)}(\theta, \Delta t/2)\right]\boldsymbol{u}_n$$
$$+ \Delta t\left[\prod_{e=1}^{N} \boldsymbol{B}^{(e)}(\theta, \Delta t/2)\right]$$
$$\times \left[\prod_{e=N}^{1} \boldsymbol{B}^{(e)}(\theta, \Delta t/2)\right]\boldsymbol{M}^{-\frac{1}{2}}\boldsymbol{f}_{n+\theta}; \tag{5.6.19}$$

which now gives second order accuracy. (The expression 'double-pass' rather than 'two-pass' contains the suggestion of doubling back which is essential for the success of the algorithm.) Hughes *et al.* also give a generalization of their EBE algorithm for use with non-linear problems.

Gilvary *et al.* (1987) make a detailed costing of the EBE method as given by Hughes *et al.* (1983a) with $\theta = 0.5$ and compare this with the costing for (a) the global Crank–Nicolson-trapezium rule algorithm and (b) an extension of this method using a consistent mass matrix with the same elements. Their motivation in forming a consistent mass EBE algorithm is (i) to show that this is possible, (ii) to investigate the errors which arise due to the lumped mass in the algorithms of Hughes *et al.* (1983). The conclusions of Gilvary *et al.* include the following:

1. For larger problems the one-pass lumped mass EBE algorithm is more economic than the global algorithm.

2. The consistent mass EBE algorithm is far more expensive than the lumped mass version but it can give much more accurate results. The numerical evidence for this comes from

(a) The linear heat conduction problem (5.6.2) solved on the standard arch-shaped region with 42 elements. For the converged solutions with decreasing time step there is 15% difference between the lumped and consistent mass versions.

(b) the linear heat conduction problem for a thick-walled tube with a temperature gradient between the inner and outer walls. For the converged solutions here there is 20% difference between the lumped and consistent mass versions. This error is a spatial discretization error and it can be reduced by refining the mesh.

The results of Gilvary *et al.* appear to indicate that the consistent mass EBE algorithm is uncompetitive for the linear first order problem.

There are exciting new possibilities for the use of element by element techniques on parallel processors. Gilvary *et al.* (1987) also discuss some aspects of these.

There are two very different approaches to the formation of an EBE solution algorithm for the second order system of dynamic equations given by Hughes *et al.* (1983b) and Ortiz *et al.* (1983). Hughes *et al.* use the Newmark algorithm on the system of equations

$$\boldsymbol{Ma} + \boldsymbol{N} = \boldsymbol{F}, \tag{5.6.20}$$

where $\boldsymbol{M}$ is the mass matrix, $\boldsymbol{a}$ the vector of accelerations, $\boldsymbol{N}$ is the vector of internal forces which can depend on the displacement, velocity and their histories, and $\boldsymbol{F}$ is the vector of external forces. At each time step the equations to be solved for the increment in the displacement $\Delta\boldsymbol{x}$ is

$$\boldsymbol{K}^*\Delta\boldsymbol{x} = \boldsymbol{R}, \tag{5.6.21}$$

where

$$\boldsymbol{K}^* = \frac{1}{\Delta t^2\beta}\{\boldsymbol{M}_{n+1} + \gamma\Delta t\boldsymbol{C}_{n+1} - \beta\Delta t^2\boldsymbol{K}_{n+1}\}, \tag{5.6.22}$$

the matrices $\boldsymbol{C}$ and $\boldsymbol{K}$ being the tangent damping and tangent stiffness matrices, respectively, and $\boldsymbol{R}$ is the residual

$$\boldsymbol{R} = \boldsymbol{F}_{n+1} - \boldsymbol{N}_{n+1} - \boldsymbol{M}_{n+1}\boldsymbol{a}_{n+1}. \tag{5.6.23}$$

Equation (5.6.21) is to be solved iteratively until $\Delta\boldsymbol{x}$ and/or $\boldsymbol{R}$ satisfy some given convergence tolerance. At each iteration an equation of the form

$$\boldsymbol{Ay} = \boldsymbol{b} \tag{5.6.24}$$

has to be solved where the matrix $\boldsymbol{A}$ is symmetric and positive definite and $\boldsymbol{A} = \sum_e \boldsymbol{A}^{(e)}$, from the sum over the elements. Equation (5.6.24) is solved by introducing an associated parabolic problem

$$\boldsymbol{W}\frac{\mathrm{d}\boldsymbol{y}}{\mathrm{d}\tau} + \boldsymbol{Ay} = H(\tau)\boldsymbol{b}, \tag{5.6.25}$$

where the matrix $\boldsymbol{W}$ can be taken as the diagonal of $\boldsymbol{A}$, $H(\tau)$ is a Heaviside function ($H(\tau)=1$ for $\tau>0$, and is zero otherwise), and τ is a non-dimensional pseudo-time. Thus a parabolic problem has been introduced and this is solved by the EBE double-pass method similar to that in the Hughes *et al.* (1983a) heat conduction EBE paper, but using the θ-method with $\theta=1$ to give fast convergence with large values of the pseudo-time-step. The EBE solution algorithm uses a 'line search', a method for which details are given in Fox (1964) p. 207, and a quasi-Newton BFGS (Broyden–Fletcher–Goldfarb–Shanno) update, details for which are given in Dennis and Moré (1977).

Ortiz *et al.* (1983) adopt a different strategy in order to put the second order system of dynamic equations into the first order form suitable for the EBE double-pass. These authors take the velocity and the stress as the unknown variables so that the equations are expressed as

$$\rho\dot{v}=\nabla\cdot\sigma+\rho b, \tag{5.6.26}$$

and

$$\dot{\sigma}=D:\nabla v \tag{5.6.27}$$

with given boundary and initial conditions, where v and σ are the velocity and stress fields, b the body forces, ρ the density and D is the elastic modulus tensor, that is, $D:\nabla v\equiv D_{ijkl}v_{k,l}$. Then the finite element space discretization gives a system of equations of the form

$$\boldsymbol{A}\dot{\boldsymbol{X}}-\boldsymbol{B}\boldsymbol{X}=\boldsymbol{F}, \tag{5.6.28}$$

where the vector $\boldsymbol{X}=[\boldsymbol{v},\boldsymbol{\sigma}]^{\mathrm{T}}$ is the vector including nodal velocities and stresses, the matrix

$$\boldsymbol{A}=\begin{bmatrix}\boldsymbol{M} & \boldsymbol{0}\\ \boldsymbol{0} & \boldsymbol{I}\end{bmatrix}$$

and the mass matrix $\boldsymbol{M}$ is taken to be diagonal. Equation (5.6.28) is then ready for the EBE double-pass algorithm to be applied. Ortiz *et al.* given numerical results obtained by applying their method to the numerical solution of a problem consisting of a mass-spring system which is linear but very stiff. The results are obtained by using exact integration on the elements so as to give an assessment of the accuracy of an EBE double-pass product formula; results using the trapezoidal formula (θ-method with $\theta = 0.5$) are similar. The displacement vector $\boldsymbol{x}_{n+1}$ is calculated from the velocity vector by the corresponding trapezoidal formula

$$\boldsymbol{x}_{n+1} = \boldsymbol{x}_n + 0.5\Delta t(\boldsymbol{v}_n + \boldsymbol{v}_{n+1}).$$

Ortiz *et al.* report that the numerical results are satisfactory with the double-pass even with time steps about 14 times greater than the critical value corresponding to the explicit Euler (θ-method with $\theta = 0$) except for the case where the initial conditions triggered off high frequency components. Then the accuracy deteriorates for time steps greater than the Euler critical value. Since the energy of the system is exactly preserved in their method the authors say that this may indicate an energy transfer between modes. Oritz *et al.* also comment that the standard first order form of the dynamic vibration equation (as given in Chapter 1) expressed with the displacement and velocity as the unknowns, is unsuitable in this context because then the solutions of the individual element equations are not energy-preserving.

5.7. Operator splitting methods

The idea of splitting the operator goes back at least as far as Gauss (1823). Gauss wrote to tell his friend Gerling about how much he enjoyed solving sets of simultaneous equations

$$\boldsymbol{Ax} = \boldsymbol{b} \tag{5.7.1}$$

not by the 'Gauss elimination method' but by the method which in the twentieth century came to be known as Southwell's Relaxation Method (Southwell 1946). This entails spotting the largest residual in the equations by eye and adjusting the values of the unknowns (relaxing them) so as to liquidate the residuals. This is a kind of splitting of the operator (in this case the operator is the matrix) but not in a systematic way. Jacobi (1844) introduced a systematic way of doing this by splitting the matrix $\boldsymbol{A}$ into

$$\boldsymbol{A} = \boldsymbol{L} + \boldsymbol{D} + \boldsymbol{U}, \tag{5.7.2}$$

where $\boldsymbol{D}$ is diagonal and the matrices $\boldsymbol{L}$ and $\boldsymbol{U}$ are strictly lower and upper triangular respectively. The solution of the eqns (5.7.1) then becomes the iteration

$$\boldsymbol{D}\boldsymbol{x}_{n+1} = -(\boldsymbol{L} + \boldsymbol{U})\boldsymbol{x}_n + \boldsymbol{b}. \tag{5.7.3}$$

Seidel (1874) discusses several methods for solving the system of equations (5.7.1) one of which is the systematic successive substitution method given by

$$(\boldsymbol{L} + \boldsymbol{D})\boldsymbol{x}_{n+1} = -\boldsymbol{U}\boldsymbol{x}_n + \boldsymbol{b}, \tag{5.7.4}$$

which is now known as the Gauss–Seidel method.

Trujillo (1977a) uses a split operator method (an adaptation of a method of Saul'yev 1964), to give an unconditionally stable explicit algorithm for the solution of the first order equation (in our notation),

$$\boldsymbol{M}\dot{\boldsymbol{x}} + \boldsymbol{K}\boldsymbol{x} = \boldsymbol{f}, \tag{5.7.5}$$

where the mass matrix $\boldsymbol{M}$ is assumed to be lumped diagonal and the symmetric stiffness matrix $\boldsymbol{K}$ is split into upper and lower triangular matrices:

$$\boldsymbol{K} = \boldsymbol{K}_{\mathrm{L}} + \boldsymbol{K}_{\mathrm{U}}, \qquad \boldsymbol{K}_{\mathrm{L}} = \boldsymbol{K}_{\mathrm{U}}^{\mathrm{T}}. \tag{5.7.6}$$

The vector $\boldsymbol{x}_n$ is updated by the formulae

$$(\boldsymbol{M} + 0.5\Delta t\boldsymbol{K}_{\mathrm{L}})\boldsymbol{x}_{n+\frac{1}{2}} = (\boldsymbol{M} - 0.5\Delta t\boldsymbol{K}_{\mathrm{U}})\boldsymbol{x}_n + 0.5\Delta t\boldsymbol{f}_n \tag{5.7.7}$$

and

$$(\boldsymbol{M} + 0.5\Delta t\boldsymbol{K}_{\mathrm{U}})\boldsymbol{x}_{n+1} = (\boldsymbol{M} - 0.5\Delta t\boldsymbol{K}_{\mathrm{L}})\boldsymbol{x}_{n+\frac{1}{2}} + 0.5\Delta t\boldsymbol{f}_n. \tag{5.7.8}$$

The method is thus like making one forward followed by one backward Gauss–Seidel pass per time step Δt and no matrix need be inverted so it is an explicit method. It also remains stable independently of the numbering of the nodes. Trujillo (1977b) presents an unconditionally stable explicit algorithm of a similar form for the structural dynamics equation

$$\boldsymbol{M}\ddot{\boldsymbol{x}} + \boldsymbol{C}\dot{\boldsymbol{x}} + \boldsymbol{K}\boldsymbol{x} = \boldsymbol{f}, \tag{5.7.9}$$

with the mass matrix $\boldsymbol{M}$ again assumed to be diagonal and $\boldsymbol{C}$ and $\boldsymbol{K}$ symmetric and positive definite. The matrices $\boldsymbol{C}$ and $\boldsymbol{K}$ are split into lower and upper triangular matrices as was the matrix $\boldsymbol{K}$ in eqns (5.7.6). Then if $\boldsymbol{x}_n$ and $\boldsymbol{v}_n$ are the approximate displacement and velocity at time $t = n\Delta t$, the solution is advanced one time step by applying the formulae

$$\left(\boldsymbol{M}+\frac{\Delta t}{2}\boldsymbol{C}_{\mathrm{L}}+\frac{\Delta t^2}{8}\boldsymbol{K}_{\mathrm{L}}\right)\boldsymbol{v}_{n+\frac{1}{2}}=\left(\boldsymbol{M}-\frac{\Delta t}{2}\boldsymbol{C}_{\mathrm{U}}-\frac{\Delta t^2}{8}\boldsymbol{K}_{\mathrm{L}}\right)\boldsymbol{v}_n$$

$$-\frac{\Delta t}{2}\boldsymbol{K}\boldsymbol{x}_n+\frac{\Delta t}{4}(\boldsymbol{f}_{n+1}+\boldsymbol{f}_n)$$

$$\boldsymbol{x}_{n+\frac{1}{2}}=\boldsymbol{x}_n+\frac{\Delta t}{4}(\boldsymbol{v}_{n+\frac{1}{2}}+\boldsymbol{v}_n)$$

$$\left(\boldsymbol{M}+\frac{\Delta t}{2}\boldsymbol{C}_{\mathrm{U}}+\frac{\Delta t^2}{8}\boldsymbol{K}_{\mathrm{U}}\right)\boldsymbol{v}_{n+1}=\left(\boldsymbol{M}-\frac{\Delta t}{2}\boldsymbol{C}_{\mathrm{L}}-\frac{\Delta t^2}{8}\boldsymbol{K}_{\mathrm{U}}\right)\boldsymbol{v}_{n+\frac{1}{2}}$$

$$-\frac{\Delta t}{2}\boldsymbol{K}\boldsymbol{x}_{n+\frac{1}{2}}+\frac{\Delta t}{4}(\boldsymbol{f}_{n+1}+\boldsymbol{f}_n)$$

$$\boldsymbol{x}_{n+1}=\boldsymbol{x}_{n+\frac{1}{2}}+\frac{\Delta t}{4}(\boldsymbol{v}_{n+1}+\boldsymbol{v}_{n+\frac{1}{2}}). \quad (5.7.10)$$

These formulae can be programmed so that the algorithm is explicit and no matrix inversion is required. By expanding in powers of the time step Δt, Trujillo shows that the accuracy of the algorithm is equivalent to that of using the (2, 0) Padé approximation of the exponential in eqn (5.3.3) with the first order form of eqn (5.7.9). But the numerical results show the accuracy is actually quite poor. In the undamped case, time steps of the order of those of the conditionally stable central difference method (§2.3) are necessary to prevent a complete distortion of the oscillatory motion. Trujillo also examines an alternate splitting but this is only conditionally stable. Park (1982) gives modifications and extensions of these methods. Park makes an improvement in the accuracy by augmenting the solution matrix with two simple diagonal matrices of the order of the local truncation error. The idea is to keep the unconditional stability at the same time as keeping the calculations effectively explicit. Park solves the same two sample problems as given in Trujillo (1977a,b) and shows much improved results.

We can also include in this section on operator splitting methods an alternating direction Galerkin technique used by Dans and Frind (1985) for the simulation of contaminant transport in complex groundwater systems. Here we have the convection–diffusion equation

$$\frac{\partial}{\partial x}\left\{D_{xx}\frac{\partial u}{\partial x}\right\}+\frac{\partial}{\partial y}\left\{D_{yx}\frac{\partial u}{\partial x}\right\}-v_x\frac{\partial u}{\partial x}+\frac{\partial}{\partial y}\left\{D_{yy}\frac{\partial u}{\partial y}\right\}$$
$$+\frac{\partial}{\partial x}\left\{D_{xy}\frac{\partial u}{\partial y}\right\}-v_y\frac{\partial u}{\partial y}-\frac{\partial u}{\partial t}=0, \quad (5.7.11)$$

where u is the concentration of the contaminant, v_x, v_y are the average groundwater velocity components and the D_{xx}, D_{xy}, D_{yy}, D_{yx} are the components of the hydrodynamic dispersion tensor (Bear 1979). When space-discretized using finite elements the terms in eqn (5.7.11) are grouped in the form

$$(\boldsymbol{L}_{xx}+\boldsymbol{L}_{yx}+\boldsymbol{L}_{yy}+\boldsymbol{L}_{xy})\boldsymbol{u}=\frac{\partial \boldsymbol{u}}{\partial t}. \quad (5.7.12)$$

The time stepping is then arranged in a typical alternating direction manner as with finite differences (Varga 1962) as

$$(\boldsymbol{L}_{xx}+\boldsymbol{L}_{yx})\boldsymbol{u}^{n+\frac{1}{2}}+(\boldsymbol{L}_{yy}+\boldsymbol{L}_{xy})\boldsymbol{u}^{n}=\frac{\boldsymbol{u}^{n+\frac{1}{2}}-\boldsymbol{u}^{n}}{\Delta t/2} \quad (5.7.13)$$

to be solved for $\boldsymbol{u}^{n+\frac{1}{2}}$, together with

$$(\boldsymbol{L}_{xx}+\boldsymbol{L}_{yx})\boldsymbol{u}^{n+\frac{1}{2}}+(\boldsymbol{L}_{yy}+\boldsymbol{L}_{xy})\boldsymbol{u}^{n+1}=\frac{\boldsymbol{u}^{n+1}-\boldsymbol{u}^{n+\frac{1}{2}}}{\Delta t/2} \quad (5.7.14)$$

to be solved for $\boldsymbol{u}^{n+1}$. If these two equations are added the effect is to take an average over one time step Δt and hence there is second order accuracy. Dans and Frind state that the cross derivative term is evaluated iteratively and with a double pass there is unconditional stability. As with the finite difference ADI (alternating direction implicit) methods stability is assured for rectangular uniform grids. Dans and Frind say that stability in a deformed mesh remains to be investigated. We note that practical users have long disregarded the mathematical limitations on ADI finite difference methods to rectangular regions and uniform mesh.

Exercises

5.1. Take the Gellert interpolation formula (5.2.10) for x in terms of the ω_1, ω_2, ω_3, ω_4, given in eqns (5.2.4), sutstitute into $a+\omega^2x=0$ and $\dot{a}+\omega^2 v=0$ and collocate at $\zeta=1$. Put $x_{n+1}=rx_n$, and $v_{n+1}=rv_n$ to obtain the stability polynomial, and show that the algorithm is stable but not C-stable. Show that the error in the factor r is $O(\Delta t^3)$.

5.2. Examine the stability for $z = i\omega$ (ω real) of the Padé approximations (0, 1), (0, 2), (1, 1), (1, 2) and (2, 1) given in Table 5.3.1.

5.3. Show that the Nørsett approximation $N_2^1(\alpha)$ for $\exp(i\omega)$, where ω is real, is stable for all ω.

5.4. Check the accuracy of the Runge–Kutta method given in eqns (5.4.13) when applied to $\dot{z} = -\lambda z$.

5.5. The Hairer (1980) method in its original form uses $b = b_1 g_1 + b_2 g_2$ in eqns (5.4.34) and (5.4.35). Find the conditions on b_1, b_2, c for the method to be (a) unconditionally stable and (b) second order accurate when applied to $\dot{y} = -\lambda y$.

5.6. Test the stability of the Trujillo method in eqns (5.7.10) on the scalar equation $m\ddot{x} + kx = 0$ by putting $K_U = K_L = \frac{1}{2}k$. Eliminate $x_{n+\frac{1}{2}}$ and $v_{n+\frac{1}{2}}$, put $v_{n+1} = rv_n$ and $x_{n+1} = rx_n$ and eliminate x_n and v_n to obtain the stability polynomial on which to use the Routh–Hurwitz conditions.

References

Addison, C. A. and Gladwell, I. (1984). Second derivative methods applied to implicit first and second order systems. *International Journal of Numerical Methods in Engineering*, **20**, 1211–1231.

Argyris, J. H., Dunne, P. C., and Angelopoulos, T. (1973a). Non-linear oscillations using the finite element technique. *Computer Methods in Applied Mechanics and Engineering*, **2**, 203–250.

Argyris, J. H., Dunne, P. C., and Angelopoulos, T. (1973b). Dynamic response by large step integration. *Earthquake Engineering and Structural Dynamics*, **2**, 185–203.

Argyris, J. H., Vaz, L. E., and Willam, K. J. (1977). Higher order methods for transient diffusion analysis. *Computational Methods in Applied Mechanics and Engineering*, **12**, 243–278.

Ashford, R. A. and Wood, W. L. (1978). Numerical integration of the motions of a tethered buoyant platform, *International Journal of Numerical Methods in Engineering*, **13**, 165–180.

Bear, J. (1979). *Hydraulics of groundwater*. McGraw-Hill.

Belytschko, T. (1983). An overview of semidiscretization and time integration procedures. In *Computational methods for transient analysis* (eds. T. Belytschko and T. J. R. Hughes), Ch. 1. North-Holland.

Belytschko, T. and Mullen, R. (1978). Stability of explicit–implicit mesh partitions in time integration. *International Journal of Numerical Methods in Engineering*, **12**, 1575–1586.

Butcher, J. C. (1964). Implicit Runge–Kutta processes. *Mathematics of Computation,* **18,** 50–64.

Brusa, L. and Nigro, L. (1980). A one-step method for direct integration of structural dynamic equations. *International Journal of Numerical Methods in Engineering,* **15,** 685–699.

Calahan, D. A. (1968). A stable accurate method of numerical integration for non-linear systems. *Proceeds of IEEE,* **56,** 744.

Conte, S. D. and de Boor, C. (1980). *Elementary numerical analysis,* McGraw-Hill.

Crank, J. and Nicolson, P. (1947). A practical method for numerical integration of solutions of partial differential equations of heat conduction type. *Proceedings of the Cambridge Philosophical Society,* **43,** 50–67.

Dahlquist, G. G. (1963). A special stability problem for linear multistep methods. *BIT*, **3,** 27–43.

Dans, A. D. and Frind, E. O. (1985). An alternating direction Galerkin technique for simulation of contaminant transport in complex groundwater systems. *Water Resources Research,* **21,** 653–664.

Dekker, K. and Verwer, J. G. (1984). *Stability of Runge–Kutta methods for stiff non-linear equations.* North-Holland.

Dennis, J. E., Jr. and Moré, J. J. (1977). Quasi-Newton methods, motivation and theory. *SIAM Review,* **19,** 46–89.

Enright, W. H. (1974). Optimal second derivative methods for stiff systems. In *Stiff differential systems* (ed. R. A. Willoughby). Plenum Press, New York.

Enright, W. H. (1980). On the efficient time integration of systems of second order equations arising in structural dynamics. *International Journal of Numerical Methods in Engineering,* **16,** 13–18.

Faddeeva, V. N. (1959). *Computational methods of linear algebra* (trans. C. D. Benster). Dover.

Fox, L. (1964). *Introduction to numerical linear algebra.* Oxford.

Gauss, C. F. (1823). Letter to Gerling. *Werke,* **9,** 278–291.

Gellert, M. (1978). A new algorithm for integration of dynamic systems. *Computers and Structures,* **9,** 401–408.

Gilvary, B., Gladwell, I., Smith, I. M., and Wong, S. W. (1987). The element by element method for linear systems. *Numerical Analysis Report no. 144*. University of Manchester, Department of Mathematics.

Hairer, E. (1980). Unconditionally stable explicit methods for parabolic equations. *Numerische Mathematik,* **35,** 57–68.

Henrici, P., (1962). *Discrete variable methods in ordinary differential equations.* Wiley.

Heun, K. (1900). Neue Methode zur approximativen Integration der

Differentialgleichungen einer unabhängigen Veränderlichen. *Zeitschrift für Mathematisch Physik,* **45,** 23–38.

Hughes, T. J. R. (1983). Analysis of transient algorithms with particular reference to stability behavior. In *Computational methods for transient analysis* (eds. T. Belytschko and T. J. R. Hughes), Ch. 2. North-Holland.

Hughes, T. J. R. and Belytschko, T. (1983). A precis of developments in computational methods for transient analysis. *Journal of Applied Mechanics,* **50,** 1033–1041.

Hughes, T. J. R. and Liu, W. K. (1978). Implicit–explicit finite elements in transient analysis: stability theory. *Journal of Applied Mechanics,* **45,** 371–374.

Hughes, T. J. R., Levit, I., and Winget, J. (1983a). Element by element implicit algorithms for heat conduction. *Journal of Engineering Mechanics,* **109,** 576–585.

Hughes, T. J. R., Levit, I., and Winget, J. (1983b). An element by element solution algorithm for problems of structural and solid mechanics. *Computational Methods in Applied Mechanics and Engineering,* **36,** 241–254.

Irons, B. M. (1970). Applications of a theorem on eigenvalues to finite element problems. *CR/132/70.* University of Wales, Department of Civil Engineering, Swansea.

Jacobi, C. G. J. (1884). Ueber eine neue Auflösungsart der bei der Methode der kleinsten Quadrate vorkommenden linearen Gleichungen. *Werke,* **3,** 467–478.

Kutta, W. (1901). Beitrag zur näherungsweisen Integration totaler Differentialgleichungen. *Zeitschrift für Mathematisch Physik,* **46,** 435–453.

Lambert, J. D. (1973). *Computational methods in ordinary differential equations.* Wiley.

Liu, W. K. and Zhang, Y. F. (1983). Improvement of mixed time implicit–explicit algorithms for thermal analysis of structures. *Computational Methods in Applied Mechanics and Engineering,* **37,** 207–223.

Liu, W. K., Belytschko, T., and Zhang, Y. F. (1984). Partitioned rational Runge–Kutta for parabolic systems. *International Journal of Numerical Methods in Engineering,* **20,** 1581–1597.

Makinson, G. J. (1968). Stable high order implicit methods for the numerical solution of systems of differential equations. *Computer Journal,* **11,** 305–310.

Neuman, S. P., Preller, C., and Narasimhan, T. N. (1982). Adaptive explicit–implicit quasi-three-dimensional model of flow and subsidence in multi-aquifer systems. *Water Resources Research,* **18,** 1551–1561.

Newmark, N. M., (1959). A method of computation for structural dynamics. *ASCE,* EM3, **85,** 67–94.

Nørsett, S. P. (1974a). One-step methods of Hermite type for numerical solution of stiff systems. *BIT,* **14,** 63–77.

Nørsett, S. P. (1974b). Semi-explicit Runge–Kutta methods. *Report Mathematics and Computation no. 6/74*. Department of Mathematics, University of Trondheim.

Ortiz, M., Pinsky, P. M., and Taylor, R. L. (1983). Unconditionally stable element by element algorithms for dynamic problems. *Computer Methods in Applied Mechanics and Engineering,* **36,** 223–229.

Padé, H. (1892). Sur la représentation approchée d'une fonction par des fractions rationnelles. *Thèse, Ann. de l'Ec. Nor.,* (3) **9,** 93pp.

Park, K. C. (1982). An improved semi-implicit method for structural dynamic analysis. *ASME Journal of Applied Mechanics,* **49,** 589–593.

Park, K. C. and Felippa, C. A. (1983). Partitioned analysis of coupled systems. In *Computational methods for transient analysis* (eds. T. Belytschko and T. J. R. Hughes), Ch. 3. North-Holland.

Richtmyer, R. D. and Morton, K. W. (1967). *Difference methods for initial value problems.* Wiley.

Runge, C. (1895). Über die numerische Auflösung von Differentialgleichungen. *Mathematische Annalen,* **46,** 167–178.

Saul'yev, V. K. (1964). *Integration of equations of parabolic type by the method of nets.* Pergamon Press.

Seidel, L. (1874). Ueber ein Verfahren, die Gleichungen, auf welche die Methode der kleinsten Quadrate führt, sowie lineare Gleichungen überhaupt, durch successive Annäherung aufzulösen. *Bayer Akademie, München,* **11,** 83–108.

Seward, W. L., Fairweather, G., and Johnson, R. L. (1984). A survey of higher order methods for the numerical integration of semi-discrete parabolic problems. *IMAJNA,* **4,** 375–425.

Siemieniuch, J. L. (1975). Properties of time-discretizations of Nørsett type for linear time-dependent partial differential equations. University of Manchester, Department of Mathematics. *Numerical Analysis Report no. 9.*

Smith, I. M. (1975). Some time-dependent soil-structure interaction problems. In *Finite elements in geomechanics* (ed. G. Gudehus), Ch. 8. Wiley.

Smith, I. M. (1977). Integration in time of diffusion and diffusion–convection equations. In *Finite elements in water resources* (eds. W. G. Gray, G. F. Pinder, and C. A. Brebbia). Pentech Press.

Smith, I. M., (1978) Transient phenomena of offshore foundations. In *Numerical methods in offshore engineering,* Ch. 15. Wiley.

Smith, I. M., Siemieniuch, J. L., and Gladwell, I. (1977). Evaluation of

Nørsett methods for integrating differential equations in time. *International Journal of Numerical and Analytical Methods in Geomechanics*, **1**, 57–74.

Sottas, G. (1984). Rational Runge–Kutta methods are not suitable for stiff systems of ODEs. *Journal of Computation and Applied Mathematics*, **10**, 169–174.

Southwell, R. V. (1946). *Relaxation methods in theoretical physics*. Clarendon Press, Oxford.

Thomas, R. M. and Gladwell, I. (1984). The methods of Gellert and of Brusa and Nigro are Padé approximant methods. *International Journal of Numerical Methods in Engineering*, **20**, 1307–1322.

Trujillo, D. M., (1975). Direct numerical integration of linear matrix differential equations using Padé approximation. *International Journal of Numerical Methods in Engineering*, **9**, 259–270.

Trujillo, D. M. (1977a). An unconditionally stable explicit algorithm for finite element heat conduction analysis. *Nuclear Engineering and Design*, **41**, 175–180.

Trujillo, D. M. (1977b). An unconditionally stable explicit algorithm for structural dynamics. *International Journal of Numerical Methods in Engineering*, **11**, 1579–1592.

Varga, R. S. (1962). *Matrix iterative analysis*. Prentice-Hall.

Wanner, G. (1987). Order stars and stability. In *The state of the art in numerical analysis* (eds. A. Iserles and M. J. D. Powell). Clarendon Press, Oxford.

Wood, W. L. (1977). On the Zienkiewicz four-time-level scheme for the numerical integration of vibration problems. *International Journal of Numerical Methods in Engineering*, **11**, 1519–1528.

Zienkiewicz, O. C. (1977). A new look at the Newmark, Houbolt and other time stepping formulas. A weighted residual approach. *Earthquake Engineering and Structural Dynamics*, **5**, 413–418.

Zienkiewicz, O. C. and Lewis, R. W. (1973). An analysis of various time-stepping schemes for initial value problems. *Earthquake Engineering and Structural Dynamics*, **1**, 407–408.

6 The energy method

6.1. Introduction

The basic idea of the Energy Method is to formulate an 'energy norm' which is a measure of the solution given by an algorithm and which must not increase. This energy norm then has two particular uses:

(1) to give conditions for stability which do not depend on the assumptions of modal decomposition which had to be made in earlier chapters;

(2) to give a criterion for the control of the time step in non-linear problems.

With the system of equations representing the dynamic vibration problem the obvious energy norm to try is that approximating the actual physical energy of the problem, i.e. the sum of the kinetic and potential energies together with the effect of the work done. We shall illustrate this with examples. The expression for the energy depends on the algorithm; some algorithms lend themselves to analysis by the energy method more easily than others.

As used by Richtmyer and Morton (1967) with various hyperbolic and parabolic finite difference schemes, the aim of the energy method is to show that this energy norm increases by a factor no greater than $1+\mathrm{O}(\Delta t)$ at each time step. This means that if u_n is the numerical solution at time $n\Delta t$ and $\|u_n\|_{\mathrm{E}}$ is the energy norm, then

$$\|u_n\|_{\mathrm{E}} \leq (1+c\Delta t)\,\|u_{n-1}\|_{\mathrm{E}}$$

for sufficiently small Δt, i.e.

$$\|u_n\|_{\mathrm{E}} \leq (1+c\Delta t)^n\,\|u_0\|_{\mathrm{E}}. \tag{6.1.1}$$

Then since we have $(1+c\Delta t)^n \leq \exp(cn\Delta t) = \exp(cT)$, which is constant for fixed $n\Delta t = T$, we know that the energy norm $\|u_n\|_{\mathrm{E}}$ is bounded and hence the solution is bounded. In the problems considered in the present text, the usual aim is simply to show that the energy norm satisfies the condition

$$\|u_n\|_{\mathrm{E}} \leq \|u_{n-1}\|_{\mathrm{E}}, \tag{6.1.2}$$

i.e. that the energy is not increasing and hence the solution is bounded. We also have to show the equivalence of this energy norm to a standard norm, taken here to be the l_2 norm given by $\|\boldsymbol{u}\|^2 = \boldsymbol{u}^{\mathrm{T}}\boldsymbol{u}$. We thus have two basic steps to the energy method and conditions on the stability may come from one or the other.

The use of energy methods with finite difference equations begins with the classic paper by Courant *et al.* (1928). The regular pattern of finite difference methods leads to a number of neat expressions for the 'energy', depending on the finite difference scheme but the construction of these energy expressions for abstract problems is often quite tricky and students are often baffled to think how it can be done.

For the θ-method applied to the linear first order parabolic system of equations

$$\boldsymbol{M}\dot{\boldsymbol{x}} + \boldsymbol{K}\boldsymbol{x} = \boldsymbol{f}(t) \tag{6.1.3}$$

the energy method is quite simple. (We note that in the context of the energy method the mass matrix $\boldsymbol{M}$ is always symmetric and positive definite and may or may not be diagonal depending on the algorithm being investigated. The stiffness matrix $\boldsymbol{K}$ is always symmetric and may be specified as positive definite or may be allowed to be positive semi-definite.) For the θ-method applied to eqn (6.1.3) we take the typical difference equation

$$\boldsymbol{M}(\boldsymbol{x}_{n+1} - \boldsymbol{x}_n) + \Delta t\boldsymbol{K}\{\theta\boldsymbol{x}_{n+1} + (1-\theta)\boldsymbol{x}_n\} = \Delta t\boldsymbol{f}_{n+\theta}, \tag{6.1.4}$$

where $\boldsymbol{f}_{n+\theta}$ is either $\boldsymbol{f}((n+\theta)\Delta t)$ or $\theta\boldsymbol{f}_{n+1} + (1-\theta)\boldsymbol{f}_n$, and write it in terms of the undivided forward difference

$$\Delta\boldsymbol{x}_n = \boldsymbol{x}_{n+1} - \boldsymbol{x}_n, \tag{6.1.5a}$$

and the mean value over the time step

$$\langle\boldsymbol{x}_n\rangle = (\boldsymbol{x}_{n+1} + \boldsymbol{x}_n)/2 \tag{6.1.5b}$$

as

$$\boldsymbol{M}\Delta\boldsymbol{x}_n + \Delta t\boldsymbol{K}\{(\theta - 0.5)\Delta\boldsymbol{x}_n + \langle\boldsymbol{x}_n\rangle\} = \Delta t\boldsymbol{f}_{n+\theta}. \tag{6.1.6}$$

Premultiplying eqn (6.1.6) by $\langle \boldsymbol{x}_n \rangle^{\mathsf{T}}$ then gives

$$\Delta \boldsymbol{x}_n^{\mathsf{T}}\{\boldsymbol{M} + (\theta - 0.5)\Delta t\boldsymbol{K}\}\langle \boldsymbol{x}_n \rangle + \Delta t \langle \boldsymbol{x}_n \rangle^{\mathsf{T}} \boldsymbol{K} \langle \boldsymbol{x}_n \rangle = \Delta t \langle \boldsymbol{x}_n \rangle^{\mathsf{T}} \boldsymbol{f}_{n+\theta}. \tag{6.1.7}$$

Thus we can identify a possible 'energy norm'

$$E_n = \tfrac{1}{2}\boldsymbol{x}_n^{\mathsf{T}}\{\boldsymbol{M} + (\theta - 0.5)\Delta t\boldsymbol{K}\}\boldsymbol{x}_n, \tag{6.1.8}$$

which is legitimate, that is, it is equivalent to the l_2 norm whose square is $\boldsymbol{x}_n^{\mathsf{T}}\boldsymbol{x}_n$, if the matrix $\{\boldsymbol{M} + (\theta - 0.5)\Delta t\boldsymbol{K}\}$ is positive definite. This is true for all values of Δt if either

(1) $\theta > 0.5$ and $\boldsymbol{K}$ is positive definite; there may be difficulties if $\boldsymbol{K}$ is only positive semi-definite and very large values of Δt are used; or

(2) $\theta = 0.5$.

We then have from eqn (6.1.7)

$$E_{n+1} = E_n - \Delta t \langle \boldsymbol{x}_n \rangle^{\mathsf{T}} \boldsymbol{K} \langle \boldsymbol{x}_n \rangle + \Delta t \langle \boldsymbol{x}_n \rangle^{\mathsf{T}} \boldsymbol{f}_{n+\theta}. \tag{6.1.9}$$

The last term on the right-hand side of eqn (6.1.9) shows the approximation to the work done by the forcing function in the time step. The stability of the algorithm is shown by applying it to the homogeneous form of equation (6.1.3) with $\boldsymbol{f} = \boldsymbol{0}$. Equation (6.1.9) shows that the energy E_n does not increase; we expect a homogeneous parabolic problem to have a decaying solution. The stability condition $\theta \geq 0.5$ in this case comes from the condition that the expression for E_n in eqn (6.1.8) is legitimate as a norm.

In §§6.2, 6.3 we discuss the energy method in connection with the Newmark algorithm when we start by using the sum of the approximations to the kinetic and potential energies and then extend this idea. In §6.4 we discuss the energy method with the SS22 algorithm which is simpler than Newmark to manipulate. In §6.5 we demonstrate the use of the energy method with non-linear problems. Non-linear problems are further considered in §6.6 with reference to the energy methods of Hughes and co-authors. Section 6.7 concludes the chapter with some more general uses of the energy method.

6.2. Newmark and the energy method

We demonstrate the energy method on a simple example of the Newmark algorithm applied to the linear system of equations

$$\boldsymbol{M}\ddot{\boldsymbol{x}} + \boldsymbol{K}\boldsymbol{x} = \boldsymbol{0}. \tag{6.2.1}$$

We can think of this as representing a finite element discretization of an elastic continuum or a structure with no friction and no external force. Hence this is a constant energy situation and by multiplying eqn (6.2.1) by the transpose of the velocity vector, $\dot{\boldsymbol{x}}^{\mathrm{T}}$, and integrating with respect to the time t, we obtain the equation

$$\tfrac{1}{2}\dot{\boldsymbol{x}}^{\mathrm{T}}\boldsymbol{M}\dot{\boldsymbol{x}} + \tfrac{1}{2}\boldsymbol{x}^{\mathrm{T}}\boldsymbol{K}\boldsymbol{x} = \tfrac{1}{2}\boldsymbol{v}_0^{\mathrm{T}}\boldsymbol{M}\boldsymbol{v}_0 + \tfrac{1}{2}\boldsymbol{x}_0^{\mathrm{T}}\boldsymbol{K}\boldsymbol{x}_0, \tag{6.2.2}$$

where $\boldsymbol{x}_0$, $\boldsymbol{v}_0$, are the initial displacement and velocity vectors respectively.

This equation suggests that we should now use the Newmark equations (Chapter 3) to obtain expressions for the change in a time step of the approximations to the kinetic energy

$$T_n = \tfrac{1}{2}\boldsymbol{v}_n^{\mathrm{T}}\boldsymbol{M}\boldsymbol{v}_n, \tag{6.2.3}$$

and the potential energy

$$U_n = \tfrac{1}{2}\boldsymbol{x}_n^{\mathrm{T}}\boldsymbol{K}\boldsymbol{x}_n.$$

For convenience the Newmark equations are written again here:

$$\boldsymbol{x}_{n+1} = \boldsymbol{x}_n + \Delta t\boldsymbol{v}_n + \Delta t^2\{\beta\boldsymbol{a}_{n+1} + (0.5-\beta)\boldsymbol{a}_n\}$$

and

$$\boldsymbol{v}_{n+1} = \boldsymbol{v}_n + \gamma\Delta t\boldsymbol{a}_{n+1} + (1-\gamma)\Delta t\boldsymbol{a}_n. \tag{6.2.4}$$

The first step is to express these two equations entirely in terms of sums and differences. Again using the notation $\Delta\boldsymbol{a}_n = \boldsymbol{a}_{n+1} - \boldsymbol{a}_n$ and $\langle\boldsymbol{a}_n\rangle = (\boldsymbol{a}_{n+1} + \boldsymbol{a}_n)/2$, we can express the second of eqns (6.2.4) as

$$\Delta\boldsymbol{v}_n = \Delta t\langle\boldsymbol{a}_n\rangle + (\gamma - 0.5)\Delta t\Delta\boldsymbol{a}_n \tag{6.2.5}$$

and taking $0.5\Delta t$ times eqn (6.2.5) minus the first of eqns (6.2.4) gives

$$\Delta t\langle\boldsymbol{v}_n\rangle = \Delta\boldsymbol{x}_n - \Delta t^2(\beta - 0.5\gamma)\Delta\boldsymbol{a}_n. \tag{6.2.6}$$

In order to obtain the change in kinetic energy in the time step we want

$$\Delta T_n = T_{n+1} - T_n = \tfrac{1}{2}(\boldsymbol{v}_n + \boldsymbol{v}_{n+1})^{\mathsf{T}}\boldsymbol{M}(\boldsymbol{v}_{n+1} - \boldsymbol{v}_n)$$

(since $\boldsymbol{M}$ is symmetric)

$$= \langle \boldsymbol{v}_n \rangle^{\mathsf{T}}\boldsymbol{M}\Delta\boldsymbol{v}_n \tag{6.2.7}$$

and for the change in potential energy in the time step we want

$$\Delta U_n = U_{n+1} - U_n = \langle \boldsymbol{x}_n \rangle^{\mathsf{T}}\boldsymbol{K}\Delta\boldsymbol{x}_n. \tag{6.2.8}$$

We thus have

$$\Delta T_n + \Delta U_n = \langle \boldsymbol{v}_n \rangle^{\mathsf{T}}\boldsymbol{M}\Delta\boldsymbol{v}_n + \langle \boldsymbol{x}_n \rangle^{\mathsf{T}}\boldsymbol{K}\Delta\boldsymbol{x}_n, \tag{6.2.9}$$

and substituting for $\Delta\boldsymbol{v}_n$ and $\langle \boldsymbol{v}_n \rangle$ from eqns (6.2.5) and (6.2.6) makes the right-hand side of eqn (6.2.9) equal to

$$\begin{aligned}
&[\Delta\boldsymbol{x}_n - \Delta t^2(\beta - 0.5\gamma)\Delta\boldsymbol{a}_n]^{\mathsf{T}}\boldsymbol{M}[\langle \boldsymbol{a}_n \rangle + (\gamma - 0.5)\Delta\boldsymbol{a}_n] \\
&\quad + \langle \boldsymbol{x}_n \rangle^{\mathsf{T}}\boldsymbol{K}\Delta\boldsymbol{x}_n \\
&= -\Delta t^2(\beta - 0.5\gamma)\langle \boldsymbol{a}_n \rangle^{\mathsf{T}}\boldsymbol{M}\Delta\boldsymbol{a}_n \\
&\quad - (\gamma - 0.5)\Delta\boldsymbol{x}_n^{\mathsf{T}}\boldsymbol{K}\Delta\boldsymbol{x}_n - \Delta t^2(\beta - 0.5\gamma) \\
&\quad \times (\gamma - 0.5)\Delta\boldsymbol{a}_n^{\mathsf{T}}\boldsymbol{M}\Delta\boldsymbol{a}_n,
\end{aligned}$$

using the Newmark collocation of the differential equation: $\boldsymbol{M}\boldsymbol{a}_j + \boldsymbol{K}\boldsymbol{x}_j = 0$, $j = n, (n+1)$. We now have the result

$$\begin{aligned}
&\Delta T_n + \Delta U_n + \Delta t^2(\beta - 0.5\gamma)\Delta A_n + (\gamma - 0.5)\Delta\boldsymbol{x}_n^{\mathsf{T}}\boldsymbol{K}\Delta\boldsymbol{x}_n \\
&\qquad + (\beta - 0.5\gamma)(\gamma - 0.5)\Delta t^2\Delta\boldsymbol{a}_n^{\mathsf{T}}\boldsymbol{M}\Delta\boldsymbol{a}_n = 0,
\end{aligned} \tag{6.2.10}$$

where

$$A_n = \tfrac{1}{2}\boldsymbol{a}_n^{\mathsf{T}}\boldsymbol{M}\boldsymbol{a}_n. \tag{6.2.11}$$

Equation (6.2.10) suggests that we take our energy norm to be

$$\hat{E}_n = \tfrac{1}{2}\boldsymbol{v}_n^{\mathsf{T}}\boldsymbol{M}\boldsymbol{v}_n + \tfrac{1}{2}\boldsymbol{x}_n^{\mathsf{T}}\boldsymbol{K}\boldsymbol{x}_n + \tfrac{1}{2}\Delta t^2(\beta - 0.5\gamma)\boldsymbol{a}_n^{\mathsf{T}}\boldsymbol{M}\boldsymbol{a}_n. \tag{6.2.12}$$

We know that the mass matrix $\boldsymbol{M}$ is positive definite and we assume here that the stiffness matrix $\boldsymbol{K}$ is also positive definite. The expression (6.2.12) is legitimate for a norm then if

(1) we have $\beta \geq 0.5\gamma$ so that the matrix $(\beta - 0.5\gamma)\boldsymbol{M}$ is positive semi-definite; and

(2) we can show that the norm $\hat{E}_n$ is equivalent to a standard norm of the solution vector.

This is a situation where it is useful to recall the different ways of regarding the solution obtained by the Newmark method described in Chapter 3. If $\beta > 0.5\gamma$, we call the solution vector $\boldsymbol{X}_n = [\boldsymbol{x}_n, \boldsymbol{v}_n, \boldsymbol{a}_n]^\mathrm{T}$ and then

$$\hat{E}_n = \tfrac{1}{2}\boldsymbol{X}_n^\mathrm{T}\boldsymbol{D}\boldsymbol{X}_n, \tag{6.2.13}$$

where $\boldsymbol{D}$ is the block diagonal matrix whose blocks are the symmetric positive definite matrices $\boldsymbol{K}$, $\boldsymbol{M}$, and $(\beta - 0.5\gamma)\Delta t^2\boldsymbol{M}$ respectively. Then we have

$$0 < \lambda_{\min}\boldsymbol{X}_n^\mathrm{T}\boldsymbol{X}_n \le \hat{E}_n = \tfrac{1}{2}\boldsymbol{X}_n^\mathrm{T}\boldsymbol{D}\boldsymbol{X}_n \le \lambda_{\max}\boldsymbol{X}_n^\mathrm{T}\boldsymbol{X}_n, \tag{6.2.14}$$

where $\lambda_{\min}$, $\lambda_{\max}$ are the minimum and maximum eigenvalues respectively of the matrix $\frac{1}{2}\boldsymbol{D}$. Thus the energy norm is equivalent to the standard l_2 norm of the solution. We note that it is important here that the stiffness matrix is positive definite and not merely positive semi-definite. If $\beta = 0.5\gamma$ then we take the solution vector as $\boldsymbol{Y}_n = [\boldsymbol{x}_n, \boldsymbol{v}_n]$ and the energy norm is now given by

$$E_n = \tfrac{1}{2}\boldsymbol{v}_n^\mathrm{T}\boldsymbol{M}\boldsymbol{v}_n + \tfrac{1}{2}\boldsymbol{x}_n^\mathrm{T}\boldsymbol{K}\boldsymbol{x}_n. \tag{6.2.15}$$

A relationship like the one in (6.2.14) can again be found with the λ's now the minimum and maximum eigenvalues of the block diagonal matrix whose blocks are the matrices $\frac{1}{2}\boldsymbol{M}$ and $\frac{1}{2}\boldsymbol{K}$. The expression for the energy in (6.2.15) is of the form we expect from the physical energy of the dynamic vibration problem. Returning to eqn (6.2.10) and the general case with $\beta > 0.5\gamma$, we have

$$\begin{aligned}\hat{E}_{n+1} = \hat{E}_n &- (\gamma - 0.5)\Delta\boldsymbol{x}_n^\mathrm{T}\boldsymbol{K}\Delta\boldsymbol{x}_n \\ &- (\beta - 0.5\gamma)(\gamma - 0.5)\Delta t^2\Delta\boldsymbol{a}_n^\mathrm{T}\boldsymbol{M}\Delta\boldsymbol{a}_n.\end{aligned} \tag{6.2.16}$$

With $\beta > 0.5\gamma$ we know that the energy norm is equivalent to a standard norm and we also need to have $\gamma \ge 0.5$ for the energy norm to be constant or decreasing, i.e. for the scheme to be stable, because then from eqn (6.2.16) we have $\hat{E}_{n+1} \le \hat{E}_n$. If $\gamma = 0.5$ we have a constant energy situation like that in the physical problem with $\hat{E}_{n+1} = \hat{E}_n$; using a value of $\gamma > 0.5$ introduces algorithmic damping and the energy norm $\hat{E}_n$ is decreasing at each step. With $\gamma = 0.5$ we have $\hat{E}_n =$ constant but if we want $E_n = T_n + U_n$ to remain constant we must take $2\beta = \gamma$ also, that is, $\beta = \frac{1}{4}$. We then have the trapezium rule equivalent given by $2\beta=\gamma=0.5$ as shown in Chapter 3. These conditions for unconditional stability are the same as those obtained in Chapter 3 but without any modal decomposition.

We now demonstrate another approach to the energy method used on the Newmark algorithm applied to the system of differential equations which retains the damping matrix $\boldsymbol{C}$, that is.

$$\boldsymbol{M}\ddot{\boldsymbol{x}}+\boldsymbol{C}\dot{\boldsymbol{x}}+\boldsymbol{K}\boldsymbol{x}=\boldsymbol{0}. \tag{6.2.17}$$

Now multiplying by the transpose of the velocity vector and integration with respect to time brings in the extra term

$$\int_0^t \dot{\boldsymbol{x}}^{\mathrm{T}}\boldsymbol{C}\dot{\boldsymbol{x}}\,\mathrm{d}t,$$

representing the work done against friction.

In order to construct the expression for the energy we multiply eqn (6.2.6) by the matrix $\boldsymbol{K}$ and substitute for the $\boldsymbol{K}\boldsymbol{x}_n$ terms from the collocated differential equation to give

$$\Delta t\boldsymbol{K}\langle\boldsymbol{v}_n\rangle+\boldsymbol{C}\Delta\boldsymbol{v}_n+\{\boldsymbol{M}+\Delta t^2(\beta-0.5\gamma)\boldsymbol{K}\}\Delta\boldsymbol{a}_n=\boldsymbol{0}.$$

Then multiplying by the appropriate terms from eqn (6.2.5) we have

$$\begin{aligned}\Delta t\langle\boldsymbol{v}_n\rangle^{\mathrm{T}}\boldsymbol{K}\Delta\boldsymbol{v}_n&+\Delta t\langle\boldsymbol{a}_n\rangle^{\mathrm{T}}\{\boldsymbol{M}+\Delta t^2(\beta-0.5\gamma)\boldsymbol{K}\}\Delta\boldsymbol{a}_n\\&+\Delta\boldsymbol{v}_n^{\mathrm{T}}\boldsymbol{C}\Delta\boldsymbol{v}_n+\Delta t(\gamma-0.5)\Delta\boldsymbol{a}_n^{\mathrm{T}}\{\boldsymbol{M}+\Delta t^2(\beta-0.5\gamma)\boldsymbol{K}\}\Delta\boldsymbol{a}_n=0.\end{aligned} \tag{6.2.18}$$

We can now introduce an energy norm

$$\overset{*}{E}_n=\tfrac{1}{2}\boldsymbol{v}_n^{\mathrm{T}}\boldsymbol{K}\boldsymbol{v}_n+\tfrac{1}{2}\boldsymbol{a}_n^{\mathrm{T}}\{\boldsymbol{M}+\Delta t^2(\beta-0.5\gamma)\boldsymbol{K}\}\boldsymbol{a}_n, \tag{6.2.19}$$

which is legitimate by a similar argument to that used for $\hat{E}_n$ if the stiffness matrix $\boldsymbol{K}$ is positive definite and $\beta\geq 0.5\gamma$. Then from eqn (6.2.18) we have

$$\begin{aligned}\overset{*}{E}_{n+1}=\overset{*}{E}_n&-\Delta\boldsymbol{v}_n^{\mathrm{T}}\boldsymbol{C}\Delta\boldsymbol{v}_n/\Delta t-(\gamma-0.5)\Delta\boldsymbol{a}_n^{\mathrm{T}}\\&\times\{\boldsymbol{M}+\Delta t^2(\beta-0.5\gamma)\boldsymbol{K}\}\Delta\boldsymbol{a}_n.\end{aligned} \tag{6.2.20}$$

Thus although the energy norm $\overset{*}{E}_n$ does not look like the physical energy we do now have the effect of the damping matrix represented and again with $2\beta \geq \gamma \geq 0.5$ and a positive semi-definite matrix $\boldsymbol{C}$ we have stability with $\overset{*}{E}_{n+1} \leq \overset{*}{E}_n$. We particularly note that these results are obtained here without the assumptions about the damping matrix (for example with Rayleigh damping) necessary when making a modal decomposition.

We can also use eqns (6.2.19) and (6.2.20) to obtain without modal decomposition the stability conditions for the popular explicit central difference method given by putting $\beta = 0$ and $\gamma = 0.5$ in the Newmark equations. From eqn (6.2.19) the norm $\overset{*}{E}_n$ is now legitimate if the matrix

$$\boldsymbol{M} - \tfrac{1}{4}\Delta t^2 \boldsymbol{K}$$

is positive definite for all values of the time step Δt. This means that we must have

$$\omega^2 \Delta t^2 < 4,$$

where ω^2 is the highest eigenvalue of $\boldsymbol{M}^{-1}\boldsymbol{K}$, which is the same result as that obtained with the scalar equation in Chapter 2.

6.3. The energy method with Newmark including natural damping and forcing terms

We now consider the system of equations representing a structure with damping proportional to the velocity and an external forcing function $\boldsymbol{f}(t)$

$$\boldsymbol{M}\ddot{\boldsymbol{x}} + \boldsymbol{C}\dot{\boldsymbol{x}} + \boldsymbol{K}\boldsymbol{x} = \boldsymbol{f}(t), \tag{6.3.1}$$

where the matrix $\boldsymbol{C}$ is symmetric and positive definite or semi-definite. Multiplying by $\dot{\boldsymbol{x}}^{\mathrm{T}}$ and integrating with respect to the time we have

$$\tfrac{1}{2}\dot{\boldsymbol{x}}^{\mathrm{T}}\boldsymbol{M}\dot{\boldsymbol{x}} + \tfrac{1}{2}\boldsymbol{x}^{\mathrm{T}}\boldsymbol{K}\boldsymbol{x} = \tfrac{1}{2}\boldsymbol{v}_0^{\mathrm{T}}\boldsymbol{M}\boldsymbol{v}_0 + \tfrac{1}{2}\boldsymbol{x}_0^{\mathrm{T}}\boldsymbol{K}\boldsymbol{x}_0 - \int_0^t \dot{\boldsymbol{x}}^{\mathrm{T}}\boldsymbol{C}\dot{\boldsymbol{x}}\,\mathrm{d}t + \int_0^t \dot{\boldsymbol{x}}^{\mathrm{T}}\boldsymbol{f}(t)\,\mathrm{d}t. \tag{6.3.2}$$

When we use the Newmark algorithm with the forcing function absent we know from eqn (6.2.20) that the natural damping will cause the energy to decrease and hence we can say that we do not need any algorithmic damping and we take the Newmark parameter γ to be equal to 0.5. The Newmark equations (6.2.5) and (6.2.6) now become:

$$\Delta \boldsymbol{v}_n = \Delta t \langle \boldsymbol{a}_n \rangle \tag{6.3.3}$$

$$\Delta t \langle \boldsymbol{v}_n \rangle = \Delta \boldsymbol{x}_n - \Delta t^2 (\beta - 0.25) \Delta \boldsymbol{a}_n \tag{6.3.4}$$

and

$$\boldsymbol{M}\boldsymbol{a}_j + \boldsymbol{C}\boldsymbol{v}_j + \boldsymbol{K}\boldsymbol{x}_j = \boldsymbol{f}_j, \qquad j = n, (n+1), \tag{6.3.5}$$

where $\boldsymbol{f}_j = \boldsymbol{f}(j\Delta t)$. Using T_n and U_n as defined in §6.2, we have:

$$\Delta T_n = \Delta \boldsymbol{v}_n^{\mathsf{T}} \boldsymbol{M} \langle \boldsymbol{v}_n \rangle \tag{6.3.6}$$

and

$$\Delta U_n = \Delta \boldsymbol{x}_n^{\mathsf{T}} \boldsymbol{K} \langle \boldsymbol{x}_n \rangle . \tag{6.3.7}$$

We can now substitute

$$\boldsymbol{K} \langle \boldsymbol{x}_n \rangle = -\boldsymbol{C} \langle \boldsymbol{v}_n \rangle - \boldsymbol{M} \langle \boldsymbol{a}_n \rangle + \langle \boldsymbol{f}_n \rangle \tag{6.3.8}$$

and we obtain, using again $A_n = \frac{1}{2} \boldsymbol{a}_n^{\mathsf{T}} \boldsymbol{M} \boldsymbol{a}_n$,

$$\Delta T_n + \Delta U_n + (\beta - 0.25) \Delta t^2 \Delta A_n = -\Delta \boldsymbol{x}_n^{\mathsf{T}} \boldsymbol{C} \langle \boldsymbol{v}_n \rangle - \Delta \boldsymbol{x}_n^{\mathsf{T}} \langle \boldsymbol{f}_n \rangle . \tag{6.3.9}$$

If $\beta > 0.25$, we can take $T_n + U_n + (\beta - 0.25)\Delta t^2 A_n$ as the energy norm and if $\beta = 0.25$ we have the approximation to the kinetic plus potential energy, $T_n + U_n$. On the right-hand side of eqn (6.3.9), we see the trapezium rule approximations to the work done in the time step against friction:

$$\int_{n\Delta t}^{(n+1)\Delta t} \dot{\boldsymbol{x}}^{\mathsf{T}} \boldsymbol{C} \dot{\boldsymbol{x}} \, \mathrm{d}t \approx \Delta \boldsymbol{x}_n^{\mathsf{T}} \boldsymbol{C} \langle \boldsymbol{v}_n \rangle \tag{6.3.10}$$

and the trapezium rule approximation to the work done by the forcing function:

$$\int_{n\Delta t}^{(n+1)\Delta t} \dot{\boldsymbol{x}}^{\mathsf{T}} \boldsymbol{f}(t) \, \mathrm{d}t \approx \Delta \boldsymbol{x}_n^{\mathsf{T}} \langle \boldsymbol{f}_n \rangle . \tag{6.3.11}$$

We know from Chapter 3 that when $\beta = \frac{1}{4}$ and $\gamma = \frac{1}{2}$ the Newmark algorithm is equivalent to the trapezium rule used on the first order form of the differential equations and hence it seems natural that the trapezium rule approximations to the integrals appear in eqns 6.3.10 and 6.3.11. This then leads us to look for trapezium rule approximations to the integrals which appear in the non-linear problems discussed in §6.5.

Fujii (1972) uses the energy method to analyse the stability of the 'Newmark-β' method (i.e. Newmark with $\gamma = 0.5$) applied to eqn (6.3.1) with the damping matrix $\boldsymbol{C} = \boldsymbol{0}$. Fujii particularly studies the effect of lumped and consistent mass matrices on the stability of the explicit scheme with $\beta = 0$ and this is discussed in §2.4.

6.4. The energy method with the SS22 algorithm

For the SS22 algorithm applied to the numerical solution of the system of equations (6.3.1), the basic equations from §4.5 are as follows:

$$\boldsymbol{x}_{n+1} = \boldsymbol{x}_n + \Delta t \boldsymbol{v}_n + \frac{\Delta t^2}{2} \boldsymbol{\alpha}_n \tag{6.4.1}$$

$$\boldsymbol{v}_{n+1} = \boldsymbol{v}_n + \Delta t \boldsymbol{\alpha}_n \tag{6.4.2}$$

and the 'weighted residual equation'

$$\boldsymbol{M\alpha}_n + \boldsymbol{C}(\boldsymbol{v}_n + \theta_1 \Delta t \boldsymbol{\alpha}_n) + \boldsymbol{K}\left(\boldsymbol{x}_n + \theta_1 \Delta t \boldsymbol{v}_n + \theta_2 \frac{\Delta t^2}{2} \boldsymbol{\alpha}_n\right) = \bar{\boldsymbol{f}}_n. \tag{6.4.3}$$

Equation (6.4.2) can be written as

$$\Delta \boldsymbol{v}_n = \Delta t \boldsymbol{\alpha}_n \tag{6.4.4}$$

and eliminating $\boldsymbol{\alpha}_n$ from eqns (6.4.1) and (6.4.2) gives

$$\Delta \boldsymbol{x}_n = \Delta t \langle \boldsymbol{v}_n \rangle. \tag{6.4.5}$$

Substituting for $\boldsymbol{\alpha}_n$ from eqn (6.4.4) into eqn (6.4.3) gives

$$\boldsymbol{M}\Delta \boldsymbol{v}_n + \Delta t \boldsymbol{C}(\boldsymbol{v}_n + \theta_1 \Delta \boldsymbol{v}_n) + \Delta t \boldsymbol{K x}_n + \Delta t^2 \theta_1 \boldsymbol{K v}_n + \theta_2 \frac{\Delta t^2}{2} \boldsymbol{K} \Delta \boldsymbol{v}_n = \Delta t \bar{\boldsymbol{f}}_n. \tag{6.4.6}$$

We next express all the $\boldsymbol{x}_n$ and $\boldsymbol{v}_n$ terms on the left-hand side of eqn (6.4.6) as sums and differences to give

$$\boldsymbol{M}\Delta \boldsymbol{v}_n + \Delta t \boldsymbol{C}\{(\theta_1 - 0.5)\Delta \boldsymbol{v}_n + \langle \boldsymbol{v}_n \rangle\} + \Delta t \boldsymbol{K}(\langle \boldsymbol{x}_n \rangle - 0.5 \Delta \boldsymbol{x}_n) + \Delta t^2 \boldsymbol{K}\{\theta_1 \langle \boldsymbol{v}_n \rangle + 0.5(\theta_2 - \theta_1)\Delta \boldsymbol{v}_n\} = \Delta t \bar{\boldsymbol{f}}_n. \tag{6.4.7}$$

Now multiplying the terms in eqn (6.4.7) by the transposes of either side of eqn (6.4.5) as appropriate, and regrouping gives

$$\langle \boldsymbol{v}_n \rangle^{\mathrm{T}} \{\boldsymbol{M} + (\theta_1 - 0.5)\Delta t \boldsymbol{C} + 0.5(\theta_2 - \theta_1)\Delta t^2 \boldsymbol{K}\} \Delta \boldsymbol{v}_n + \langle \boldsymbol{x}_n \rangle^{\mathrm{T}} \boldsymbol{K} \Delta \boldsymbol{x}_n$$
$$= -\Delta t \langle \boldsymbol{v}_n \rangle^{\mathrm{T}} \boldsymbol{C} \langle \boldsymbol{v}_n \rangle - (\theta_1 - 0.5) \Delta \boldsymbol{x}_n^{\mathrm{T}} \boldsymbol{K} \Delta \boldsymbol{x}_n + \Delta \boldsymbol{x}_n^{\mathrm{T}} \bar{\boldsymbol{f}}_n. \quad (6.4.8)$$

The energy method with SS22 is thus much simpler than with Newmark; it gives directly the energy norm to be used as

$$E_n = \tfrac{1}{2} \boldsymbol{v}_n^{\mathrm{T}} \{\boldsymbol{M} + (\theta_1 - 0.5)\Delta t \boldsymbol{C} + 0.5(\theta_2 - \theta_1)\Delta t^2 \boldsymbol{K}\} \boldsymbol{v}_n + \tfrac{1}{2} \boldsymbol{x}_n^{\mathrm{T}} \boldsymbol{K} \boldsymbol{x}_n. \quad (6.4.9)$$

We have to suppose that the matrices $\boldsymbol{M}$ and $\boldsymbol{K}$ are positive definite and the damping matrix $\boldsymbol{C}$ is positive semi-definite. The conditions on the parameters, $\theta_2 \geq \theta_1 \geq 0.5$ are sufficient to make the energy norm legitimate for all values of the time step Δt. Equation (6.4.8) can be written as

$$E_{n+1} = E_n - \Delta t \langle \boldsymbol{v}_n \rangle^{\mathrm{T}} \boldsymbol{C} \langle \boldsymbol{v}_n \rangle - (\theta_1 - 0.5) \times \Delta \boldsymbol{x}_n^{\mathrm{T}} \boldsymbol{K} \Delta \boldsymbol{x}_n + \Delta \boldsymbol{x}_n^{\mathrm{T}} \bar{\boldsymbol{f}}_n. \quad (6.4.10)$$

Hence with zero external force the energy norm is decreasing either from the effect of the natural damping or from the effect of the algorithmic damping introduced by taking $\theta_1 > 0.5$. The stability conditions are the same as those obtained from the modal analysis in Chapter 4 but here there is no assumption about the damping matrix apart from its being positive semi-definite.

6.5. The energy method with Newmark and SS22 used on non-linear problems

Belytschko and Schoeberle (1975) use the 'Newmark-β' method (Newmark with the parameter $\gamma = 0.5$) for a non-linear structural dynamics problem. Their proof of unconditional stability with $\beta = 0.25$ using the energy method applies only to non-linear material properties but they state that numerical results show unconditional stability with geometric non-linearities also. Using the notation of this text, the strain-displacement relation is given by

$$\boldsymbol{\epsilon} = \boldsymbol{B} \boldsymbol{L}^{(e)} \boldsymbol{x}, \quad (6.5.1)$$

where $\boldsymbol{L}^{(e)}$ is the connectivity matrix from the finite element space discretization (see Zienkiewicz 1977). The stresses are non-linear functions of the strains given by

$$\boldsymbol{\sigma} = \boldsymbol{\sigma}(\boldsymbol{\epsilon}). \tag{6.5.2}$$

The system of equations to be solved is expressed as

$$\boldsymbol{M}\boldsymbol{a} + \boldsymbol{P} = \boldsymbol{f}, \tag{6.5.3}$$

where $\boldsymbol{M}$ is the mass matrix, $\boldsymbol{a}$ is the vector of nodal accelerations, $\boldsymbol{f}$ is the vector of external nodal forces and $\boldsymbol{P}$ is the vector of internal nodal forces given by

$$\boldsymbol{P} = \sum_e \boldsymbol{L}^{(e)\mathrm{T}} \int_{V^{(e)}} \boldsymbol{B}^{\mathrm{T}}\boldsymbol{\sigma}\, \mathrm{d}V, \tag{6.5.4}$$

where the summation is over the elements and $V^{(e)}$ is the volume of element e.

The kinetic energy at time $t = n\Delta t$ is taken as

$$T_n = \tfrac{1}{2}\boldsymbol{v}_n^{\mathrm{T}}\boldsymbol{M}\boldsymbol{v}_n \tag{6.5.5}$$

and the internal energy is defined as generated by the formula

$$U_{n+1} = U_n + \sum_e \int_{V^{(e)}} \Delta\boldsymbol{\epsilon}_n^{\mathrm{T}}\langle \boldsymbol{\sigma}_n \rangle\, \mathrm{d}V$$

$$= U_n + \sum_e \Delta\boldsymbol{x}_n^{\mathrm{T}}\boldsymbol{L}^{(e)\mathrm{T}} \int_{V^{(e)}} \boldsymbol{B}^{\mathrm{T}}\langle \boldsymbol{\sigma}_n \rangle\, \mathrm{d}V,$$

using eqn (6.5.1), i.e.

$$U_{n+1} = U_n + \Delta\boldsymbol{x}_n^{\mathrm{T}}\langle \boldsymbol{P}_n \rangle \tag{6.5.6}$$

from eqn (6.5.4). Belytschko and Schoeberle start their examples with a given velocity and zero displacement and hence they suppose that the initial internal energy is zero. But we can have a situation where the formula for the internal energy is known and the initial conditions can consist of a given displacement and zero velocity, for example as in Wood and Oduor (1988). Hence, more generally we can say $U_0 =$ the initial potential energy in the system.

From §6.3 the basic Newmark equations with $\gamma = 0.5$ may be written in the form

$$\Delta \boldsymbol{x}_n = \Delta t\langle \boldsymbol{v}_n \rangle + \Delta t^2(\beta - 0.25)\Delta \boldsymbol{a}_n \tag{6.5.7}$$

and

$$\Delta \boldsymbol{v}_n = \Delta t\langle \boldsymbol{a}_n \rangle. \tag{6.5.8}$$

Writing eqn (6.5.3) in incremental form we have

$$\boldsymbol{M}\Delta \boldsymbol{a}_n + \Delta \boldsymbol{P}_n = \Delta \boldsymbol{f}_n \tag{6.5.9}$$

and substituting for $\Delta \boldsymbol{a}_n$ from eqn (6.5.7) gives

$$(\boldsymbol{M} + \beta\Delta t^2 \boldsymbol{K})\Delta \boldsymbol{x}_n = \boldsymbol{M}(\Delta t \boldsymbol{v}_n + 0.5\Delta t^2 \boldsymbol{a}_n) + \beta\Delta t^2 \Delta \boldsymbol{f}_n \tag{6.5.10}$$

as the equation to be solved at each time step for the increment in the displacements, where the tangent stiffness matrix $\boldsymbol{K}$ given by the relationship

$$\Delta \boldsymbol{P}_n = \boldsymbol{K}\Delta \boldsymbol{x}_n \tag{6.5.11}$$

has been introduced for use in the non-linear equation solver. The increment $\Delta \boldsymbol{x}_n$ is obtained from eqn (6.5.10) and then $\boldsymbol{x}_{n+1} = \boldsymbol{x}_n + \Delta \boldsymbol{x}_n$ is calculated and hence the new values of the strains, stresses and internal nodal forces. The new acceleration $\boldsymbol{a}_{n+1}$ is then found from eqn (6.5.7) and the equation of motion (6.5.3) checked at $(n+1)\Delta t$ for the out of balance or residual error force given by

$$\boldsymbol{f}^{\text{err}}_{n+1} = \boldsymbol{f}_{n+1} - \boldsymbol{P}_{n+1} - \boldsymbol{M}\boldsymbol{a}_{n+1}. \tag{6.5.12}$$

The iteration is continued until

$$|\Delta \boldsymbol{x}^{\text{T}}_n(\boldsymbol{f}^{\text{err}}_{n+1} + \boldsymbol{f}^{\text{err}}_n)| < \epsilon(U_n + T_n), \qquad n \geq 2 \tag{6.5.13}$$

where ϵ is some small number. Belytschko and Schoeberle show that with the criterion (6.5.13) for the convergence of the iteration this algorithm is stable with the parameter $\beta = 0.25$. Hughes (1977) points out that if we change the criterion (6.5.13) to

$$|\Delta \boldsymbol{x}^{\text{T}}_n(\boldsymbol{f}^{\text{err}}_{n+1} + \boldsymbol{f}^{\text{err}}_n)| < \epsilon(U_n + T_n + (\beta - 0.25)\Delta t^2 A_n), \tag{6.5.14}$$

introducing the extra term $A_n = \frac{1}{2}\boldsymbol{a}^{\text{T}}_n \boldsymbol{M}\boldsymbol{a}_n$ as in §6.3, we can show stability with the parameter β remaining. The quantity

$$E_n = T_n + U_n + (\beta - 0.25)\Delta t^2 A_n \tag{6.5.15}$$

is legitimate as an energy norm if the internal energy U_n (as defined in eqn (6.5.6)) is positive definite and if $\beta \geq 0.25$. Defining the external work done on the system as

$$W_{n+1} = W_n + \Delta \boldsymbol{x}_n^{\mathrm{T}} \langle \boldsymbol{f}_n \rangle, \qquad W_0 = 0, \tag{6.5.16}$$

and interpreting (6.5.14) as

$$\Delta \boldsymbol{x}_n^{\mathrm{T}} (\boldsymbol{f}_{n+1}^{\mathrm{err}} + \boldsymbol{f}_n^{\mathrm{err}}) = \pm \epsilon' E_n, \tag{6.5.17}$$

where $0 < \epsilon' < \epsilon$, the left-hand side of eqn (6.5.17) is equal to

$$\begin{aligned}
-2\Delta \boldsymbol{x}_n^{\mathrm{T}} \{ \boldsymbol{M} \langle \boldsymbol{a}_n \rangle + \langle \boldsymbol{P}_n \rangle - \langle \boldsymbol{f}_n \rangle \} & \\
&= -2\{ \Delta t \langle \boldsymbol{v}_n \rangle + \Delta t^2 (\beta - 0.25) \Delta \boldsymbol{a}_n \}^{\mathrm{T}} \\
&\quad \times \boldsymbol{M} \langle \boldsymbol{a}_n \rangle - 2\Delta \boldsymbol{x}_n^{\mathrm{T}} \langle \boldsymbol{P}_n \rangle + 2\Delta \boldsymbol{x}_n^{\mathrm{T}} \langle \boldsymbol{f}_n \rangle,
\end{aligned}$$

using eqn (6.5.7),

$$\begin{aligned}
= -2\{ & \langle \boldsymbol{v}_n \rangle^{\mathrm{T}} \boldsymbol{M} \Delta \boldsymbol{v}_n + \Delta \boldsymbol{x}_n^{\mathrm{T}} \langle \boldsymbol{P}_n \rangle \\
& + \Delta t^2 (\beta - 0.25) \Delta \boldsymbol{a}_n^{\mathrm{T}} \boldsymbol{M} \langle \boldsymbol{a}_n \rangle - \Delta \boldsymbol{x}_n^{\mathrm{T}} \langle \boldsymbol{f}_n \rangle \},
\end{aligned}$$

using eqn (6.5.8). Thus, substituting the last expression into eqn (6.5.17) and using the energy definition (6.5.15), we have

$$E_{n+1} - E_n - \Delta W_n = \pm \frac{\epsilon'}{2} E_n$$

and hence

$$E_{n+1} = \left(1 \pm \frac{\epsilon'}{2}\right) E_n + \Delta W_n < (1 + \epsilon) E_n + \Delta W_n. \tag{6.5.18}$$

We can examine the conditions for stability by putting $\boldsymbol{f} \equiv \boldsymbol{0}$ in eqn (6.5.3) and from (6.5.18) we have $E_n < (1+\epsilon)^n E_0 < e^{n\epsilon} E_0$, i.e. the energy is bounded for suitably small ϵ and so we have stability in the energy norm given by eqn (6.5.15) provided that $\beta > 0.25$.

The numerical results given by Belytschko and Schoeberle (1975) using the convergence criterion (6.5.13) are obtained with the non-linear equation (6.5.10) solved with the tangent stiffness matrix kept fixed during the iterations, i.e. with a fixed tangent method not Newton–Raphson. A consistent mass matrix is used. The examples are problems with non-linear elastic–plastic material. Stable results are obtained with $\beta = 0.25$ and $\beta = 0.375$; with $\beta = 0.125$ the authors report that the energy started deviating immediately and later the iterative procedure failed to converge. Other interesting points are that with one of the problems (a clamped ring), (a) the procedure failed with a very small time step, this is ascribed as probably due to round-off error, and (b) with the larger time steps a very large number of iterations are required in the first time step and failure to iterate sufficiently causes significant changes in the solution.

Wood and Oduor (1988) discuss the problem of a mass–spring system represented by the scalar equation

$$\ddot{x} + P(x) = 0 \tag{6.5.19}$$

including a hardening spring with

$$P(x) = s_1 x + s_2 x^3, \qquad s_1, s_2 > 0, \tag{6.5.20}$$

potential energy

$$U(x) = \tfrac{1}{2} s_1 x^2 + \tfrac{1}{4} s_2 x^4 \tag{6.5.21}$$

and a softening spring with

$$P(x) = s \tanh x, \qquad s > 0, \tag{6.5.22}$$

potential energy

$$U(x) = s \ln\{\cosh x\}. \tag{6.5.23}$$

Figures 6.5.1 to 6.5.4 show the energy ratio E_n/E_0 (with the potential energy calculated using the exact $U(x)$), for Newmark with $2\beta = \gamma = 0.5$, and SS22 with $\theta_1 = \theta_2 = 0.5$ applied to these two problems. There are various ways of implementing the SS22 algorithm for non-linear problems which are discussed in Chapter 7. The results here are obtained by using the standard equations given in §4.5 together with

$$\alpha_n + P(x_n) + \left(\theta_1 \Delta t \dot{x}_n + \frac{\Delta t^2}{2} \theta_2 \alpha_n\right) P'(\bar{x}_n) = 0, \tag{6.5.24}$$

where $\bar{x}_n = \frac{1}{2}(x_n + x_{n+1})$.

The parameters for the hardening spring are taken as $s_1 = 1$, $s_2 = 10$. Figure 6.5.1 shows that the Newmark algorithm remains stable for $\Delta t = 0.01$, 0.1, 0.5. Figure 6.5.2 shows that the SS22 algorithm as in eqn (6.5.24) is stable for $\Delta t = 0.01$ and 0.1, but becomes unstable for $\Delta t = 0.2$.

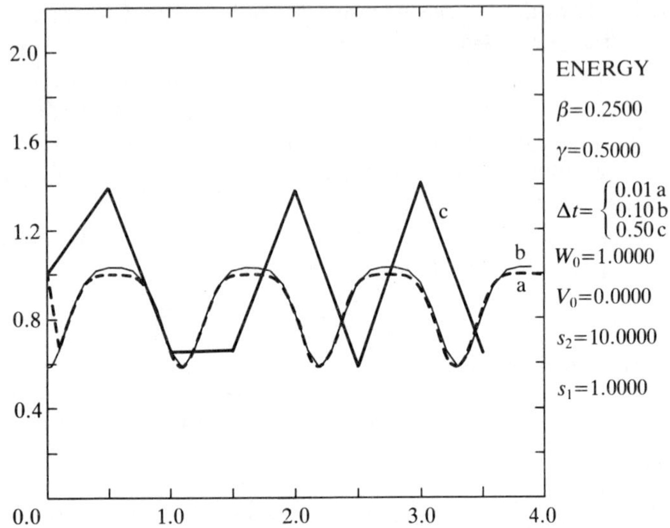

Fig. 6.5.1. Energy ratios with Newmark and the hardening spring. Newmark method. Case: $W'' + s_1 W + s_2 W^3 = 0$.

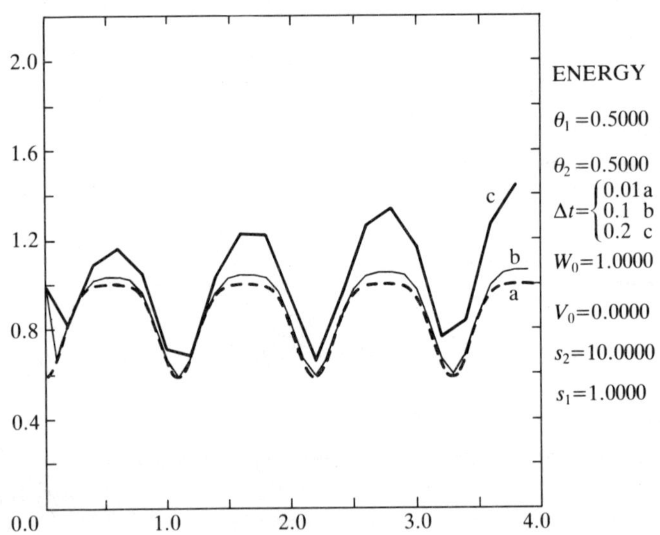

Fig. 6.5.2. Energy ratios with SS22 and the hardening spring. SS22 method. Case: $W'' + s_1 W + s_2 W^3 = 0$.

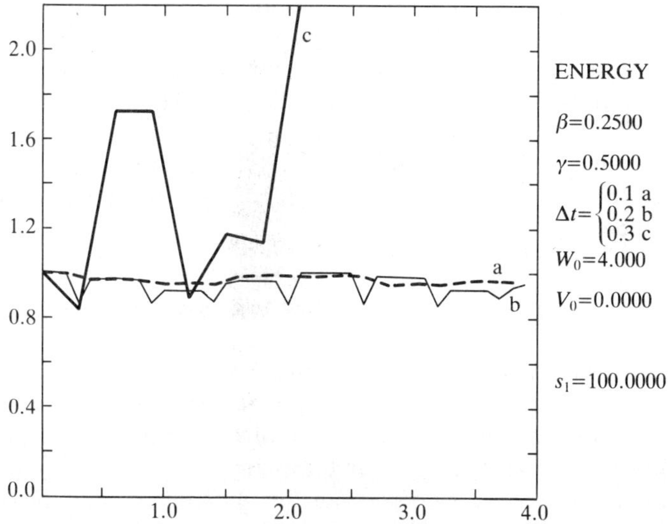

Fig. 6.5.3. Energy ratios with Newmark and the hardening spring. Newmark method. Case: $W'' + s_1 \tanh W = 0$.

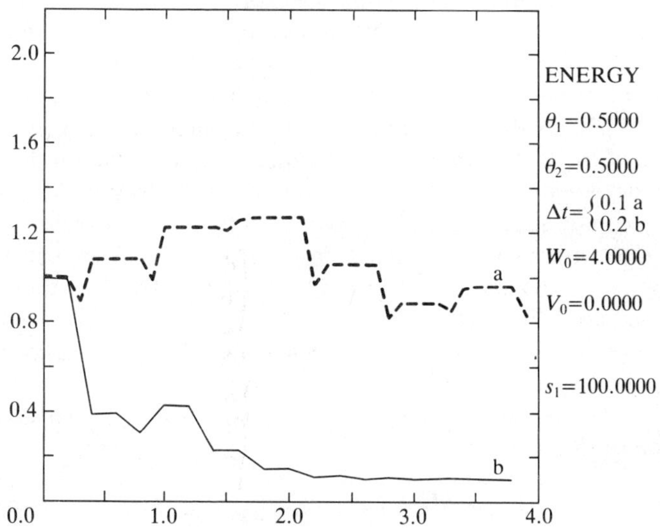

Fig. 6.5.4. Energy ratios with SS22 and the hardening spring. SS22 method. Case: $W'' + s_1 \tanh W = 0$.

The parameter for the softening spring is taken as $s_1 = 100$. Figure 6.5.3 shows that the Newmark algorithm now becomes unstable for $\Delta t = 0.3$. Figure 6.5.4 shows that the SS22 algorithm as in eqn (6.5.24) is stable but damped. These results are compatible with those reported in Wood and Oduor (1988) which are obtained with the same softening spring but $s_1 = 100$ and $s_2 = 1000$ for the hardening spring and stability judged from the amplitude of the oscillation.

6.6. The energy methods of Hughes and co-authors

There is much discussion in the literature about how effective the trapezium rule equivalent Newmark algorithm with $2\beta = \gamma = 0.5$ (also known as the average acceleration method) is for the numerical solution of non-linear forms of the dynamic vibration equation. Hughes (1976) points out the apparently contradictory statements from Belytschko and Schoeberle (1975) and from Park (1975). As described in §6.5, Belytschko and Schoeberle report satisfactory results from using the Newmark-trapezium rule on problems with non-linear material properties. They use a fixed tangent method for solving the non-linear equation at each time step (the method of calculation of the tangent stiffness matrix is unspecified) and a convergence criterion on the residual or out-of-balance of the equation to be satisfied. Park (1975 and later 1977), on the other hand, gives the unsatisfactory performance of the trapezium rule for the first order form of non-linear dynamic equations as the motivation for introducing the Park time-stepping scheme (an A-stable combination of Gear two-step and Gear three-step discussed in §4.11). Park does not use energy methods so further reference to the Park method is deferred until Chapter 7.

Hughes postulates a smooth scalar-valued function $U(\boldsymbol{x})$ for the potential energy, where $\boldsymbol{x}$ is the vector of nodal displacements. The system of dynamic vibration equations for a structure is then of the form

$$\boldsymbol{M}\ddot{\boldsymbol{x}} + \boldsymbol{C}\dot{\boldsymbol{x}} + \boldsymbol{K}(\boldsymbol{x}) = \boldsymbol{f} \tag{6.6.1}$$

where $\boldsymbol{K}(\boldsymbol{x})$ is the vector whose jth component is $\partial U/\partial x_j$. The tangent stiffness matrix is then given by $\boldsymbol{DK}(\boldsymbol{x})$ the matrix whose

(i, j)th entry is

$$\partial^2 U / \partial x_i \, \partial x_j. \tag{6.6.2}$$

Hughes takes the potential energy of the approximate solution as $U(\boldsymbol{x}_n)$ and takes two Taylor series with remainder term type expansions for $U(\boldsymbol{x}_{n+1})$ and $U(\boldsymbol{x}_n)$ to give

$$U(\boldsymbol{x}_{n+1}) = U(\boldsymbol{x}_n) + \Delta\boldsymbol{x}_n^{\mathrm{T}}\boldsymbol{K}(\boldsymbol{x}_n) + \tfrac{1}{2}\Delta\boldsymbol{x}_n^{\mathrm{T}}\boldsymbol{DK}(\boldsymbol{x}_n^{(1)})\Delta\boldsymbol{x}_n \tag{6.6.3}$$

and

$$U(\boldsymbol{x}_n) = U(\boldsymbol{x}_{n+1}) - \Delta\boldsymbol{x}_n^{\mathrm{T}}\boldsymbol{K}(\boldsymbol{x}_{n+1}) + \tfrac{1}{2}\Delta\boldsymbol{x}_n^{\mathrm{T}}\boldsymbol{DK}(\boldsymbol{x}_n^{(2)})\Delta\boldsymbol{x}_n, \tag{6.6.4}$$

where the values $\boldsymbol{x}_n^{(1)}, \boldsymbol{x}_n^{(2)}$ in the remainder terms are intermediate between $\boldsymbol{x}_n$ and $\boldsymbol{x}_{n+1}$. Then subtracting eqn (6.6.4) from (6.6.3) gives

$$\Delta U(\boldsymbol{x}_n) = \Delta\boldsymbol{x}_n^{\mathrm{T}}\langle \boldsymbol{K}(\boldsymbol{x}_n)\rangle + \Delta\boldsymbol{x}_n^{\mathrm{T}}\boldsymbol{S}_n\Delta\boldsymbol{x}_n, \tag{6.6.5}$$

where the matrix $\boldsymbol{S}_n = \tfrac{1}{4}[\boldsymbol{DK}(\boldsymbol{x}_n^{(1)}) - \boldsymbol{DK}(\boldsymbol{x}_n^{(2)})]$ gives a measure of the non-linearity of the problem. If we now take the energy of the approximate solution as the sum of the terms representing the kinetic and potential energies

$$E_n = \tfrac{1}{2}\boldsymbol{v}_n^{\mathrm{T}}\boldsymbol{M}\boldsymbol{v}_n + U(\boldsymbol{x}_n) \tag{6.6.6}$$

then using eqn (6.6.5) gives

$$\Delta E_n = \Delta\boldsymbol{v}_n^{\mathrm{T}}\boldsymbol{M}\langle \boldsymbol{v}_n\rangle + \Delta\boldsymbol{x}_n^{\mathrm{T}}\langle \boldsymbol{K}(\boldsymbol{x}_n)\rangle + \Delta\boldsymbol{x}_n^{\mathrm{T}}\boldsymbol{S}_n\Delta\boldsymbol{x}_n. \tag{6.6.7}$$

Substituting from the Newmark algorithm equations with $2\beta = \gamma = 0.5$ given by eqns (6.5.7) and (6.5.8), i.e.

$$\Delta\boldsymbol{x}_n = \Delta t\langle \boldsymbol{v}_n\rangle \quad \text{and} \quad \Delta\boldsymbol{v}_n = \Delta t\langle \boldsymbol{a}_n\rangle, \tag{6.6.8}$$

and have

$$\Delta E_n = \Delta\boldsymbol{x}_n^{\mathrm{T}}\{\boldsymbol{M}\langle \boldsymbol{a}_n\rangle + \langle \boldsymbol{K}(\boldsymbol{x}_n)\rangle\} + \Delta\boldsymbol{x}_n^{\mathrm{T}}\boldsymbol{S}_n\Delta\boldsymbol{x}_n. \tag{6.6.9}$$

Now the non-linear dynamic vibration equation (6.6.1) cannot be satisfied exactly; we suppose it is satisfied with a residual $\boldsymbol{f}_n^{\mathrm{err}}$ which is governed by a convergence criterion and which is given by

$$\boldsymbol{f}_n^{\mathrm{err}} = \boldsymbol{M}\boldsymbol{a}_n + \boldsymbol{C}\boldsymbol{v}_n + \boldsymbol{K}(\boldsymbol{x}_n) - \boldsymbol{f}_n. \tag{6.6.10}$$

Then substituting from eqn (6.6.10) into the energy balance

equation (6.6.9) we have

$$\Delta E_n = \Delta \boldsymbol{x}_n^{\mathrm{T}}\{-\boldsymbol{C}\langle \boldsymbol{v}_n\rangle + \langle \boldsymbol{f}_n\rangle + \langle \boldsymbol{f}_n^{\mathrm{err}}\rangle\} + \Delta \boldsymbol{x}_n^{\mathrm{T}}\boldsymbol{S}_n\Delta \boldsymbol{x}_n,$$

i.e.

$$\Delta E_n = -\Delta t\langle \boldsymbol{v}_n\rangle^{\mathrm{T}}\boldsymbol{C}\langle \boldsymbol{v}_n\rangle + \Delta \boldsymbol{x}_n^{\mathrm{T}}\langle \boldsymbol{f}_n\rangle + \Delta \boldsymbol{x}_n^{\mathrm{T}}\langle \boldsymbol{f}_n^{\mathrm{err}}\rangle + \Delta \boldsymbol{x}_n^{\mathrm{T}}\boldsymbol{S}_n\Delta \boldsymbol{x}_n. \tag{6.6.11}$$

On the right-hand side of eqn (6.6.11) we have the energy loss due to the friction, the work done by the forcing function, the effect of the error in the solution of the non-linear equation and the measure of the non-linearity respectively. Considering first the measure of the non-linearity, we can take as examples the problems studied in Wood and Oduor (1988). One of these had a single degree of freedom with a mass on a hardening spring governed by the stress–strain rule given in eqn (6.5.20). Using Hughes notation this is

$$K(x) = s_1 x + s_2 x^3, \qquad s_1 > 0, \qquad s_2 \geq 0. \tag{6.6.12}$$

Thus here the potential energy

$$U(x) = s_1 x^2/2 + s_2 x^4/4 \tag{6.6.13}$$

and the tangent stiffness

$$DK(x) = s_1 + 3s_2 x^2. \tag{6.6.14}$$

The expansions corresponding to those in eqns (6.6.3) and (6.6.4) are

$$U(x_{n+1}) = U(x_n) + \Delta x_n(s_1 x_n + s_2 x_n^3) + \tfrac{1}{2}\Delta x_n^2\{s_1 + s_2(3x_n^2 + 2x_n\Delta x_n + \tfrac{1}{2}\Delta x_n^2)\} \tag{6.6.15}$$

and

$$U(x_n) = U(x_{n+1}) - \Delta x_n(s_1 x_{n+1} + s_2 x_{n+1}^3) + \tfrac{1}{2}\Delta x_n^2\{s_1 + s_2(3x_n^2 + 4x_n\Delta x_n + \tfrac{3}{2}\Delta x_n^2)\}. \tag{6.6.16}$$

From these last two equations we can identify $x_n^{(1)}$ and $x_n^{(2)}$ if we wish. Subtracting eqn (6.6.16) from eqn (6.6.15) then gives

$$\Delta U(x_n) = \Delta x_n\langle K(x_n)\rangle + \Delta x_n^2 S_n,$$

where

$$S_n = -s_2\Delta x_n^3(x_n + x_{n+1})/4. \tag{6.6.17}$$

At points where $(x_n + x_{n+1}) > 0$ and $\Delta x_n < 0$ this term will be positive and at points where $(x_n + x_{n+1}) < 0$ and $\Delta x_n < 0$ it will be negative and as the solution proceeds through the cycle the effect will tend to cancel out. Also in Wood and Oduor (1988) there is consideration of a problem with a softening spring given by

$$K(x) = s \tanh(x).$$

Supposing that x is small enough to take $\tanh(x) \approx x - x^3/3$, we can see that this simply has the effect of changing the sign of s_2 in eqn (6.6.17) and hence with the softening spring we have the same result with opposite signs. With reasonable values of the time step (remembering the engineer's rule of thumb of at least ten time steps per period) the non-linearity effects represented by the S_n terms should not accumulate. We also note that the S_n term effect corresponds to the error that Park (1975) ascribes to the use of past derivatives. Park is discussing an equation of the form

$$\ddot{x} = f(x)$$

and takes $\Omega^2(x) = \partial f / \partial x$ which corresponds to $DK(x)$ in eqn (6.6.2). Then S_n in eqn (6.6.5) corresponds to the change in $\Omega^2(x)$ and it would seem that the errors due to the use of past derivatives might also change sign and cancel out. The pathological examples given in these papers and observed in the Wood and Oduor investigation do appear to occur mainly with time steps larger than one-tenth of the period. A program for the solution of a practical problem can include a procedure for reducing the size of the time step if danger signals occur such as untoward increase in energy and/or failure of the iteration to converge.

The effect of the error in the solution of the non-linear equation (assuming that the iterative scheme for the solution of this is convergent) is more worrying. Even if the convergence criterion is linked to the time step so that the result from the Belytschko and Schoeberle approach given in inequality (6.5.18) with $W_n = 0$ becomes

$$E_n < (1 + \epsilon \Delta t)^n E_0 < \exp(\epsilon T) E_0$$

for $n\Delta t = T$, fixed, this is not satisfactory because there should not be any growth in the energy.

Hughes *et al.* (1978) propose a modification of the Newmark trapezium rule which does exactly conserve energy for non-linear elastodynamics in the absence of external forces. The problem is represented by an equation like eqn (6.6.1) but with no damping term:

$$\boldsymbol{M}\ddot{\boldsymbol{x}} + \boldsymbol{K}(\boldsymbol{x}) = \boldsymbol{f}, \tag{6.6.18}$$

where again $\boldsymbol{K}(\boldsymbol{x}) = \{\partial U/\partial x_j\}$, $U(\boldsymbol{x})$ being the scalar potential energy function assumed non-negative and such that if $U(\boldsymbol{x})$ is bounded so is $\|\boldsymbol{x}\|$. Collocating the differential equation (6.6.18) at $t = (n+1)\Delta t$ gives

$$\boldsymbol{M}\boldsymbol{a}_{n+1} + \boldsymbol{K}(\boldsymbol{x}_{n+1}) = \boldsymbol{f}_{n+1}, \tag{6.6.19}$$

and substituting from the Newmark equation

$$\boldsymbol{a}_{n+1} = \frac{4}{\Delta t^2}(\boldsymbol{x}_{n+1} - \boldsymbol{x}_n - \Delta t\boldsymbol{v}_n) - \boldsymbol{a}_n$$

gives the result

$$\frac{4}{\Delta t^2}\boldsymbol{M}\boldsymbol{x}_{n+1} + \boldsymbol{K}(\boldsymbol{x}_{n+1}) = \boldsymbol{f}_{n+1} + \boldsymbol{M}\left\{\boldsymbol{a}_n + \frac{4}{\Delta t^2}(\boldsymbol{x}_n + \Delta t\boldsymbol{v}_n)\right\}. \tag{6.6.20}$$

A functional is then formed such that eqn (6.6.20) is the Euler or Euler–Lagrange equation of this functional. Courant and Hilbert (1953) refer to these equations as Euler equations even when the Lagrange multipliers are included but in the context of numerical solutions the label 'Euler' on its own is inevitably associated with the explicit forward difference formula.) The required functional is

$$F(\boldsymbol{x}_{n+1}) = \frac{2}{\Delta t^2}\boldsymbol{x}_{n+1}^{\mathrm{T}}\boldsymbol{M}\boldsymbol{x}_{n+1} + U(\boldsymbol{x}_{n+1}) - \boldsymbol{x}_{n+1}^{\mathrm{T}}$$
$$\times \{\boldsymbol{f}_{n+1} + \boldsymbol{M}(\boldsymbol{a}_n + 4(\boldsymbol{x}_n + \Delta t\boldsymbol{v}_n)/\Delta t^2\}. \tag{6.6.21}$$

Differentiating $F(\boldsymbol{x}_{n+1})$ with respect to the components of $\boldsymbol{x}_{n+1}$ and equating the result to zero gives eqn (6.6.20) and thus this equation is the condition for a stationary variation of the functional.

The Newmark trapezium rule algorithm is to be modified so that, if E_n is defined as in eqn (6.6.6), we have the energy balance equation

$$\Delta E_n = \Delta \boldsymbol{x}_n^{\mathrm{T}} \langle \boldsymbol{f}_n \rangle. \tag{6.6.22}$$

This is achieved by introducing another functional

$$G(\boldsymbol{x}_{n+1}) = E\left(\boldsymbol{x}_{n+1}, \frac{2}{\Delta t}\Delta \boldsymbol{x}_n - \boldsymbol{v}_n\right) - E_n - \Delta \boldsymbol{x}_n^{\mathrm{T}} \langle \boldsymbol{f}_n \rangle. \tag{6.6.23}$$

The two functionals are then combined together as

$$F(\boldsymbol{x}_{n+1}) + \lambda G(\boldsymbol{x}_{n+1}), \tag{6.6.24}$$

where λ is a Lagrange multiplier. (Courant and Hilbert (1953) discuss the theoretical concepts of Lagrange multipliers and Johnson and Riess (1977) have some discussion on their use in numerical work.) The conditions for a stationary variation of the functional (6.6.24) are given by differentiating with respect to λ, i.e.

$$G(\boldsymbol{x}_{n+1}) = 0 \tag{6.6.25}$$

and differentiating with respect to the components of $\boldsymbol{x}_{n+1}$ which produces the equation

$$(1+\lambda)\left\{\frac{4}{\Delta t^2}\boldsymbol{M}\boldsymbol{x}_{n+1} + \boldsymbol{K}(\boldsymbol{x}_{n+1})\right\} - \boldsymbol{f}_{n+1} - \lambda \langle \boldsymbol{f}_n \rangle$$
$$- \boldsymbol{M}\left[\boldsymbol{a}_n + \frac{4}{\Delta t^2}\{(1+\lambda)\boldsymbol{x}_n + \Delta t(2+\lambda)\boldsymbol{v}_n/2\}\right] = \boldsymbol{0}. \tag{6.6.26}$$

Equations (6.6.25) and (6.6.26) then have to be solved for λ and $\boldsymbol{x}_{n+1}$. Hughes *et al.* employ a modified Newton–Raphson procedure with the Jacobian matrix only updated when the convergence rate of the iterates is too slow. Their program requires only one additional forward reduction/back substitution compared wih a similar implementation of the straight Newmark-trapezium rule algorithm. The numerical results for problems with non-linear elastic materials with no external forcing function show that the modified algorithm exactly preserves the energy for the larger time steps which give rise to very erratic behaviour from the unmodified algorithm.

6.7. Energy methods—more general algorithms

In this section we discuss some other algorithms using the energy method in different ways. Oden and Fost (1973) consider the problem of the longitudinal motion of a rod such that x is the axial coordinate and the displacement at x at time t is $u(x, t)$. The motion is set off by some initial conditions and there are no body forces or forces at the ends. The energy $E(t)$ at time t is the sum of the kinetic energy

$$K = \frac{\rho_0 A_0}{2} \int_0^L \dot{u}^2 \, \mathrm{d}x \tag{6.7.1}$$

and the potential energy

$$H = A_0 \int_0^L W \, \mathrm{d}x, \tag{6.7.2}$$

where ρ_0, A_0, L are the initial density, cross-sectional area, and length of the rod respectively. The strain energy function W is a function of $\partial u/\partial x$, denoted by u_x, only. The authors say that this represents a non-linear elastic effect which is very common in continuum mechanics. Then if

$$\frac{\mathrm{d}W(u_x)}{\mathrm{d}u_x} \equiv T(u_x) \tag{6.7.3}$$

the equation of the motion of the rod is

$$\rho_0 A_0 \frac{\partial^2 u}{\partial t^2} - A\left\{\frac{\mathrm{d}T(u_x)}{\mathrm{d}u_x}\right\} \frac{\partial^2 u}{\partial x^2} = 0. \tag{6.7.4}$$

Then, multiplying eqn (6.7.4) by $\partial u/\partial t$ and integrating with respect to x from $x = 0$ to L gives

$$\rho_0 A_0 [\ddot{u}, \dot{u}] + A_0 \{T(u_x), \dot{u}_x\} = 0, \tag{6.7.5}$$

where the inner product

$$[u, v] \equiv \int_0^L uv \, \mathrm{d}x. \tag{6.7.6}$$

Equation (6.7.5) is the statement that the kinetic plus potential energy does not change with time, that is, it is constant. This equation, (6.7.5), is then approximated by piecewise linear finite elements in space and central differences for the time derivatives. The scheme is considered to be stable in energy if the finite element approximation, E_h, to the energy satisfies

$$E_h(n\Delta t) \le C^{(n)} E_h((n-1)\Delta t), \tag{6.7.7}$$

where $C^{(n)} > 0$, for $n = 1, 2, \ldots, R$, say. Then with a consistent mass matrix the scheme is shown to be stable in energy in this sense if

$$\frac{h}{\Delta t} > \sqrt{6}C_{\max}^{(n)}, \tag{6.7.8}$$

where h is the minimum element length and $C_{\max}^{(n)}$ is the maximum wave speed in the rod at time $t = n\Delta t$, given by

$$C_{\max}^{(n)} = \max_x [T'(U_x^{(n)})/\rho_0]^{\frac{1}{2}},$$

where $U_x^{(n)}$ is the finite element approximation to u_x at time $t = n\Delta t$ and $T'(u_x) = \mathrm{d}T/\mathrm{d}u_x$. With a lumped mass matrix the $\sqrt{6}$ on the right-hand side of inequality (6.7.8) is replaced by $\sqrt{2}$. This is a similar result to that obtained by Fujii (1972), that lumping the mass matrix increases the stability limit on the time step by a factor of $\sqrt{3}$. Oden and Fost also prove convergence for their scheme applied to the non-homogeneous form of eqn (6.7.4) provided the relevant stability condition is satisfied. The lumped and consistent mass approximations have the same rate of convergence. It is interesting to note that Belytschko and Schoeberle (1975) comment that the restrictions on the types of problem in the papers of Fujii (1972) and Oden and Fost (1973) were probably included to allow the proof of convergence, this being more difficult to prove than stability. Neither of these papers gives numerical results.

In the Trujillo (1977) operator-splitting method described in §5.7 for use with the linear dynamic vibration equation, the author attempts to establish stability using an energy method. With $E_n = \boldsymbol{v}_n^{\mathrm{T}}\boldsymbol{M}\boldsymbol{v}_n + \boldsymbol{x}_n^{\mathrm{T}}\boldsymbol{K}\boldsymbol{x}_n$, he obtains, without giving the details, the result

$$E_{n+1} - E_n + \frac{2}{\Delta t}\Delta\boldsymbol{x}_n^{\mathrm{T}}\boldsymbol{C}\Delta\boldsymbol{x}_n + \frac{\Delta t}{8}\Delta\boldsymbol{v}_n^{\mathrm{T}}\boldsymbol{C}\Delta\boldsymbol{v}_n + 2\Delta\boldsymbol{x}_n^{\mathrm{T}}(\boldsymbol{C}_{\mathrm{U}} - \boldsymbol{C}_{\mathrm{L}})\langle\boldsymbol{v}_n\rangle = 0 \tag{6.7.9}$$

for the case with zero forcing function, where $\boldsymbol{C}_{\mathrm{U}}$ and $\boldsymbol{C}_{\mathrm{L}}$ are the upper and lower triangular matrix splittings of the damping matrix in this operator splitting algorithm (the equations are given in §5.7). Equation (6.7.9) shows that the algorithm is unconditionally stable, i.e. the energy E_n is preserved if there is zero natural damping, but the last term on the left-hand side does not lend itself to any conclusion when natural damping is present. Numerical examples show that the algorithm is stable with and without natural damping but the accuracy is disappointing because time steps of the same order as those of the conditionally stable central difference method are required for reasonable results. An alternative operator splitting gives better accuracy but only conditional stability.

Hamdan and Patamapongs (1985) describe a non-linear elastic one-dimensional problem using the central difference in time but with approximations introduced in the non-linear stiffness terms such that the algorithm expresses a discrete conservation law. The problem is given as the motion of a finite slab represented by the equation

$$\rho\ddot{u} - \mathrm{D}\{\bar{t}(1 + \mathrm{D}u)\} = f, \tag{6.7.10}$$

where u is the displacement perpendicular to the slab. The stress

$$\bar{t} = E_1\gamma + E_2\gamma^2, \tag{6.7.11}$$

γ is the Green–St Venant strain:

$$\gamma = \mathrm{D}u + \tfrac{1}{2}(\mathrm{D}u)^2 \tag{6.7.12}$$

and $\mathrm{D} \equiv \mathrm{d}/\mathrm{d}x$. The authors state that their method is most efficient when only terms of third degree or less are retained in the non-linear expression. The method then becomes a linearization technique given by

$$(\mathrm{D}u_n)^2 = \tfrac{1}{3}\mathrm{D}u_n[\mathrm{D}u_{n+1} + \mathrm{D}u_n + \mathrm{D}u_{n-1}], \tag{6.7.13}$$

$$(\mathrm{D}u_n)^3 = \tfrac{1}{2}\mathrm{D}u_n^2[\mathrm{D}u_{n+1} + \mathrm{D}u_{n-1}]. \tag{6.7.14}$$

Hamdan and Patamapongs show that for the homogeneous problem energy is conserved, i.e. the stability is unconditional, with the algorithm given by

$$\begin{aligned} &\left\langle \frac{\rho}{\Delta t^2}(U^{n+1} - 2U^n + U^{n+1}), V \right\rangle \\ &\quad + \langle [E_1\mathrm{D}^*U^n + (\tfrac{3}{2}E_1 + E_2)\mathrm{D}U^n(\mathrm{D}U^{n+1} + \mathrm{D}U^n + \mathrm{D}U^{n-1})/3 \\ &\quad + (\tfrac{1}{2}E_1 + 2E_2)(\mathrm{D}U^N)^2(\mathrm{D}U^{n+1} + \mathrm{D}U^{n-1})/2], \mathrm{D}V \rangle \\ &= \langle f^n, V \rangle, \end{aligned} \tag{6.7.15}$$

where $\langle \, . \, , . \, \rangle$ represents the space inner product,

$$D^*U^n = rDU^n + (1-r)(DU^{n+1} + DU^{n-1})/2, \qquad 0 < r < 0.5,$$

V is any of the basis functions of the finite element space approximation, and U^n is the finite element space approximation at time $t = n\Delta t$. The algorithm with $r = 1$ is shown to be conditionally stable; stable numerical results are demonstrated with $r = 1/3$.

References

Belytschko, T. and Schoeberle, D. F. (1975). On the unconditional stability of an implicit algorithm for non-linear structural dynamics. *ASME Journal of Applied Mechanics*, **17**, 865–869.

Courant, R., Friedrichs, K., and Lewy, H. (1928). Über die partiellen Differenzengleichungen der Mathematischen Physik. *Mathematische Annale*, **100**, 32–74. English translation (1967). On partial differential equations of mathematical physics. *IBM Journal*, **11**, 215–234.

Courant, R. and Hilbert, D. (1953). *Methods of mathematical physics*, Vol. 1. Interscience.

Fujii, H. (1972). Finite element schemes: stability and convergence. In *Advances in computational methods in structural mechanics and design* (ed. J. R. Oden *et al.*). U.A.H. Press, Huntsville, Alabama.

Hamdan, S. M. and Patamapongs, N. (1985). Stability and convergence of some finite element algorithms for nonlinear elastodynamics. *International Journal of Numerical Methods in Engineering*, **21**, 975–999.

Hughes, T. J. R. (1976). Stability, convergence and growth and decay of energy of the average acceleration method in non-linear structural dynamics. *Computers and Structures*, **6**, 313–324.

Hughes, T. J. R. (1977). A note on the stability of Newmark's algorithm in non-linear structural dynamics. *International Journal of Numerical Methods in Engineering*, **11**, 383–386.

Hughes, T. J. R. (1983). In *Computational methods for transient analysis* (eds. T. Belytschko, and T. J. R. Hughes), Ch. 2. North-Holland.

Hughes, T. J. R. and Liu, W. K. (1978). Implicit–explicit finite elements in transient analysis: stability theory. *ASME Journal of Applied Mechanics*, **45**, 371–374.

Hughes, T. J. R., Caughey, T. K., and Liu, W. K., (1978). Finite element methods for non-linear elastodynamics which conserve energy. *ASME Journal of Applied Mechanics*, **45**, 366–370.

Johnson, L. W. and Riess, R. D. (1977). *Numerical analysis*. Addison Wesley.

Oden, J. R. and Fost, R. B. (1973). Convergence, accuracy and stability of finite element approximations of a class of non-linear hyperbolic equations. *International Journal of Numerical Methods in Engineering*, **6**, 357–365.

Park, K. C. (1975). Evaluating time integration methods for non-linear dynamics analysis. In *Finite element analysis of transient non-linear structural behavior* (ed. T. Belytschko). *ASME Symposia Series AMD*, Vol. 14, 35–58.

Richtmyer, R. D. and Morton, K. W. (1967). *Difference methods for initial value problems*. Interscience.

Trujillo, D. M. (1977). An unconditionally stable explicit algorithm for structural dynamics, *International Journal of Numerical Methods in Engineering*, **11**, 1579–1592.

Wood, W. L. and Oduor, M. E. (1988). Stability properties of some algorithms for the solution of nonlinear dynamic vibration equations. *Communications in Applied Numerical Methods*, **4**, 205–212.

Zienkiewicz, O. C. (1977). *The finite element method*. McGraw-Hill.

7 Non-linear equations

7.1. Introduction

In this chapter we discuss methods currently used by practical engineers and scientists to obtain the numerical solution of non-linear equations. These, when space-discretized, will typically have the form of the second order system

$$\boldsymbol{M}\ddot{\boldsymbol{x}} + \boldsymbol{C}\dot{\boldsymbol{x}} + \boldsymbol{P}(\boldsymbol{x}) = \boldsymbol{f}(t), \tag{7.1.1}$$

which can represent the vibration of a structure with some non-linear stress–strain and damping characteristics, or of the first order system

$$\boldsymbol{C}\dot{\boldsymbol{x}} + \boldsymbol{P}(\boldsymbol{x}) = \boldsymbol{f}(t), \tag{7.1.2}$$

which can represent, for example, non-linear heat conduction or diffusion or flow in a porous medium. In eqn (7.1.2) the mass or capacitance matrix $\boldsymbol{C}$ may also be dependent on the solution (Chapter 1).

The methods vary from the very simple explicit time-stepping schemes with a lumped mass matrix using the central difference algorithm with the second order system or the Euler forward time-stepping algorithm with the first order system to the sophisticated implicit schemes using Newton–Raphson, Quasi-Newton or gradient methods. If we think of this as a spectrum of methods with simple explicit at one end and sophisticated Newton type methods at the other, then in between we can put the methods which use some kind of *ad hoc* linearization to form an iteration and then the quasi-Newton and gradient methods. The implicit methods are all essentially predictor–corrector methods since some formula is used to start the iteration and the new values for the end of the time step are corrected until convergence is judged to have been reached. The starting value is crucial for the optimization of the convergence. The convergence criterion is sometimes based on a norm of the change in

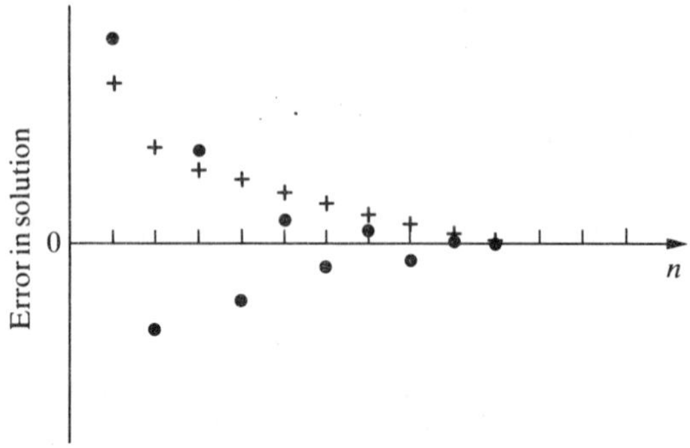

Fig. 7.1.1. Overdamped (+) and underdamped (•) iterative solutions.

the solution from one iteration to the next and sometimes on the residual or out of balance in the equation to be satisfied, and sometimes both. Those with experience of using Gauss–Seidel iteration for the solution of systems of simultaneous equations will recall its notorious slowness to converge even when implemented with an acceleration parameter as Successive Over-Relaxation (SOR). Thus there could be a very small change from one iteration to the next when still far from the solution. We can visualize this as a kind of overdamped situation as illustrated in Fig. 7.1.1. When we have an underdamped situation which can show itself by producing oscillations in the solutions of successive iterations (also illustrated for one degree of freedom in Fig. 7.1.1), it is safe to stop the iterations when the maximum change in the solution between successive iterations is less than some suitable tolerance.

The possible unreliability of the method based on stopping the iteration when the residual is less than some tolerance is illustrated by an example from approximation theory. Suppose we want to find the best fit polynomial $\sum_{j=1}^{n} a_j x^j$ to a given function $f(x)$ on the interval $[0, 1]$, with $f(0) = 0$. We want to minimize

$$I = \int_0^1 \left\{ f(x) - \sum_{j=1}^{n} a_j x^j \right\}^2 \mathrm{d}x$$

with respect to the parameters $a_1, a_2, \ldots, a_n$. We equate to zero the partial derivatives of I with respect to each a_j, $j = 1, 2, \ldots, n$.

For $n = 4$ we obtain the set of n linear equations as follows

$$\begin{aligned}\tfrac{1}{2}a_1 + \tfrac{1}{3}a_2 + \tfrac{1}{4}a_3 + \tfrac{1}{5}a_4 &= b_1\\ \tfrac{1}{3}a_1 + \tfrac{1}{4}a_2 + \tfrac{1}{5}a_3 + \tfrac{1}{6}a_4 &= b_2\\ \tfrac{1}{4}a_1 + \tfrac{1}{5}a_2 + \tfrac{1}{6}a_3 + \tfrac{1}{7}a_4 &= b_3\\ \tfrac{1}{5}a_1 + \tfrac{1}{6}a_2 + \tfrac{1}{7}a_3 + \tfrac{1}{8}a_4 &= b_4\end{aligned} \tag{7.1.3}$$

where $b_j = \int_0^1 f(x)x^j \, dx$, $j = 1, 2, 3, 4$. The type of matrix in eqns (7.1.3) is called the Hilbert matrix. For $b^{\mathrm{T}} = [1, 0.7227, 0.5697, 0.4714]$, the exact solution of eqns (7.1.3) is

$$a^{\mathrm{T}} = [1.162, \ -0.234, \ 2.436, \ -0.56] \tag{7.1.4}$$

(this is from an example given by Fox and Mayers 1968). Now suppose in the course of some iteration we have the solution

$$a^{\mathrm{T}} = [0.702, \ 3.046, \ -3.904, \ 3.05]. \tag{7.1.5}$$

Substituting the values from (7.1.5) into the eqns (7.1.3) we find that the residuals are given by $r^{\mathrm{T}} = [3 \times 10^{-4}, 3 \times 10^{-4}, 5 \times 10^{-5}, 2 \times 10^{-4}]$. Thus if the residual tolerance is set to 5×10^{-4}, for instance, we might be led to accept solution (7.1.5) as correct but comparing it with (7.1.4) we can see that it does not even have one figure accuracy. The effect is due to the large condition number of the Hilbert matrix which increases rapidly with n. This is an extreme example of an ill-conditioned problem but it does show that residuals cannot be absolutely relied upon as a check for when convergence has been obtained.

The rest of this chapter is arranged in sections covering the different approaches to the solution of non-linear time dependent problems with finite element space discretization.

Section 7.2 gives the simplest explicit methods for non-linear first and second order equations. Section 7.3 describes methods which are essentially predictor–corrector type algorithms adapted for use with the particular problem being solved and hence referred to here as '*ad hoc*' iterative methods. The recurrent question of whether lumped or distributed mass matrices should be used is examined in detail in §7.5 with references to sources where further details can be found.

Methods for dealing directly with non-linear second order systems are discussed in §7.6. There are some differing opinions on the question of optimum starting conditions considered in

§7.7. In §7.8 we consider criteria for time-step adjustment; the control of stability in this way is often the practical answer when unconditional stability cannot be guaranteed. The criterion can be based on an estimate of the error or a condition of non-increasing energy (Chapter 6) or mass balance or a combination of these.

In §7.9 we discuss the recent revival of interest in the Conjugate Gradient Method in the form of the Preconditioned Conjugate Gradient Method for the solution of the system of equations encountered at each time step. This section concludes with a reference to the Dynamic Relaxation Method of Otter (1965) which was disparaged by mathematicians but popular with engineers who like the idea of being able to visualize the iterative method as a damped oscillation to a steady state. This method is still in use.

7.2. Explicit methods, first and second order systems

First order equations

For the first order system of equations

$$\boldsymbol{C}\dot{\boldsymbol{u}} + \boldsymbol{P}(\boldsymbol{u}) = \boldsymbol{f}(t), \tag{7.2.1}$$

the simplest explicit method is the Euler forward time-stepping algorithm which is also the θ-method with $\theta = 0$:

$$\boldsymbol{C}\boldsymbol{u}_{n+1} = \boldsymbol{C}\boldsymbol{u}_n - \Delta t(\boldsymbol{P}(\boldsymbol{u}_n) - \boldsymbol{f}_n). \tag{7.2.2}$$

With a lumped mass matrix $\boldsymbol{C}$, evaluated as $\boldsymbol{C}(\boldsymbol{u}_n)$ if it is non-linear, there is no matrix to invert and the programming is extremely simple. The method is only conditionally stable as with the linear case (§2.6), where if $\boldsymbol{P}(\boldsymbol{x}) = \boldsymbol{K}\boldsymbol{x}$ and λ is the maximum eigenvalue of $\boldsymbol{C}^{-1}\boldsymbol{K}$, for stability the time step is limited by the rule $\lambda\Delta t < 2$, but many practical users accept this for the sake of the simplicity of the method. Some form of error estimate can be used to control the size of time step. The popularity of this method is not reflected in the literature because it does not inspire the kind of papers which get published.

Another explicit algorithm which has been popular is Lees' algorithm (Lees 1966) discussed in §4.11. For the non-linear equation

$$\boldsymbol{C}\dot{\boldsymbol{u}} + \boldsymbol{K}(\boldsymbol{u})\boldsymbol{u} = \boldsymbol{f}(t), \tag{7.2.3}$$

this algorithm is written as

$$\frac{1}{2\Delta t}\boldsymbol{C}(\boldsymbol{u}_{n+1}-\boldsymbol{u}_{n-1})+\tfrac{1}{3}\boldsymbol{K}(\boldsymbol{u}_n)(\boldsymbol{u}_{n+1}+\boldsymbol{u}_n+\boldsymbol{u}_{n-1})=\boldsymbol{f}_n. \quad (7.2.4)$$

As explained in §4.11, this algorithm is unconditionally stable but liable to give trouble with spurious oscillations.

Culham and Varga (1971) discuss the numerical solution of the type of non-linear equations representing the flow of gas in a porous medium which when space-discretized is the same as eqn (7.2.3). They compare several kinds of space discretization which are not relevant here and they also compare the performance of four time-stepping schemes:

1. The backward time-stepping algorithm

   $$(\boldsymbol{C}_{n+1}+\Delta t\boldsymbol{K}_{n+1})\boldsymbol{u}_{n+1}=\boldsymbol{C}_{n+1}\boldsymbol{u}_n+\Delta t\boldsymbol{f}_{n+1}, \quad (7.2.5)$$

   where $\boldsymbol{C}_{n+1}$, $\boldsymbol{K}_{n+1}$ signify that the matrices are evaluated as $\boldsymbol{C}(\boldsymbol{u}_{n+1})$, $\boldsymbol{K}(\boldsymbol{u}_{n+1})$ respectively. This is an implicit scheme requiring iteration and is mentioned again in §7.3.

2. The Crank–Nicolson type scheme with the matrices and the source term all averaged

   $$\left(\boldsymbol{C}_{n+\frac{1}{2}}+\frac{\Delta t}{2}\boldsymbol{K}_{n+\frac{1}{2}}\right)\boldsymbol{u}_{n+1}=\left(\boldsymbol{C}_{n+\frac{1}{2}}-\frac{\Delta t}{2}\boldsymbol{K}_{n+\frac{1}{2}}\right)\boldsymbol{u}_n + \Delta t\boldsymbol{f}_{n+\frac{1}{2}}. \quad (7.2.6)$$

3. Lees' algorithm as in eqn (7.2.4).

4. A scheme which Culham and Varga call a Modified Backward Time Approximation, which is like scheme (1) but with the matrices evaluated at level $n\Delta t$

   $$(\boldsymbol{C}_n+\Delta t\boldsymbol{K}_n)\boldsymbol{u}_{n+1}=\boldsymbol{C}_n\boldsymbol{u}_n+\Delta t\boldsymbol{f}_{n+1}. \quad (7.2.7)$$

   This is another explicit scheme.

Culham and Varga take a problem with one space dimension for which they have an exact solution and compare the cost efficiency of these time stepping schemes. For each type of space approximation, they find that the first and fourth methods are superior to the other two is in the sense that the second order accuracy of the Crank–Nicolson and Lees algorithms only shows when the time step is very much reduced. Culham and Varga come to the conclusion that when numerical solutions with relative errors of approximately 1.0 to 7.5 per cent are required the first order backward and modified backward schemes are superior. The second order Crank–Nicolson type method (2) is better for solutions where relative errors less than 5×10^{-3} per cent are required. Lees' algorithm is rated unsatisfactory because although second order the magnitude of the error is large compared to those of the other schemes except for very small time steps. As with other users of Lees' algorithm, Culham and Varga also find that it produces spurious oscillations.

Second order equations

The most obvious and most popular explicit algorithm for the second order system of equations (7.1.1) is that given by using Newmark with $\gamma = 0.5$ and $\beta = 0$, otherwise known as the central difference algorithm. This algorithm applied to linear problems is discussed in Chapter 2. Applied to eqn (7.1.1) we take

$$\boldsymbol{x}_{n+1} = \boldsymbol{x}_n + \Delta t \boldsymbol{v}_n + \frac{\Delta t^2}{2}\boldsymbol{a}_n \tag{7.2.8}$$

$$\boldsymbol{v}_{n+1} = \boldsymbol{v}_n + \Delta t(\boldsymbol{a}_n + \boldsymbol{a}_{n+1})/2 \tag{7.2.9}$$

and substitute for $\boldsymbol{x}_{n+1}$ and $\boldsymbol{v}_{n+1}$ into

$$\boldsymbol{M}\boldsymbol{a}_{n+1} + \boldsymbol{C}\boldsymbol{v}_{n+1} + \boldsymbol{P}(\boldsymbol{x}_{n+1}) = \boldsymbol{f}_{n+1} \equiv \boldsymbol{f}((n+1)\Delta t), \tag{7.2.10}$$

where $\boldsymbol{x}_n$, $\boldsymbol{v}_n$, $\boldsymbol{a}_n$ are the approximate displacement, velocity, and acceleration respectively at time $t = n\Delta t$. The new value of the acceleration $\boldsymbol{a}_{n+1}$ is then given by

$$\left(\boldsymbol{M} + \frac{\Delta t}{2}\boldsymbol{C}\right)\boldsymbol{a}_{n+1} = -\frac{\Delta t}{2}\boldsymbol{C}\boldsymbol{a}_n - \boldsymbol{P}\left(\boldsymbol{x}_n + \Delta t\boldsymbol{v}_n + \frac{\Delta t^2}{2}\boldsymbol{a}_n\right) + \boldsymbol{f}_n. \tag{7.2.11}$$

If the mass matrix $\boldsymbol{M}$ and the damping matrix $\boldsymbol{C}$ are lumped, this is another very simple algorithm to program. Several examples of the use of the central difference algorithm have been discussed in §2.4 in connection with the effect of mass lumping in producing some increase in the limit on the time step for stability. Oden and Fost (1973) which is mentioned there is actually discussing a non-linear problem representing a wide class of hyperelastic bodies including both physically non-linear materials and finite amplitude motion by the equation $u_{tt} - F(u_x) = f(x, t)$.

There are several slightly different forms of the central difference method which appear in the literature, equivalent to eqns (7.2.8) to (7.2.11) but differently expressed. For example, Park (1977) works with equations which can be expressed as

$$\boldsymbol{v}_{n+\frac{1}{2}} = \boldsymbol{v}_{n-\frac{1}{2}} + \Delta t \boldsymbol{a}_n \tag{7.2.12}$$

and

$$\boldsymbol{x}_{n+1} = \boldsymbol{x}_n + \Delta t \boldsymbol{v}_{n+\frac{1}{2}}, \tag{7.2.13}$$

which with

$$\boldsymbol{v}_n = (\boldsymbol{v}_{n+\frac{1}{2}} + \boldsymbol{v}_{n-\frac{1}{2}})/2 \tag{7.2.14}$$

comes to the same result as in the previous equations. McNamara (1974) works with increments using the notation $\Delta \boldsymbol{x}_{t+\Delta t}$ for $(\boldsymbol{x}_{n+1} - \boldsymbol{x}_n)$ in the notation of the present text. Then he expresses the central difference operator in the form

$$\Delta \boldsymbol{x}_{t+\Delta t} = \Delta \boldsymbol{x}_t + \Delta t^2 \ddot{\boldsymbol{x}}_t, \tag{7.2.15}$$

which is the same as

$$\boldsymbol{a}_n = (\boldsymbol{x}_{n+1} - 2\boldsymbol{x}_n + \boldsymbol{x}_{n-1})/\Delta t^2,$$

the central finite difference expression for the acceleration which is used in the central difference algorithm.

Belytschko *et al.* (1976) use direct calculation of the internal nodal forces without forming the stiffness matrix in a program for large-scale non-linear structure–media interaction (plane or axi-symmetric). They have zero damping and use the central difference method with a lumped mass matrix. The time step limit is given as

$$\Delta t \le \min \alpha \Delta l / c_{\mathrm{L}}, \tag{7.2.16}$$

where Δl is an element dimension and

$$c_{\mathrm{L}}^2 = \frac{3k(1-\nu)}{\rho(1+\nu)},$$

where k, ν, and ρ are the bulk modulus, Poisson's ratio and density respectively. Thus $\Delta l / c_{\mathrm{L}}$ is the time for the pressure wave to cross the element. The parameter α is a reduction factor, $0.2 < \alpha < 0.9$ which the authors say is necessary because of the destabilizing effects of round-off and the possibility of rapidly varying material properties. Energy balance (Chapter 6) is used to check the stability. Belytschko *et al.* remark that energy generated by an instability may be dissipated as the material becomes plastic, then larger time steps are permitted and stability is recaptured. The authors call this 'arrested stability' and say that this may cause errors of 10 to 100% which can, however, be detected by energy checks. ('Hour-glass' instability, caused by using one-point Gauss quadrature on bilinear elements, is also made apparent by an energy balance check.)

Zienkiewicz *et al.* (1986) use the explicit method SS22 with $\theta_2 = 0$ for the solution of a dynamic problem with the displacement, strain and stress independently approximated.

There are a number of instances where comparison has been made between the use of the central difference algorithm and an iterative method necessary for the solution of an implicit time stepping scheme for a non-linear dynamic problem. The iterative methods generally used in this context are those known as 'tangent stiffness' or 'pseudo-load' and these are described in §7.4. We must emphasise that there is no evidence to suggest that one method is suitable for all types of problem. The optimum method is problem-dependent. We have to remember that the properties of the algorithms proved for their application to linear equations do not automatically carry over to the non-linear equations. Also the one-degree-of-freedom problems frequently discussed in the literature may be helpful to some extent but they are not sufficient to determine the best methods because they do not include the stiffness effect from the high-frequency modes produced by the finite element space discretization. Another point to note is that some problems must have small time steps because of the rapid change of the forcing function.

McNamara looks at the performance of Newmark ($2\beta = \gamma = 0.5$), Houbolt (see §4.7) and the central difference method for the solution of a non-linear elastic problem of structural dynamics. He finds that with the Newmark and Houbolt algorithms used iteratively, convergence is only obtained when the time step is limited to the stability limit of the central difference operator, hence the performance of the central difference operator compares very favourably. He recommends that whatever the algorithm being used the time step should always be less than 10^{-2}

times the period T of the solution, a time step of $5 \times 10^{-3}\, T$ is reasonable. Then with the implicit methods the non-linear stiffness need only be updated every ten time steps and iteration is not necessary.

Belytschko *et al.* (1975) discuss the use of the central difference operator in the solution of large displacement problems. They observe that empirical evidence suggests that the stability limit

$$\lambda_{max}\Delta t \leq 2, \tag{7.2.17}$$

proved for linear problems still holds for a non-linear continuum but for structural problems the critical time step appears to be smaller. Belytschko *et al.* (1975) say that problems where the behaviour at the wave front is of engineering importance and high frequencies dominate are best solved by the explicit central difference algorithm. These problems include shock response from explosions or impact. Other problems are called inertial problems. These are problems where low frequencies dominate and they include seismic response and large deformations of elasto-plastic material structures with ramp or step loading. The rise time and duration of the forcing function compared with the time for a wave to traverse the structure is used in the identification here. If the rise time and duration exceed several traversal times, the problem is usually of inertial type and should be solved by an implicit algorithm. But the cost-efficiency of the two approaches is also dependent on the topology of the finite element mesh and the computer to be used. Belytschko *et al.* (1975) also have some comments on the spurious oscillations which frequently appear in the numerical solution. Damping these by algorithmic damping, for example using Newmark with $\gamma > 0.5$, is undesirable because this also damps out the oscillations which should be preserved. Belytschko *et al.* suggest removing the spurious oscillations by a filtering process which is a kind of averaging. We note that tacit averaging out of the oscillations produced by using too large elements in the spatial discretization has long been practised.

Park (1977) prefers the central difference method to other explicit multi-step methods because of the trouble with spurious solutions from methods which have stability polynomials of higher order than two. The results obtained by Park agree with those of McNamara (1974) to the effect that the restriction on the time step for convergence with implicit methods makes the central difference preferable.

Park and Underwood (1980) consider the situation where only low accuracy is required for the solution of the non-linear dynamic vibration equation (7.1.1). They claim that estimates of the local truncation error with the central difference algorithm or the difference between predictor and corrector are acceptable when the local relative error has to be kept to about 10^{-5} but they are unreliable and/or too expensive when the accuracy required is only a few per cent. They propose a criterion based on an 'apparent frequency' to estimate the highest frequency 'noise' in the system in any time step. This is calculated by forming what they call the perturbed apparent frequency

$$(\omega_{\mathrm{PA}}^2)_i = \frac{|\Delta a_{n,i}|}{|\Delta x_{n,i}|}, \tag{7.2.18}$$

where x and a denote displacement and acceleration as usual and $\Delta x_{n,i} = x_{n+1,i} - x_{n,i}$, the change in the displacement at the ith node and similarly for $\Delta a_{n,i}$. Then the time step is adjusted to make $\Delta t(\omega_{\mathrm{PA}})_{\max} \leq 2$ for stability. Underwood and Park (1980) gives numerical examples to illustrate the use of this criterion.

Kujawski and Desai (1984) propose an explicit method for the non-linear equation $\boldsymbol{M}\ddot{\boldsymbol{x}} + \boldsymbol{K}(\boldsymbol{x})\boldsymbol{x} = \boldsymbol{f}$ which is like the two-step form of Newmark with $\gamma = 0.5$ but uses an averaged extrapolated form of the stiffness matrix

$$\boldsymbol{K}_\theta = \boldsymbol{K}(\boldsymbol{x}_\theta), \tag{7.2.19}$$

where

$$\boldsymbol{x}_\theta = (1-\theta_1-\theta_2)\boldsymbol{x}_n + \theta_1\boldsymbol{x}_{n-1} + \theta_2\boldsymbol{x}_{n-2}. \tag{7.2.20}$$

The algorithm is given by the recurrence relation

$$\begin{aligned}(\boldsymbol{M} + \beta\Delta t^2\boldsymbol{K}_\theta)\boldsymbol{x}_{n+1} = \Delta t^2\boldsymbol{F}_n + \{2\boldsymbol{M} - (1-2\beta)\Delta t^2\boldsymbol{K}_\theta\}\boldsymbol{x}_n \\ - (\boldsymbol{M} + \beta\Delta t^2\boldsymbol{K}_\theta)\boldsymbol{x}_{n-1},\end{aligned} \tag{7.2.21}$$

where $\boldsymbol{F}_n = \beta\boldsymbol{f}_{n+1} + (1-2\beta)\boldsymbol{f}_n + \beta\boldsymbol{f}_{n-1}$, that is, the forcing function is averaged in the same pattern as the $\boldsymbol{Kx}$ terms. For the scalar equation $\ddot{x} + \omega^2 x = 0$, putting $x_{n+1} = rx_n$, the stability polynomial of the algorithm (7.2.21) is given by

$$(1+\beta\omega^2\Delta t^2)r^2 + \{-2 + (1-2\beta)\omega^2\Delta t^2\}r + (1+\beta\omega^2\Delta t^2) = 0. \tag{7.2.22}$$

We can use the Routh–Hurwitz conditions (Appendix 1) on this quadratic or simply note that the product of the roots is unity hence we want them to be complex. Either way the condition for stability is

$$(1-4\beta)\omega^2\Delta t^2 \leq 4, \tag{7.2.23}$$

so that (provided $\omega^2 > 0$) there is unconditional stability for $\beta \geq \frac{1}{4}$ as with Newmark. Kujawski and Desai recommend using $\beta = \frac{1}{4}$ and $\theta_1 = \theta_2 = 0$ so that the stiffness matrix is computed as $\boldsymbol{K}_\theta = \boldsymbol{K}(\boldsymbol{x}_n)$ using the value at the end of the previous time step. They obtain numerical results which compare favourably with standard Newmark with the non-linear equations solved by Newton–Raphson. This idea of using a single averaged value of the non-linear stiffness corresponds to the generalized midpoint method for first order equations which is discussed in §7.3. An extrapolated value of the argument of the stiffness matrix is used by Kujawski and Desai in order to construct an explicit scheme but there is the further possibility of including $\boldsymbol{x}_{n+1}$ in the argument and using iteration on the resultant one-step form of the algorithm.

7.3. *Ad hoc* iteration for first order systems

The general form of equation we want to consider in this section is given by

$$\boldsymbol{C}(\boldsymbol{u})\dot{\boldsymbol{u}} + \boldsymbol{K}(\boldsymbol{u})\boldsymbol{u} = \boldsymbol{f}(t). \tag{7.3.1}$$

As explained in chapter 1 this equation can represent the spatially discretized numerical model of an aquifer, either vertically integrated up to the water table or using Richards' equation for saturated–unsaturated subsurface flow, or a problem of non-linear heat conduction or diffusion. The matrix $\boldsymbol{C}(\boldsymbol{u})$ is the non-linear mass matrix which is positive definite and may or may not be taken as diagonal. The matrix $\boldsymbol{K}(\boldsymbol{u})$ is the non-linear stiffness matrix type which is positive definite or positive semi-definite depending on the boundary conditions. With a complicated non-linear problem it is usual in practice to keep to a simple time-stepping scheme. But even with the θ-method, which is so simple in the linear case, it is not immediately obvious how best to handle the non-linear mass and stiffness matrices. In order to investigate the stability, we look first at a homogeneous scalar equation of the same form as (7.3.1).

$$\dot{u} + \lambda(u)u = 0, \tag{7.3.2}$$

where we assume that $\lambda(u) > 0$ (corresponding to the stiffness matrix remaining positive definite). The θ-method as given in Chapter 2 applied to eqn (7.3.2) gives the formula

$$\frac{1}{\Delta t}(u_{n+1} - u_n) = -\{\theta\lambda_{n+1}u_{n+1} + (1-\theta)\lambda_n u_n\}, \tag{7.3.3}$$

writing $\lambda_n = \lambda(u_n)$, where $0 \leq \theta \leq 1$. This gives the recurrence relation

$$u_{n+1} = Au_n,$$

where

$$A = \frac{\{1-(1-\theta)\Delta t\lambda_n\}}{(1+\theta\Delta t\lambda_{n+1})}. \tag{7.3.4}$$

For stability we want the modulus of the amplification factor A to stay less than or equal to unity, hence we require

$$|1-(1-\theta)\Delta t\lambda_n| \leq 1 + \theta\Delta t\lambda_{n+1},$$

i.e.

$$\Delta t\{(1-\theta)\lambda_n - \theta\lambda_{n+1}\} \leq 2. \tag{7.3.5}$$

For the fully implicit scheme with $\theta = 1$ there is no limitation on the time step for stability. For any other value of θ there is a limitation whenever the left-hand side of (7.3.5) is positive, i.e. when $(1-\theta)\lambda_n > \theta\lambda_{n+1}$. In particular for the trapezium rule equivalent, Crank–Nicolson, with $\theta = 0.5$, the algorithm is liable to become unstable whenever $\lambda_n > \lambda_{n+1}$ unless the time step is restricted by the formula

$$\Delta t \leq 4/(\lambda_n - \lambda_{n+1}). \tag{7.3.6}$$

This difficulty is removed if instead of using the θ-method in the form (7.3.3) the formula is adjusted so that λ appears in it evaluated at one level instead of two. Thus we can write

$$\frac{1}{\Delta t}(u_{n+1} - u_n) = -\lambda(u_{n+\theta})\{\theta u_{n+1} + (1-\theta)u_n\}, \tag{7.3.7}$$

where $u_{n+\theta} = \theta u_{n+1} + (1-\theta)u_n$. The amplification factor is now

$$A = \frac{1-(1-\theta)\Delta t \lambda_{n+\theta}}{1+\theta \Delta t \lambda_{n+\theta}} \tag{7.3.8}$$

and the stability condition gives

$$(1-2\theta)\lambda_{n+\theta}\Delta t \leq 2. \tag{7.3.9}$$

Thus we are in the same position as with the linear case that with $\theta \geq 0.5$ we have unconditional stability. In particular with $\theta = 0.5$ we have the midpoint rule which when applied to eqn (7.3.2) produces the formula

$$\frac{1}{\Delta t}(u_{n+1} - u_n) = -\lambda(\bar{u}_n)\bar{u}_n, \tag{7.3.10}$$

where $\bar{u}_n = (u_n + u_{n+1})/2$. Gourlay (1970) pointed out the advantage of the midpoint rule in this context. Hughes (1977) did the same thing with the generalized midpoint rule with reference to non-linear heat conduction. Dahlquist (1975) refers to the midpoint rule as the one-leg twin of the Crank–Nicolson trapezium rule form as in eqn (7.3.7) with $\theta = 0.5$.

We note from eqn (7.3.8) that whatever averaging formula is used for the non-linear term the only value of θ that guarantees that the amplification factor stays positive is $\theta = 1$. For any other value we are liable to get oscillations when $\lambda \Delta t > 1/(1-\theta)$ and if we are also retaining $\theta \geq 0.5$ for unconditional stability, the right hand side of this inequality is least when $\theta = 0.5$.

We now return to the equations representing groundwater flow because these are useful in illustrating various points in connection with the solution of non-linear first order systems. Extending their introduction in Chapter 1, these equations can be taken as the following:

1. Equations applying to the saturated zone only which is a system of equations of the form

$$n_e \frac{\partial h}{\partial t} = \frac{\partial}{\partial x}\left\{K_x(h-\eta)\frac{\partial h}{\partial x}\right\} + \frac{\partial}{\partial y}\left\{K_y(h-\eta)\frac{\partial h}{\partial y}\right\} + Q \tag{7.3.11}$$

(the Boussinesq equation), where h is the height of the water table and η the height of the (impervious) bottom of the aquifer above datum, n_e is the effective porosity, x and y are horizontal coordinates, K_x, K_y are the hydraulic conductivity components in the x and y directions respectively and Q represents a source or sink term. Water engineers frequently linearize eqn (7.3.11) by taking averaged values of $K_x(h-\eta)$, $K_y(h-\eta)$ which are then called the transmissivities T_x, T_y, as, for example, in Bear (1979), Connorton and Wood (1983). Without this extra assumption, the space dicretization gives the non-linear form

$$\boldsymbol{M}\dot{\boldsymbol{h}} + \boldsymbol{K}(\boldsymbol{h})\boldsymbol{h} = \boldsymbol{f}(t), \tag{7.3.12}$$

where $\boldsymbol{M}$ is the standard mass matrix, $\boldsymbol{h}$ is the vector of nodal values of the height of the water table and $\boldsymbol{K}(\boldsymbol{h})$ is now the non-linear stiffness type matrix whose entries depend on $\boldsymbol{h}$.

2. Alternatively the subsurface flow can be modelled using Richards' equation (Richards 1931) for the saturated and unsaturated flow in the three-dimensional region Ω up to ground level. This can be written as an equation for $\Phi(x, y, z, t)$ the total hydraulic potential as follows

$$C(\Psi)\frac{\partial \Phi}{\partial t} = \frac{\partial}{\partial x}\left\{K_x(\Psi)\frac{\partial \Phi}{\partial x}\right\} + \frac{\partial}{\partial y}\left\{K_y(\Psi)\frac{\partial \Phi}{\partial y}\right\} + \frac{\partial}{\partial z}\left\{K_z(\Psi)\frac{\partial \Phi}{\partial z}\right\} + Q, \tag{7.3.13}$$

where x, y are horizontal coordinates and z is height above datum. The moisture potential Ψ is related to the hydraulic potential by the formula $\Psi = \Phi - z$, $C(\Psi)$ is the specific moisture capacity, $K_x(\Psi)$, $K_y(\Psi)$, $K_z(\Psi)$ are the components of the hydraulic conductivity, and Q is the source or sink term. We can take the source or sink term Q situated at (x_s, y_s, z_s) as $Q = Q\delta(x - x_s, y - y_s, z - z_s)$, using the Dirac delta function (Stakgold 1967), where Q is the strength and the dimensions of the source or sink are small compared with the size of the finite elements.

With any model of groundwater flow we must have starting conditions. There are some very different opinions about these which are discussed separately in §7.5. We also have boundary conditions which must be either

(1) Dirichlet, where the values of the height of the water table in eqn (7.3.12) or the total hydraulic potential Φ in eqn

(7.3.15) are known on part of the boundary Γ_1, also referred to as a 'known head' boundary; or

(2) a prescribed flux on the rest of the boundary Γ_2, where we have equations

$$K_x \frac{\partial \Phi}{\partial x} n_1 + K_y \frac{\partial \Phi}{\partial y} n_2 = v, \tag{7.3.14}$$

where (n_1, n_2) is the unit outward normal on the boundary Γ_2, or the equivalent in three dimensions. If $v = 0$, eqn (7.3.14) represents the Neumann no-flow condition.

When space discretized, the Richards' equation for the saturated–unsaturated flow problem produces the non-linear system

$$\boldsymbol{M}(\boldsymbol{\Phi})\dot{\boldsymbol{\Phi}} + \boldsymbol{K}(\boldsymbol{\Phi})\boldsymbol{\Phi} = \boldsymbol{f}. \tag{7.3.15}$$

The matrix $\boldsymbol{M}(\boldsymbol{\Phi})$ has the form of the standard mass matrix with entries

$$m_{ij} = \iiint_{\Omega} C(\Psi) N_i N_j \, \mathrm{d}x \, \mathrm{d}y \, \mathrm{d}z, \tag{7.3.16}$$

where N_j are the basis functions. In the context of groundwater flow the matrix $\boldsymbol{M}$ is known as the capacitance matrix. The matrix $\boldsymbol{K}(\boldsymbol{\Phi})$ is the standard stiffness matrix type with entries

$$k_{ij} = \iiint_{\Omega} \left(K_x \frac{\partial N_i}{\partial x} \frac{\partial N_j}{\partial x} + K_y \frac{\partial N_i}{\partial y} \frac{\partial N_j}{\partial y} + K_z \frac{\partial N_i}{\partial z} \frac{\partial N_j}{\partial z} \right) \mathrm{d}x \, \mathrm{d}y \, \mathrm{d}z. \tag{7.3.17}$$

The matrix $\boldsymbol{M}$ contains the non-linear term $C(\Psi)$ and the stiffness matrix $\boldsymbol{K}$ contains the non-linear terms $K_x(\Psi)$, $K_y(\Psi)$, $K_z(\Psi)$ which all have to be calculated from curves fitted to the data which allow for smooth transfer between the unsaturated and saturated regions.

The vector $\boldsymbol{f}$ in eqn (7.3.15) includes the Dirichlet boundary conditions from the known head boundary as well as the data from the source or sink terms. It is of course possible to reduce the number of equations to be solved by some averaging in the direction of one of the horizontal coordinates.

The two approaches to modelling groundwater flow thus lead to systems of equations (7.3.12) and (7.3.15) which each have the same pattern as that of eqn (7.3.1) Equation (7.3.1) could also represent non-linear heat conduction or other types of parabolic problem.

We consider variations on the θ-method time-stepping scheme applied to the numerical solution of the system of eqns (7.3.15) which may be written as

$$(\boldsymbol{M}+\theta\Delta t\boldsymbol{K})\boldsymbol{\Phi}_{n+1}=\{\boldsymbol{M}-(1-\theta)\Delta t\boldsymbol{K}\}\boldsymbol{\Phi}_n+\Delta t\{\theta\boldsymbol{f}_{n+1}+(1-\theta)\boldsymbol{f}_n\}. \tag{7.3.18}$$

A decision has to be made about how to deal with the non-linear terms in the matrices $\boldsymbol{M}$ and $\boldsymbol{K}$ which depend on the solution vector $\boldsymbol{\Phi}$. This means that for each time step there has to be an iterative solution of eqn (7.3.18). The iteration is continued until some criterion is satisfied as discussed in §7.1. The iteration process is then said to have converged. It seems relevant here to discuss the use of the word 'convergence' as it is frequently used in two different senses without qualification and this can be confusing.

The mathematics literature puts much emphasis on the convergence of the numerical solution to the true solution of the differential equation as the element size and the time step tend to zero. Strang and Fix (1973), Chapter 7, for example, gives the error in the solution of eqn (7.3.13) by the time-stepping scheme in eqn (7.3.18) as

$$\mathrm{O}(h^k)+\mathrm{O}(\Delta t) \quad \text{for } \theta\neq 0.5 \tag{7.3.19}$$

and

$$\mathrm{O}(h^k)+\mathrm{O}(\Delta t^2) \quad \text{for } \theta=0.5, \tag{7.3.20}$$

where h is a measure of the size of the elements and k is the order of the complete polynomial in the finite element basis functions (for example, $k=2$ for linear triangles or four-node bilinear quadrilaterals). From (7.3.19) and (7.3.20), we have the sense in which the numerical solution converges to the true solution as the finite element size h and the time-step size Δt tend to zero. But in practice we usually have to work with elements which are not particularly small in order to keep the number N of unknown nodal coordinates of manageable size. If N is increased too much we not only have a larger matrix to invert but also possible trouble with the condition number. From eqn (7.3.18) we can see that with $\theta \geq 0.5$ for unconditional stability and the rather large values of the time step Δt which may consequently be used, the matrix to be inverted becomes dominated by the term with the stiffness matrix which is notorious for becoming ill-conditioned as its order increases. A further consideration is that it may not be possible to use a value of the time step sufficiently small to justify writing the corresponding error term as '$O(\Delta t)$' or '$O(\Delta t^2)$' which necessarily refer to what happens as Δt tends to zero.

The other sense of the word 'convergence' which needs to be made specific when it is used is that of the convergence of the iteration on the time step which is necessary to solve the non-linear equation (7.3.18) where the matrices $\boldsymbol{M}$ and $\boldsymbol{K}$ depend on $\boldsymbol{\Phi}$.

We now look at three ways of tackling the non-linear equation (7.3.18) each expressed for convenience in the notation employed in this section.:

1. Neuman (1973) in a paper on the solution of saturated–unsaturated flow by finite elements uses the equivalent of

$$\{\boldsymbol{M}(\boldsymbol{\Phi}_{n+\frac{1}{2}}) + 0.5\Delta t_n \boldsymbol{K}(\boldsymbol{\Phi}_{n+\frac{1}{2}})\}\boldsymbol{\Phi}_{n+1}$$
$$= \{\boldsymbol{M}(\boldsymbol{\Phi}_{n+\frac{1}{2}}) - 0.5\Delta t_n \boldsymbol{K}(\boldsymbol{\Phi}_{n+\frac{1}{2}})\}\boldsymbol{\Phi}_n + \Delta t_n \boldsymbol{f}_{n+\frac{1}{2}}, \quad (7.3.21)$$

where the time step is given as Δt_n to make allowance for change from one time step to the next. If the equation were linear this would be the same as the $\theta = 0.5$, Crank–Nicolson method except for the last term on the right hand side which has $\boldsymbol{f}_{n+\frac{1}{2}}$ instead of $0.5(\boldsymbol{f}_n + \boldsymbol{f}_{n+1})$.

Neuman is considering either a plane flow problem or an axisymmetric problem and using linear triangles or bilinear quadrilaterals.

In order to start the iteration on the time step the value of $\boldsymbol{\Phi}_{n+\frac{1}{2}}$ is predicted by linear extrapolation from previous values as

$$\boldsymbol{\Phi}_{n+\frac{1}{2}} = \boldsymbol{\Phi}_n + \frac{\Delta t_n}{2\Delta t_{n-1}}(\boldsymbol{\Phi}_n - \boldsymbol{\Phi}_{n-1}). \quad (7.3.22)$$

This is not available on the first time step but it has to be accepted that there will be more iterations there. Then if $\boldsymbol{\Phi}^k_{n+1}$ is the value of $\boldsymbol{\Phi}_{n+1}$ given by the kth iteration, the iterative process is continued by replacing

$$\boldsymbol{\Phi}_{n+\frac{1}{2}} \quad \text{by} \quad 0.5(\boldsymbol{\Phi}_n + \boldsymbol{\Phi}^k_{n+1}). \tag{7.3.23}$$

2. Frind and Verge (1978) use a form of eqn (7.3.18) equivalent to

$$\{\boldsymbol{M}(\boldsymbol{\Phi}_{n+\frac{1}{2}}) + \Delta t_n \boldsymbol{K}(\boldsymbol{\Phi}_{n+\frac{1}{2}})\}\boldsymbol{\Phi}_{n+1} = \boldsymbol{M}(\boldsymbol{\Phi}_{n+\frac{1}{2}})\boldsymbol{\Phi}_n + \Delta t_n \boldsymbol{f}_{n+\frac{1}{2}}. \tag{7.3.24}$$

If the equation were linear this would be the $\theta = 1$ form of the θ-method except that the last term on the right-hand side would then be $\boldsymbol{f}_{n+1}$. The iteration is started by taking $\boldsymbol{\Phi}_{n+\frac{1}{2}}$ extrapolated as in eqn (7.3.22) and continued as in (7.3.23). Frind and Verge are considering a three-dimensional problem using isoparametric bricks with which any side can be either linear, quadratic, or cubic.

3. The Institute of Hydrology Distributed Model (IHDM) (Beven *et al.* 1987), uses for the solution of Richards' equation the same scheme as that used by Frind and Verge (1978) given by eqns (7.3.24) and (7.3.22). The iteration is accelerated by an adaptation of the Cooley method (Cooley 1983) which works extremely efficiently for this problem. The equations are arranged so that they are solved for the change in the value of the pressure $\boldsymbol{\Phi}_{n+1}$ from one iteration to the next, that is for

$$\Delta\boldsymbol{\Phi}^k_{n+1} = \boldsymbol{\Phi}^k_{n+1} - \boldsymbol{\Phi}^{k-1}_{n+1}. \tag{7.3.25}$$

Then the method takes

$$\boldsymbol{\Phi}^k_{n+1} = \omega_k \Delta\boldsymbol{\Phi}^k_{n+1} + \boldsymbol{\Phi}^{k-1}_{n+1}, \tag{7.3.26}$$

where ω_k is a damping parameter. This method depends essentially on the assumption that these changes in nodal pressure will oscillate in sign from one iteration to the next so that the convergence will be accelerated by damping the oscillation, that is, by under-relaxation. If Δ_k is the maximum change in nodal pressures at the kth iteration then we put

(1)
$$\begin{aligned} s &= 1, \quad \text{for} \quad k = 0 \\ s &= \Delta_{k+1}/(\omega_k \Delta_k), \quad \text{for} \quad k > 0, \end{aligned} \tag{7.3.27}$$

where ω_k is the Cooley relaxation parameter from the previous iteration;

(2)
$$\hat{\omega} = (3+s)/(3+|s|), \quad \text{when } s > -1, \qquad \hat{\omega} = 0.5/|s|, \quad \text{otherwise,} \tag{7.3.28}$$

(3) If $\hat{\omega}\,|\Delta_{k+1}|$ is greater than $e_{\max}$, the specified maximum change in the potential allowed in an iteration, then

$$\omega_{k+1} = e_{\max}/|\Delta_{k+1}|,$$

otherwise (7.3.29)

$$\omega_{k+1} = \hat{\omega}.$$

The iteration is stopped when $|\Delta_{k+1}|$ is less than some prescribed tolerance. We can see that this is necessarily an underrelaxation situation. (We recall that the successive over relaxation (SOR) method for solving systems of simultaneous equations is notorious for producing small changes in the solution when still a long way from convergence.)

Frind and Verge (1978) remark that the backward difference scheme using the non-linear matrices evaluated with the average values of the nodal potentials as in eqn (7.3.24) tends to be resistant to spurious oscillations. In the experience of the author of this book these spurious oscillations or ripples appear when a scheme of the form (7.3.21) is used and are due to the Crank–Nicolson effect producing negative eigenvalues in the amplification matrix as described in Chapter 2. This effect is also illustrated in the simple scalar example in eqn (7.3.2). When the implicit scheme (7.3.24) is used these spurious ripples disappear, whether the non-linear terms are evaluated at the averaged or at the end value of the nodal coordinates.

In general with the iterative methods discussed in this chapter it is understood that the system of equations is solved at each iteration by a standard Gauss or Cholesky package. There can be difficulties when the number of equations is very large and/or the matrix to be inverted is ill-conditioned. In §7.9 we describe the preconditioned conjugate gradient method which, with element-by-element implementation, is an exciting new method for dealing with these problems. Here we note a proposal by Gellert (1985) to solve the vertically integrated unconfined groundwater flow equation (7.3.11) by a scheme corresponding to eqn (7.3.12) time stepped by the formula

$$\{\boldsymbol{M} + \theta\Delta t_{n+1}\boldsymbol{K}(\bar{\boldsymbol{h}})\}\boldsymbol{h}_{n+1} = \{\boldsymbol{M} - (1-\theta)\Delta t_{n+1}\boldsymbol{K}(\bar{\boldsymbol{h}})\}\boldsymbol{h}_n + \bar{\boldsymbol{f}}, \tag{7.3.30}$$

where the bar denotes averaged values. Gellert takes a lumped mass matrix and, splitting the stiffness matrix into diagonal and non-diagonal terms, solves the system by accelerated Jacobi iteration (Varga 1962). This avoids the inversion of non-diagonal matrices. Conditions for relating the size of time step to the number of iterations necessary for numerical stability are examined in detail.

The groundwater flow problems naturally extend to the analysis of aquifer contaminant transport. When the contaminant is of the same density as the groundwater, the velocities may be obtained from using Darcy's equation on the solution of eqn (7.3.12) or (7.3.15) to substitute into the convection–diffusion equation for the contaminant given by Bear (1979, p. 251) as

$$\frac{\partial(nc)}{\partial t} = \nabla \cdot nD \cdot \nabla c - \nabla \cdot qc, \tag{7.3.31}$$

where c = concentration of contaminant, n = the volumetric fraction of the liquid phase, D is the diffusion tensor and q is the vector of velocity components of the groundwater (see Bear 1979 for details.) When space discretized by finite elements the result is an equation of the form

$$\boldsymbol{M}^* \frac{\partial \boldsymbol{c}}{\partial t} + \boldsymbol{K}^*\boldsymbol{c} = \boldsymbol{f}^*, \tag{7.3.32}$$

where $\boldsymbol{M}^*$ is of mass matrix type but $\boldsymbol{K}^*$, because of the presence of the convection terms in eqn (7.3.31), is no longer symmetric. Difficulties can arise when the effect of convection is large compared with that of the diffusion. In the differential equation this is represented by the Reynolds number. The reader is again referred to Bear (1979) for details; most groundwater flows are such that the viscous forces predominate and the Reynolds number is less than 10. In the numerical solution, the effect of the diffusion and convection terms in the matrix K^* is measured by the 'mesh Peclet number',

$$P_e = |q|\,\Delta x/D, \tag{7.3.33}$$

where Δx is a measure of the size of the element and $|q|$ and D are the local magnitudes of the velocity and diffusion respectively. Yeh (1985) reports some interesting experiments on the solution of the linear equation for the concentration, eqn (7.3.32), by the fully implicit θ-method ($\theta = 1$). Yeh compares Successive Under-Relaxation (SUR), Gauss–Seidel, and Successive Over-Relaxation (SOR), (for details see Varga 1962), with Peclet numbers $P_e = 0.05$, 1.0, 50, ∞, and concludes that SUR and Gauss–Seidel converge for all Peclet numbers but SOR will only converge for $P_e \ll 1$. Yeh uses a distributed mass matrix.

Mariño (1981) also solves the subsurface flow problem using finite element space discretization on the saturated–unsaturated flow equation (7.3.13) in two dimensions in elevation (that is, no change in one horizontal direction). The Darcy equations for the velocity components may be written as

$$q_x = -K_x \frac{\partial h}{\partial x}, \qquad q_y = -K_y \frac{\partial h}{\partial y}, \tag{7.3.34}$$

where $h =$ the hydraulic head obtained from the groundwater flow equation, and K_x, K_y are the effective components of the hydraulic conductivity in the x and y directions respectively. Mariño solves the eqns (7.3.34) weakly by expressing the q_x and q_y in terms of nodal values and the same basis functions as are used for the hydraulic head, h. Then the nodal values of the velocity components are available for use in the equation for the movement of the contaminant. Mariño is using linear triangle elements.

In this connection we note that with any equation of the form of the vertically integrated flow in an unconfined aquifer as in eqn (7.3.11) or the saturated–unsaturated flow equation (7.3.13), if the basis functions are N_j, the program has to calculate $\partial N_j/\partial x$, $\partial N_j/\partial y, \ldots$, at the numerical quadrature points in order to form the stiffness matrix $\boldsymbol{K}$. Hence only another line or two of program is needed to calculate the values of the velocities at the quadrature points:

$$q_x = -\sum_j h_j K_x \frac{\partial N_j}{\partial x}, \ldots$$

or (7.3.35)

$$q_x = -\sum_j \Phi_j K_x \frac{\partial N_j}{\partial x}, \ldots$$

when the solution for h or Φ has been obtained. Thus the components of velocity at the quadrature points are available ready to calculate the convection terms in the matrix $\boldsymbol{K}^*$ in the concentration equation (7.3.32).

With the eight-node element, picking up the components of velocity at the two-by-two Gauss points also has the advantage of higher accuracy because these are the Barlow 'stress points' (Barlow 1976). This idea is used in Matthews and Wood (1987).

Another example of an iterative scheme is given in Griffiths (1978) for the purpose of solving the vertically integrated flow in an unconfined aquifer represented by eqn (7.3.11) incorporating the effect of a seepage face. With the bottom of the aquifer level and taken as datum the equation becomes

$$n_e \frac{\partial h}{\partial t} = \frac{\partial}{\partial x}\left\{K_x \frac{\partial}{\partial x}\left(\frac{h^2}{2}\right)\right\} + \frac{\partial}{\partial y}\left\{K_y \frac{\partial}{\partial y}\left(\frac{h^2}{2}\right)\right\} + Q. \quad (7.3.36)$$

Griffiths takes $\Psi = h^2/2$ so that eqn (7.3.36) becomes

$$n_e \frac{\partial h}{\partial t} = \frac{\partial}{\partial x} K_x \frac{\partial \Psi}{\partial x} + \frac{\partial}{\partial y} K_y \frac{\partial \Psi}{\partial y} + Q. \quad (7.3.37)$$

Then h and Ψ are approximated with the same set of basis functions using eight-node quadrilaterals. The time-stepping scheme is arranged so that the solution on each time step is given by a predictor–corrector form with

predictor: $$\boldsymbol{h}_{n+1}^{\mathrm{P}} = \boldsymbol{h}_n - \Delta t \boldsymbol{M}^{-1}(\boldsymbol{K}\boldsymbol{\Psi}_n - \boldsymbol{f}_n) \quad (7.3.38)$$

where the matrices $\boldsymbol{M}$ and $\boldsymbol{K}$ are the (distributed) mass and stiffness matrices respectively and $\boldsymbol{f}_n$ contains the known boundary conditions

corrector: starting with $\boldsymbol{h}_{n+1}^{\circ} = \boldsymbol{h}_{n+1}^{\mathrm{P}}$,

$$\Psi_{n+1}^{i+1} = \tfrac{1}{2}(h_{n+1}^{i})^2 \quad \text{at nodes not on the seepage face,}$$

$$\Psi_{n+1}^{i+1} = \frac{s^2}{2} \quad \text{at nodes on the seepage face,} \quad (7.3.39)$$

where s is the water level at the seepage face, and

$$\boldsymbol{h}_{n+1}^{i+1} = \boldsymbol{h}_n - \frac{\Delta t}{2}\boldsymbol{M}^{-1}(\boldsymbol{K}\boldsymbol{\Psi}_{n+1}^{i+1} - \hat{\boldsymbol{f}}_{n+1} + \boldsymbol{K}\boldsymbol{\Psi}_n - \boldsymbol{f}_n), \quad (7.3.40)$$

where $\hat{f}_{n+1}$ includes the boundary flux due to the seepage as estimated from the last iteration. This is then intended to be iterated to convergence.

In the original Griffiths (1978) dissertation there is a wrong sign for the flux normal to the seepage face which in the program caused the water to pile up there and prevented convergence. With this corrected, the program has been used successfully for modelling large-scale groundwater flow. In particular it is the program used in Matthews and Wood (1987).

In the literature on the numerical solution of these groundwater flow problems, there are differing opinions on whether the mass matrix should be lumped or not. This question, which has implications for other kinds of parabolic problems, is discussed in the next section.

7.4. Lumped versus distributed mass matrices in subsurface flow

A mathematician is likely to feel instinctively that the mass matrix in eqn (7.3.21) or (7.3.24) should be distributed because this is more accurate. But it is clearly easier to program the numerical solution with a lumped mass matrix. If N_j are the basis functions and the lumping is done by simply adding the terms in a row into the diagonal, these diagonal entries for the saturated–unsaturated flow equation (7.3.13), for example, are

$$\sum_i \iint_V CN_iN_j \,\mathrm{d}V = \iint_V \sum_i CN_iN_j \,\mathrm{d}V = \iint_V CN_j \,\mathrm{d}V, \quad (7.4.1)$$

where V is the region occupied by the flow, because the sum of the basis functions is everywhere equal to unity, that is,

$$\sum_i N_i = 1.$$

This formula is safe when linear triangles or bilinear quadrilaterals are being used but not for higher order elements. For example, for six-node triangles, the vertex basis functions in terms of area coordinates L_1, L_2, L_3 (Zienkiewicz 1977) are

$$\Phi_j = 2L_j^2 - L_j, \quad j = 1, 2, 3.$$

With the useful formula

$$\iint_{\Delta} L_1^a L_2^b L_3^c \, dx \, dy = \frac{2a!b!c!}{(a+b+c+2)!} \Delta, \tag{7.4.2}$$

where Δ is the area of the triangle, we have

$$\int_{\Delta} \Phi_j \, dx \, dy = 0.$$

Hence this formula will not do for the lumped mass matrix here. For higher order elements some simple form of positive weighting must be used.

It might also seem reasonable to suppose that less cpu time would be needed with a lumped mass matrix because there are only the diagonal terms to update and add in to the matrix to be inverted at each iteration. But the reasons given for lumping are less straightforward than these. Neuman (1973) appears to have formed a paradigm for many authors and researchers. Neuman advocates the use of a lumped mass matrix, saying that experience shows that this is necessary 'for the numerical method to converge in the case of unsaturated flow... A much more stable solution is obtained by...' taking the mass matrix as lumped. Presumably Neuman is referring to the convergence of the iteration on the time step here but he has previously referred to the convergence in the L_2 norm of the finite element solution to the true solution at a particular time as the number of nodes tends to infinity. The use of the word 'convergence' unqualified may lead to confusion. Neuman then uses the lumped mass matrix for both saturated and unsaturated flow which also makes it easier for him to employ both triangular and quadrilateral elements. The time-stepping scheme used by Neuman is described in §7.3 eqns (7.3.21) to (7.3.23).

Neuman does not say how many iterations are necessary on average nor what criterion is used for convergence of the iteration. But he does say that near quadratic convergence can be obtained in many cases and the method is superior to Newton–Raphson which requires a complicated matrix and a close starting value. Neuman also considers the possibility of the physical conditions of the subsurface flow being such that the diagonal terms of the lumped mass matrix might vanish and then he advocates using the backward difference formula as in eqn (7.3.24).

Mariño (1981) cites Neuman (1973) as the reason for using the lumped mass matrix for both the groundwater flow and in the equation for the movement of contaminant.

Narasimhan (1978), who is also in favour of lumping the mass matrix, quotes Neuman (1973) as generally having trouble with the convergence of the iteration when using a distributed mass matrix, whereas Neuman only reported difficulty in the unsaturated flow. Narasimhan also gives as evidence in favour of lumping the mass matrix the example of Fujii (1973) who, Narasimhan says, showed that when the consistent mass matrix is used the time step must not be too small nor too large for the maximum principle to hold. Actually what Fujii showed was that the maximum principle holds if but not only if the time step is between certain limits.

Frind and Verge (1978) investigate the use of the lumped mass matrix for the purpose of economizing on computer time. They report excellent results obtained using the lumped mass matrix in the unsaturated region but not in the saturated region. (They also investigated the effect of reduced integration as suggested by Zienkiewicz *et al.* (1971) for the general analysis of plates and shells and found that this is not to be recommended for flow problems.)

Cooley (1983) also quotes Fujii (1973) as evidence that lumping the mass matrix improves stability. This is so, as shown in §2.4, when a conditionally stable time-stepping scheme is used. But Cooley is using the fully implicit scheme corresponding to

$$\frac{1}{\Delta t_n}\boldsymbol{M}_{n+1}(\boldsymbol{\Phi}_{n+1}-\boldsymbol{\Phi}_n)+\boldsymbol{K}_{n+1}\boldsymbol{\Phi}_{n+1}=\boldsymbol{Q}_{n+1}, \tag{7.4.3}$$

which is unconditionally stable.

Further examples of opinions on whether the mass matrix should or should not be lumped include the following. Hromadka and Guymon (1980) have a one space dimensional problem which is a numerical model of an unsaturated horizontal homogeneous column of soil. They are primarily looking at ways of modelling the soil water diffusivity but they also experiment with different forms of the mass/capacitance matrix and conclude that for the non-linear problem the distributed mass matrix together with diffusivity coefficients evaluated as the mean parabolic value of the state variable gives the best results. Huyakorn and Pinder (1983) in a textbook on computational methods in subsurface flow again refer to Neuman (1973) and say that experience shows that a more stable but usually less accurate solution is produced by using the lumped mass matrix.

It seems relevant to recall here that Gresho *et al.* (1978) report tests on numerical solutions with lumped and distributed mass matrices for one- and two-dimensional linear convection–diffusion equations (like the linear form of eqn (7.3.31)). They use very accurate time-stepping schemes so that the effect of the space discretization is highlighted and conclude that the solutions with the distributed mass matrices are greatly superior. They find the same result with pure convection. An eminently quotable comment from this paper is 'partial justification for mass lumping might be that virtually all finite difference methods use it (another point of view, of course, would tend to shun mass lumping for this same reason)'.

The Institute of Hydrology Distributed Model (IHDM) (Beven *et al.* 1987) for saturated–unsaturated groundwater flow, as described in §7.3, uses a backward difference on the time step with averaged values for arguments of the mass and stiffness matrices. Wood and Calver (1990) report experiments to compare the performance of this program with lumped and distributed mass matrices. The particular subsurface solution considered in these experiments deals with saturated–unsaturated flow in a vertical section of a hill slope with variable width. At the downslope face outflow can take place into a channel. Figure 7.4.1 (from Calver and Wood 1989) shows the hill slope and the kind of finite element meshes used in the experiments. The meshes chosen are from those used in Calver and Wood (1989), listed in Table 7.4.1. The simulation includes a period of 25 hours without rainfall followed by 5 hours rainfall at 4 mm per hour followed by another 25 hours without rainfall. The experiments include cases where the proportion of the area which is saturated varies from 20 to 80% approximately, because of the statements in the literature reported above that this could make a significant difference to the performance of the program depending on whether the mass matrix is lumped or not. Care has to be

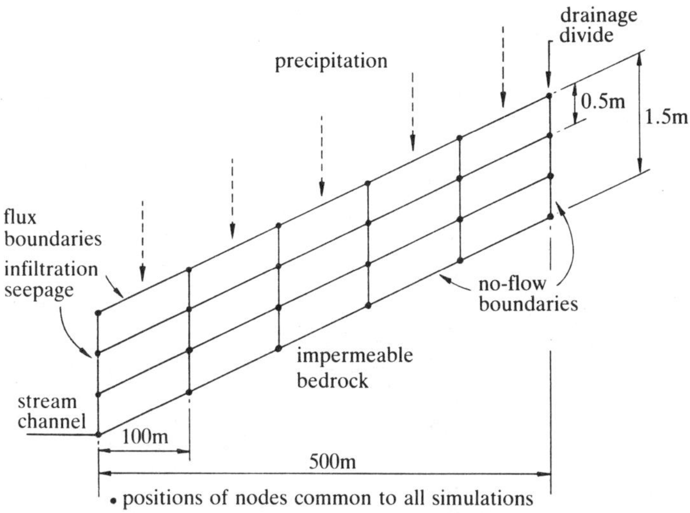

Fig. 7.4.1. Hill-slope elements. (From Calver and Wood 1989.)

taken not to involve overland flow. The conclusions from these experiments are that

(1) lumping makes no difference to the rate of convergence of the iteration;

(2) with the distributed mass matrix the cpu time is only very slightly greater;

(3) the difference in the solution values for the potential is within the accuracy justified by observational error;

(4) there is a significant difference in the values for the flow at known head boundaries which are calculated by the Lynch method.

The following is a brief explanation of the method of Lynch (1984) which was proposed in connection with mass conservation in groundwater flow but is actually of very general application. We first take a non-dimensionalized example with one space dimension

$$\frac{\partial \Phi}{\partial t}=\frac{\partial^2 \Phi}{\partial x^2} \tag{7.4.4}$$

Table 7.4.1. Finite element grid specifications for hill-slope problem (from Calver and Wood 1989).

Run	Horizontal dimension of elements Δx(m)	Vertical dimension of elements Δz(m)	$\Delta x/\Delta z$	Subsurface calculation time step Δt (h)	Total number of nodes	cpu time (s)	Total number of iterations	Mass balance (m^3)
1a	0.5	0.5	1	0.5	4004	1407	253	0.00530
2a	100	0.5	200	0.5	24	9	230	−0.00053
3a	20	0.1	200	0.5	416	258	252	0.01107
4a	50	0.5	100	0.5	44	15	229	0.00161
5a	25	0.25	100	0.5	147	64	249	0.01033
6a	10	0.1	100	0.5	816	523	253	0.00945
7a	10	0.5	20	0.5	204	77	252	0.00610
8a	5	0.25	20	0.5	707	314	251	0.00869
9a	2	0.1	20	0.5	4016	2327	253	0.00969
10a	5	0.5	10	0.5	404	145	254	0.00467
11a	2.5	0.25	10	0.5	1407	629	252	0.00920
12a	100, 10*	0.5	200, 20	0.5	60	20	226	0.00354
13a	**variable			0.5	342	205	226	0.00442

1b	0.5	0.5	1	0.1	4004	5182	853	0.02501
2b	100	0.5	200	0.1	24	34	841	0.01552
3b	20	0.1	200	0.1	416	855	838	0.02433
4b	50	0.5	100	0.1	44	57	839	0.02219
5b	25	0.25	100	0.1	147	216	840	0.02136
6b	10	0.1	100	0.1	816	1718	836	0.02454
7b	10	0.5	20	0.1	204	246	831	0.02668
8b	5	0.25	20	0.1	707	1017	857	0.02050
9b	2	0.1	20	0.1	4016	8182	867	0.02284
10b	5	0.5	10	0.1	404	533	839	0.02671
11b	2.5	0.25	10	0.1	1407	2160	868	0.02064
12b	100,10*	0.5	200, 20	0.1	60	78	830	0.02763
13b	**variable			0.1	342	730	811	0.02516

* Lower fifth of hillslope has the finer mesh.

** Depths (m) of node rows from surface: 0.00, 0.05, 0.10, 0.15, 0.20, 0.30, 0.40, 0.50, 0.60, 0.70, 0.80, 0.90, 1.00, 1.10, 1.20, 1.30, 1.40, 1.50.

Distances (m) of node columns from divide: 0, 100, 200, 300, 400, 410, 420, 430, 440, 450, 460, 470, 480, 490, 492, 494, 496, 498, 500.

on $0 \leq x \leq 1$, $t > 0$, with given initial conditions and $\Phi(0, t) = a$, $\partial\Phi/\partial x = b$ at $x = 1$, for $t \geq 0$, that is, there is no jump in values at the start. We divide the interval $[0, 1]$ into N equal linear elements of length h and form the system of ordinary differential equations in the usual way, initially disregarding the Dirichlet boundary condition at $x = 0$, so that we have $(N + 1)$ equations of which the first is

$$\frac{h}{6}(2\dot{\Phi}_0 + \dot{\Phi}_1) + \frac{1}{h}(\Phi_0 - \Phi_1) + \left[\frac{\partial\Phi}{\partial x}\right]_{x=0} = 0. \tag{7.4.5}$$

We do not know the value of $\partial\Phi/\partial x$ at $x = 0$ because this is a Dirichlet node where we are given that $\Phi_0 = a$. Hence we follow the usual routine of substituting $\Phi_0 = a$ in the other N equations and solving for the other nodal values by some time-stepping scheme. Having done this at a particular time step we can substitute back into eqn (7.4.5) to obtain the gradient at $x = 0$ which is given by

$$\left(\frac{\partial\Phi}{\partial x}\right)_{x=0} = -\frac{h}{6}(2\dot{\Phi}_0 + \dot{\Phi}_1) + \frac{1}{h}(\Phi_1 - \Phi_0) \tag{7.4.6}$$

with the distributed mass matrix, or

$$\left(\frac{\partial\Phi}{\partial x}\right)_{x=0} = -\frac{h}{2}\dot{\Phi}_0 + \frac{1}{h}(\Phi_1 - \Phi_0) \tag{7.4.7}$$

with the lumped mass matrix. What Lynch (1984) is looking at is the difference between what he calls the 'conventional practice' of computing the flow across the boundary by simply taking the finite element gradient there which gives

$$\left(\frac{\partial\Phi}{\partial x}\right)_{x=0} = \frac{1}{h}(\Phi_1 - \Phi_0) \tag{7.4.8}$$

and allowing for the other terms on the right-hand side of eqn (7.4.6) (Lynch is using the distributed mass matrix). These other terms represent the storage terms which the 'conventional practice' would omit. If the Dirichlet value $\Phi(0) = a$ is constant so that $\dot{\Phi}_0 = 0$, then the difference between the results with and without the mass storage term is just the same as the difference between the results with the distributed and lumped mass matrices given by eqns (7.4.6) and (7.4.7).

We can look at the result of using the Lynch method in higher dimensions by considering the weak form of eqn (7.3.13) using two space dimensions x and z, and supposing $K_x = K_z = K$. Corresponding to a node j with basis function N_j we have the equation

$$\iint_V CN_j \sum_i N_i \dot{\Phi}_i \,\mathrm{d}V + \iint_V K\Big\{\nabla N_j \cdot \sum_i \Phi_i \nabla N_i\Big\}\,\mathrm{d}V$$

$$- \int_S KN_j \sum_i \Phi_i \nabla N_i \cdot \boldsymbol{n}\,\mathrm{d}S = \iint_V N_j Q \,\mathrm{d}V, \quad (7.4.9)$$

where S is the boundary of the flow region V and $\boldsymbol{n}$ is the unit vector of the outward normal. Then either

(1) j is an interior node and the boundary integral is not present; or

(2) j is a boundary node where the flux $\nabla\Phi \cdot \boldsymbol{n}$ is known and can be substituted; or

(3) j is a Dirichlet node.

In case (3) the equation is kept separately and when the nodal values of Φ are found from the rest of the equations the flux term can be found by substituting into eqn (7.4.9) for each node j on the Dirichlet boundary. We can again see from eqn (7.4.9) that with a constant fixed head at the boundary point j and a lumped mass matrix the storage term will be absent. In general the use of the lumped mass in connection with the calculation of the boundary flow is ignoring the contribution to the information about the change in storage within the element of the terms $\mathrm{d}\Phi_i/\mathrm{d}t$ where i is the label of any node with which the boundary node j is coupled. But there is also the effect on the boundary flow of the difference between the values of the hydraulic potential Φ given by the solutions with the lumped and distributed mass matrices. Whether this term or the storage term makes the greater contribution can depend on the mesh.

The uniform grid with $\Delta x = \Delta z = 0.5\,m$ (4004 nodes) with a time step of 0.5 hour (listed as run 1a in Table 7.4.1), using a

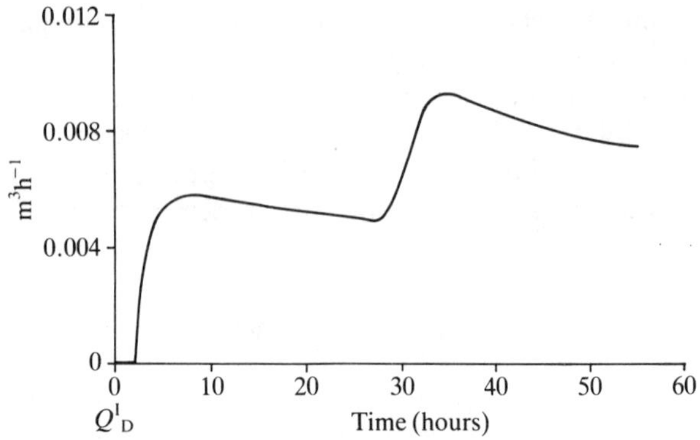

Fig. 7.4.2. Hydrograph for run ID.

distributed mass matrix, which is found to give a good performance in Calver and Wood (1989), is used as a baseline run from which the other runs are measured. Figure 7.4.2 shows the hydrograph from this run now referred to as run ID. Figure 7.4.3 shows the error in run IL, which has the same mesh and conditions as run ID but with a lumped mass matrix. With this fine mesh the difference is small. The experiments reported in Wood and Calver (1990) with various meshes and hydrological conditions show that in general the error in the discharge is greater with the lumped mass matrix. Figures 7.4.4 and 7.4.5 show a typical result for the errors from using another mesh, mesh A (8a in Table 7.4.1) with distributed and lumped mass

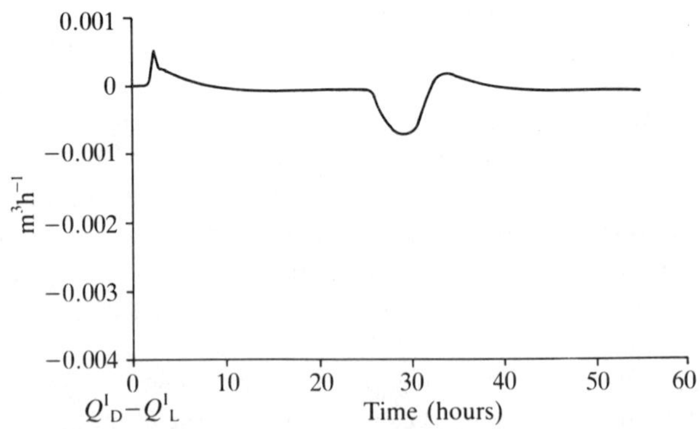

Fig. 7.4.3. Error in discharge IL.

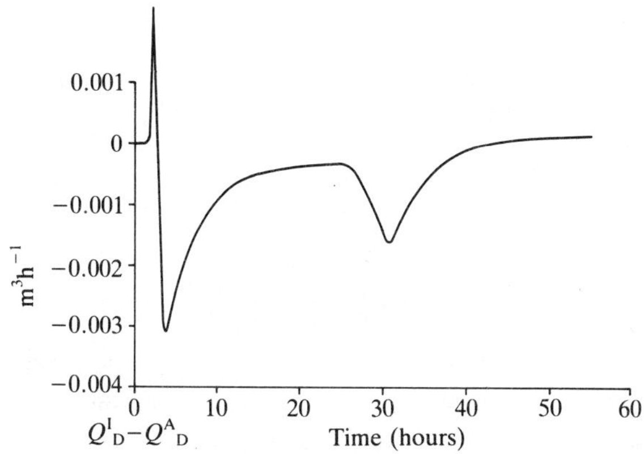

Fig. 7.4.4. Error in discharge AD.

matrices respectively. The graphs start with zero error because run-in time is allowed. Mesh A with $\Delta x = 5\,m$, $\Delta z = 0.25\,m$ may seem to be a reasonable choice of mesh for the long thin region shown in Fig. 7.4.1. The difference between the solution values for the potential Φ given by runs AD and AL appears to be within the accuracy justified by observational error. But with the parallelogram elements as shown in Fig. 7.4.1, the element

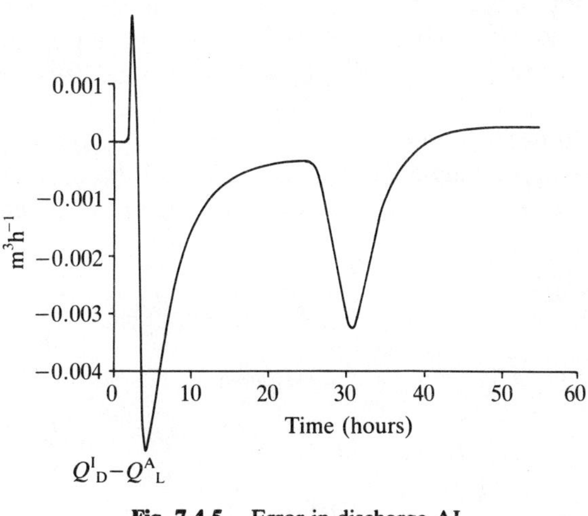

Fig. 7.4.5. Error in discharge AL.

stiffness matrix has terms as follows:

$$k_{12} = K\left\{\frac{\Delta z}{6\Delta x} - \frac{\Delta x}{3\Delta z}(1+t^2)\right\},$$

$$k_{12} = K\left\{-\frac{\Delta z}{3\Delta x} + \frac{\Delta x}{6\Delta z}(1+t^2)\right\}, \tag{7.4.10}$$

$$k_{14} = K\left\{\frac{\Delta z}{6\Delta x} - \frac{\Delta x}{6\Delta z}(1+t^2) + \frac{t}{2}\right\}$$

where $t = \tan\theta$ and θ is the angle of the slope. The ratio $\Delta x : \Delta z = 20:1$ in mesh A has the effect of magnifying the difference between the two solutions given by the lumped and distributed mass matrices and this is responsible for the greater error shown by the lumped mass matrix result in Fig. 7.4.5; the storage term effect is smaller. Further details are given in Wood and Calver (1990).

7.5. Newton or quasi-Newton method for first order non-linear equations

We start this section by giving a brief description of the Newton or Newton–Raphson method for solving a system of non-linear equations. (The two labels seem to be used without distinction.) The Newton method for the solution of a single non-linear equation is given in all the textbooks on elementary numerical analysis; see, for example, Conte and De Boor (1980), Phillips and Taylor (1973). Suppose we have a system of two non-linear equations in two unknowns: $g_1(x_1, x_2) = 0$, $g_2(x_1, x_2) = 0$. Then if we have an approximate solution (x_1^i, x_2^i), and the exact solution is $(x_1^i + \epsilon_1, x_2^i + \epsilon_2)$, we assume we can expand the functions g_1 and g_2 to give

$$\begin{aligned} g_1(x_1^i + \epsilon_1, x_2^i + \epsilon_2) &= g_1(x_1^i, x_2^i) + \epsilon_1 \frac{\partial g_1}{\partial x_1} + \epsilon_2 \frac{\partial g_1}{\partial x_2} \\ &\quad + \mathrm{O}(\|\boldsymbol{\epsilon}\|^2) = 0 \\ g_2(x_1^i + \epsilon_1, x_2^i + \epsilon_2) &= g_2(x_1^i, x_2^i) + \epsilon_1 \frac{\partial g_2}{\partial x_1} + \epsilon_2 \frac{\partial g_2}{\partial x_2} \\ &\quad + \mathrm{O}(\|\boldsymbol{\epsilon}\|^2) = 0 \end{aligned} \tag{7.5.1}$$

where $\boldsymbol{\epsilon}$ is the vector $[\epsilon_1, \epsilon_2]^T$ and $\|\boldsymbol{\epsilon}\|$ is the l_2 vector norm. Neglecting the second order terms in eqns (7.5.1) we can obtain the correction $\boldsymbol{\epsilon}^i$ to the approximation solution $\boldsymbol{x}^i$ from the equations

$$\boldsymbol{g}(\boldsymbol{x}^i) + J(\boldsymbol{x}^i)\boldsymbol{\epsilon}^i = \mathbf{0}, \tag{7.5.2}$$

where $\boldsymbol{J}(\boldsymbol{x}^i)$ is the Jacobian matrix whose (j, k)th entry is $\dfrac{\partial g_j}{\partial x_k}$ evaluated at $\boldsymbol{x} = \boldsymbol{x}^i$. We take the next approximation as $\boldsymbol{x}^{i+1} = \boldsymbol{x}^i + \boldsymbol{\epsilon}^i$ and continue until $\boldsymbol{g}(\boldsymbol{x}^i)$ is small enough to satisfy some convergence criterion. For the method to work the Jacobian matrix has to exist and to be invertible, that is non-singular. As with the single scalar non-linear equation, the starting value for the iteration must be near enough or the method will not converge. When the Newton method works it works very well and gives quadratic convergence with

$$\|\boldsymbol{x}^{i+1} - \boldsymbol{x}\| \leq c\, \|\boldsymbol{x}^i - \boldsymbol{x}\|^2 \tag{7.5.3}$$

for some constant c, where $\boldsymbol{x}$ is the exact solution. Apart from the necessity of a good starting value the most important disadvantage for the solution of a large system of equations is the cost of obtaining the Jacobian matrix and updating it at every iteration. Where this cost is excessive there are variants of Newton's method such as

(1) 'fixed tangent', where the Jacobian matrix is obtained but kept constant as $\boldsymbol{J}(\boldsymbol{x}^0)$ during the iteration; or

(2) 'secant', where the Jacobian matrix is approximated by using the previous value.

We can illustrate this on the example in eqns (7.5.1) by taking, for instance,

$$\frac{\partial g_1}{\partial x_1}(x_1, x_2) \approx \{g_1(x_1^i, x_2^i) - g_1(x_1^{i-1}, x_2^i)\}/(x_1^i - x_1^{i-1}). \tag{7.5.4}$$

Henrici (1962) gives the theoretical background to Newton's method. The classic paper by Dennis and Moré (1977) on Quasi-Newton methods gives the motivation and theory for the modifications.

Dennis and Moré comment on the importance of having more than one criterion for stopping the iteration. For example, $\|\boldsymbol{x}^{i+1}-\boldsymbol{x}^{i}\| \leq c_1 \|\boldsymbol{x}^{i}\|$ combined with one on the residual, $\|\boldsymbol{g}(\boldsymbol{x}^{i})\| \leq c_2$, for some convenient tolerances c_1 and c_2, using the notation of the example above. The danger of relying on one of these only is illustrated in §7.1. Many users simply rely on the criterion based on the change in the solution.

We now consider the idea of using Newton or Quasi-Newton on the solution of eqns (7.3.21) or (7.3.24) which represent time-stepping schemes for the first order equation typified by the saturated–unsaturated groundwater flow equation (7.3.15), included again here for convenience:

$$\boldsymbol{M}(\boldsymbol{\Phi})\dot{\boldsymbol{\Phi}} + \boldsymbol{K}(\boldsymbol{\Phi})\boldsymbol{\Phi} = \boldsymbol{f}. \tag{7.5.5}$$

The non-linear entries in the mass and stiffness matrices here are obtained from empirical relationships and it would be difficult to differentiate to obtain the Jacobian matrix. The pattern of the terms is that of 'non-linear matrix times vector'. Thus this type of equation suggests the approach described in §7.3 where a kind of *ad hoc* iteration is evolved, updating the coefficient matrix and solving for the correction at each iteration.

On the other hand, the Boussinesq equation (7.3.11), which gives the non-linear form

$$\boldsymbol{M}\dot{\boldsymbol{h}} + \boldsymbol{K}(\boldsymbol{h})\boldsymbol{h} = \boldsymbol{f} \tag{7.5.6}$$

has a mass matrix which does not change and a stiffness matrix of a form which is amenable to Newton's method. Looking at the non-linear term in detail, we can see that it comes from the expressions

$$\frac{\partial}{\partial x} K_x h \frac{\partial h}{\partial x}, \qquad \frac{\partial}{\partial y} K_y h \frac{\partial h}{\partial y}$$

in eqn (7.3.11). With the substitution of h in terms of the basis functions N_l, multiplication by N_j and integration by parts gives terms of the form

$$\sum_l \sum_k h_1 h_k \iint_{\Omega_j} N_l \frac{\partial N_k}{\partial x}\frac{\partial N_j}{\partial x}\,\mathrm{d}\Omega, \ldots, \tag{7.5.7}$$

where the summations are each over the nodes coupled to node j in the neighbourhood Ω_j of node j. Yoon and Yeh (1975) show that this is particularly simple with a one-dimensional problem with equal linear elements. The terms combine so that in the typical jth equation the non-linear entries produce the expression:

$$\tfrac{1}{2}(-h_{j-1}^2 + 2h_j^2 - h_{j+1}^2),$$

where the nodes are numbered consecutively. Consequently the stiffness matrix $K(h)$ in eqn (7.5.6) is tridiagonal with typical entries

$$K_{ij} = h_i/\Delta x, \quad \text{if } i = j,$$
$$K_{ij} = -h_j/(2\Delta x), \quad \text{if } |i-j| = 1,$$

and $K_{ij} = 0$, otherwise, where Δx is the length of the elements. Thus with the time-stepping scheme

$$\frac{1}{\Delta t}\boldsymbol{M}(\boldsymbol{h}_{n+1} - \boldsymbol{h}_n) + \tfrac{1}{2}\{\boldsymbol{K}(\boldsymbol{h}_{n+1})\boldsymbol{h}_{n+1} + \boldsymbol{K}(\boldsymbol{h}_n)\boldsymbol{h}_n\} = \bar{\boldsymbol{f}}, \quad (7.5.8)$$

the Jacobian matrix for the Newton method is readily available as

$$\boldsymbol{J} = \boldsymbol{M} + \Delta t\boldsymbol{K}(\boldsymbol{h}_{n+1}). \quad (7.5.9)$$

The stopping criterion used by Yoon and Yeh is based on the sum of the differences in an iteration

$$\sum_{j=1}^{N} |h_j^{i+1} - h_j^i| \le 10^{-5},$$

rather than on the residual.

With the two-dimensional problem where we have the non-linear terms as in eqn (7.5.7), it is possible to compromise by making the lumping approximation

$$\sum_k h_k N_k \approx h_j \sum_k N_k = h_j, \quad (7.5.10)$$

where the summation is over the neighbourhood of node j. The effect is that each entry in the jth row of the matrix $K(h)$ contains h_j and the Jacobian matrix is fairly simple to form and use.

Anderson and Zienkiewicz (1974) in a paper on spontaneous ignition have an equation of the form

$$\boldsymbol{M}\dot{\boldsymbol{\Phi}} + \boldsymbol{K}\boldsymbol{\Phi} - \delta\boldsymbol{F} = \boldsymbol{0}, \tag{7.5.11}$$

where $\boldsymbol{\Phi}$ is the non-dimensionalized temperature vector, δ is a parameter, the matrices $\boldsymbol{M}$ and $\boldsymbol{K}$ depend on $\boldsymbol{\Phi}$, and $\boldsymbol{F}$ in particular depends on $\exp(\Phi)$.

The time stepping scheme is taken as

$$\left(\frac{2}{\Delta t}\boldsymbol{M} + \boldsymbol{K}\right)\boldsymbol{\Phi}_{n+1} - \delta\boldsymbol{F}_{n+1} = \left(\frac{2}{\Delta t}\boldsymbol{M} - \boldsymbol{K}\right)\boldsymbol{\Phi}_{n} + \delta\boldsymbol{F}_{n}. \tag{7.5.12}$$

This is then solved by Newton's method with the assumption that the time step is small enough for the mass and stiffness matrices not to vary significantly within the time step so that the non-linearity is concentrated in the $\boldsymbol{F}_{n+1}$ term. The Jacobian matrix is taken to have components

$$\frac{2}{\Delta t}M_{ij} + K_{ij} - \delta \sum_{e} \int_{\Omega_e} N_i N_j \exp(\Phi)\, \mathrm{d}\Omega, \tag{7.5.13}$$

where N_i are the basis functions and e and Ω_e are the element number and area respectively. Anderson and Zienkiewicz state that the Newton method, in spite of the extra computation per iteration, is more efficient than a method of successive approximation. The cost-effectiveness of the Newton method here evidently depends on the particular form of the non-linearity.

7.6. The non-linear second order dynamic vibration equation

We consider here a dynamic vibration problem which after space discretization is represented by the system of equations

$$\boldsymbol{M}\ddot{\boldsymbol{x}} + \boldsymbol{C}\dot{\boldsymbol{x}} + \boldsymbol{P}(\boldsymbol{x}) = \boldsymbol{f}(t), \tag{7.6.1}$$

where $\boldsymbol{x}, \dot{\boldsymbol{x}}, \ddot{\boldsymbol{x}}$ are the nodal displacements, velocities and accelerations respectively, $\boldsymbol{M}$ and $\boldsymbol{C}$ are the mass and damping matrices and $\boldsymbol{P}(\boldsymbol{x})$ represents the internal forces of the structure. For linear elastic behaviour $\boldsymbol{P}(\boldsymbol{x}) = \boldsymbol{K}\boldsymbol{x}$ where $\boldsymbol{K}$ is the usual stiffness matrix, but for non-linear behaviour with large deformations and/or plasticity, $\boldsymbol{P}(\boldsymbol{x})$ usually has to be calculated from an expression of the type

$$P = \int_V B^{\mathrm{T}} \sigma \, \mathrm{d}V, \qquad (7.6.2)$$

where σ is the vector of the non-linear stresses, B is the matrix which expresses the strains in terms of the nodal displacements and V is the region occupied by the structure. Details of the stress analysis background to eqn (7.6.2) are given in Zienkiewicz (1977).

We look first at the motivation given by Park (1975) for introducing Park's method (the three-step second order method described in §4.11). Park is discussing the solution by the Newmark algorithm with $2\beta = \gamma = 0.5$ (the trapezium rule equivalent), of the non-linear scalar equation of the form

$$m\ddot{x} + k(x)x = 0 \qquad (7.6.3)$$

with no damping or forcing term. The Newmark equations used here are

$$\begin{aligned} x_{n+1} &= x_n + \Delta t v_n + \frac{\Delta t^2}{4}(a_n + a_{n+1}), \\ v_{n+1} &= v_n + \frac{\Delta t}{2}(a_n + a_{n+1}). \end{aligned} \qquad (7.6.4)$$

Using the Newmark collocation at $t = t_n$, $t_{n+1} = t_n + \Delta t$, means substituting $a_n = -k(x_n)x_n/m$, $a_{n+1} = -k(x_{n+1})x_{n+1}/m$ into the first of eqns (7.6.4) which gives

$$m(x_n - x_{n+1} + \Delta t v_n) - \frac{\Delta t^2}{4}\{k(x_n)x_n + k(x_{n+1})x_{n+1}\} = 0. \qquad (7.6.5)$$

In order to eliminate the v_n term we have to use another time step and the result is the equivalent two-step scheme in the displacements

$$m(x_{n+1} - 2x_n + x_{n-1}) + \frac{\Delta t^2}{4} \times \{k(x_{n+1})x_{n+1} + 2k(x_n)x_n + k(x_{n-1})x_{n-1}\} = 0. \qquad (7.6.6)$$

Park claims to show that 'local instability' is evident in the algorithm given by eqn (7.6.6) because of the three $k(x)x$ terms.

He prefers the Houbolt algorithm (§4.7) which uses backward differences for the derivatives. We note that the objection to the three non-linear terms in eqn (7.6.6) could also be answered by using a method analogous to the central difference method discussed in §7.3 whose advantage for first order problems was pointed out by Gourlay (1970). The central difference method suggests taking the numerical solution of $m\ddot{x} + P(x) = 0$ as given by

$$m(x_{n+1} - 2x_n + x_{n-1}) + P(\bar{x}_n) = 0,$$

where $\bar{x}_n = (x_{n+1} + 2x_n + x_{n-1})/4$, or $(x_{n+1} + x_{n-1})/2$. This is essentially what happens with the single-step SS22 algorithm applied to non-linear problems, which uses an average value, α_n, of the acceleration over the time step and an average value of the argument in $P(x)$. This method is discussed later.

Park (1975) uses the backward difference approximation for the acceleration given by

$$a_{n+1} = \frac{1}{\Delta t^2}(2x_{n+1} - 5x_n + 4x_n - x_{n-2}) \tag{7.6.7}$$

the algorithm for the numerical solution of eqn (7.6.3) becomes

$$m(2x_{n+1} - 5x_n + 4x_{n-1} - x_{n-2}) + \Delta t^2 k(x_{n+1})x_{n+1} = 0. \tag{7.6.8}$$

With, for example, $k(x)x = k_1x + k_2x^3$, the solution of the recurrence relation for very large $\Delta t^2 k/m$ is simply $x = 0$ (showing the excessive damping of the Houbolt method). Park concludes that backward differencing is therefore advisable and Park's method corresponds to a linear combination of the Gear two-step and three-step backward difference methods for the first order form of the dynamic vibration equation. The 'local instability' evidence produced by Park is obtained by making the trial solution '$x_{n+1} = rx_n$'. But we have to be careful here to distinguish between the component of the solution of a linear recurrence relation denoted by r^n, where r appears in the assumption made by substituting the trial solution $x_{n+1} = rx_n$ and is complex and the actual solution for the displacement x_n itself which is real. It is not in general possible to put trial solutions of the form $x_{n+1} = rx_n$ into a non-linear recurrence relation.

Suppose we have a cubically hardening spring with equation

$$m\ddot{x} + k_1x + k_2x^3 = 0. \tag{7.6.9}$$

Corresponding to the Newmark algorithm in eqn (7.6.6) we now have

$$4(x_{n+1}-2x_n+x_{n-1})+\Omega_1^2(x_{n+1}+2x_n+x_{n-1})$$
$$+\Omega_2^2(x_{n+1}^3+2x_n^3+x_{n-1}^3)=0, \quad (7.6.10)$$

where $\Omega_1^2=\Delta t^2k_1/m$ and $\Omega_2^2=\Delta t^2k_2/m$. By inspection, the recurrence relation (7.6.10) for very large values of Ω_1^2 and/or Ω_2^2 has a solution $x_n=(-1)^na$, where a is constant. Thus the algorithm is stable and this suggests that it is unconditionally stable. Wood and Oduor (1988) describe experiments with the Newmark method applied in its single-step form which is equivalent to the two-step given by eqn (7.6.6). These numerical experiments show that the solution remains stable in the sense that the amplitude remains bounded though it may be inaccurate. Figure 7.6.1 (from Wood and Oduor 1988) shows the stable solution given by Newmark for eqn (7.6.9) with a time step which is approximately 0.16 of the true period.

For the non-linear elastic softening spring represented by the

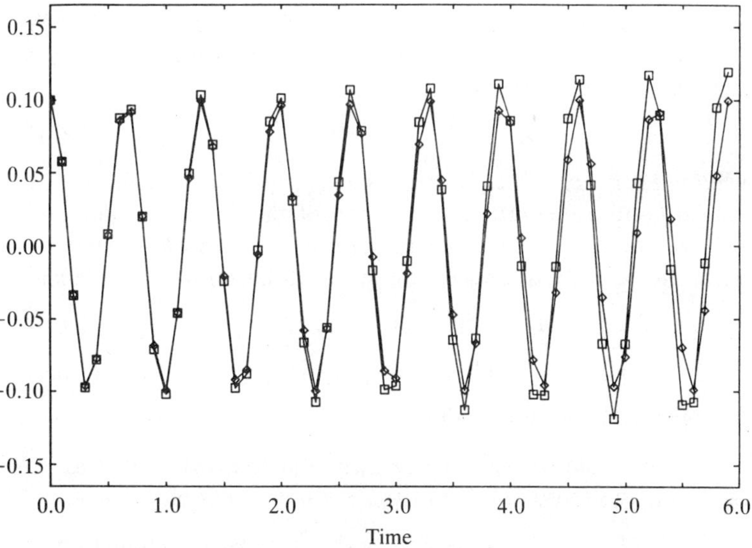

Fig. 7.6.1. Hardening spring with Newmark and SS22a, ($\Phi=0.5$). $\ddot{u}+100u(1+10u^2)=0$, $u_0=0.1$, $v_0=0$, $\Delta t=0.10$. Newmark ($\gamma=0.5$, $\beta=0.25$) and SS22 ($\theta_1=\theta_2=0.5$) with $\hat{u}_n=\frac{1}{2}(u_n+u_{n+1})$.

equation

$$\ddot{x} + k \tanh x = 0, \qquad k > 0, \tag{7.6.11}$$

the recurrence relation corresponding to eqn (7.6.6) is now

$$4(x_{n+1} - 2x_n + x_{n-1}) + \Omega^2\{\tanh(x_{n+1}) + 2\tanh(x_n) + \tanh(x_{n-1})\} = 0, \tag{7.6.12}$$

where $\Omega^2 = \Delta t^2 k/m$. Now for large values of Ω^2 we can see by inspection that there is a solution $x_n = (-a)^n$, where a is some constant and a and n are large enough for

$$\tanh(a^{n-1}) \approx 1$$

(for example, $\tanh(4) = 0.99934$, to five figures). This suggests that the algorithm is only conditionally stable here. The numerical experiments reported in Wood and Oduor (1988) again using the single step form of this Newmark algorithm, found instability with this example when $k = 100$, $\Delta t = 0.3$ and starting conditions $x_0 = 4$, $v_0 = 0$. This instability is also demonstrated in Fig. 6.5.3 which shows the effect on the energy.

The one-step time-stepping schemes are the most convenient and popular for the numerical integration of eqn (7.6.1). We consider first Newmark where the basic idea is to collocate so that we wish to satisfy the equation

$$\boldsymbol{M}\boldsymbol{a}_{n+1} + \boldsymbol{C}\boldsymbol{v}_{n+1} + \boldsymbol{P}(\boldsymbol{x}_{n+1}) = \boldsymbol{f}_{n+1}, \tag{7.6.13}$$

where $\boldsymbol{x}_{n+1}$, $\boldsymbol{v}_{n+1}$, $\boldsymbol{a}_{n+1}$ are the approximate displacement, velocity and acceleration at the end of the $(n+1)$th time step, and we know their values at the beginning of this time step. We write the basic Newmark equations again here for convenience

$$\boldsymbol{x}_{n+1} = \boldsymbol{x}_n + \Delta t\boldsymbol{v}_n + (\tfrac{1}{2} - \beta)\Delta t^2\boldsymbol{a}_n + \beta\Delta t^2\boldsymbol{a}_{n+1} \tag{7.6.14}$$

$$\boldsymbol{v}_{n+1} = \boldsymbol{v}_n + (1-\gamma)\Delta t\boldsymbol{a}_n + \gamma\Delta t\boldsymbol{a}_{n+1}. \tag{7.6.15}$$

We observe that we can either:

1. Substitute for $\boldsymbol{x}_{n+1}$ and $\boldsymbol{v}_{n+1}$ in eqn (7.6.13) so that $\boldsymbol{a}_{n+1}$ is the unknown to be obtained from the non-linear equation

$$(\boldsymbol{M} + \gamma\Delta t\boldsymbol{C})\boldsymbol{a}_{n+1} + \boldsymbol{P}(\tilde{\boldsymbol{x}}_n + \beta\Delta t^2\boldsymbol{a}_{n+1}) - \boldsymbol{F}_{n+1} = \boldsymbol{0}, \tag{7.6.16}$$

where $\tilde{\boldsymbol{x}}_n = \boldsymbol{x}_n + \Delta t\boldsymbol{v}_n + (\tfrac{1}{2} - \beta)\Delta t^2\boldsymbol{a}_n$ is known, and $\boldsymbol{F}_{n+1} = \boldsymbol{f}(t_{n+1}) - \boldsymbol{C}\{\boldsymbol{v}_n + (1-\gamma)\Delta t\boldsymbol{a}_n\}$ is also known.

Then the Newton method applied to eqn (7.6.16) regarded as a non-linear equation in $\boldsymbol{a}_{n+1}$ has the Jacobian matrix

$$\boldsymbol{J}_a^i = \boldsymbol{M} + \gamma \Delta t \boldsymbol{C} + \beta \Delta t^2 \boldsymbol{K}(\tilde{\boldsymbol{x}}_n + \beta \Delta t^2 \boldsymbol{a}_{n+1}^i), \qquad (7.6.17)$$

where $\{K(x)\}_{ij} = \mathrm{d}P_i/\mathrm{d}x_j$, corresponding to the ith iterate and the iteration given by

$$\begin{aligned}\boldsymbol{a}_{n+1}^{i+1} &= \boldsymbol{a}_{n+1}^i - (\boldsymbol{J}_a^i)^{-1} \\ &\times \{(\boldsymbol{M} + \gamma \Delta t \boldsymbol{C})\boldsymbol{a}_{n+1}^i + \boldsymbol{P}(\tilde{\boldsymbol{x}}_n + \beta \Delta t^2 \boldsymbol{a}_{n+1}^i) - \boldsymbol{F}_{n+1}\}. \qquad (7.6.18)\end{aligned}$$

2. Alternatively we can make the non-linear equation to be solved an equation in $\boldsymbol{x}_{n+1}$. This is achieved by substituting for $\boldsymbol{a}_{n+1}$ in terms of $\boldsymbol{x}_n$, $\boldsymbol{v}_n$, $\boldsymbol{a}_n$, and $\boldsymbol{x}_{n+1}$ from eqn (7.6.14):

$$\beta \Delta t^2 \boldsymbol{a}_{n+1} = \boldsymbol{x}_{n+1} - \boldsymbol{x}_n - \Delta t \boldsymbol{v}_n - (\tfrac{1}{2} - \beta)\Delta t^2 \boldsymbol{a}_n \qquad (7.6.19)$$

and obtaining $\boldsymbol{v}_{n+1}$ likewise from eqns (7.6.15) and (7.6.19) to give

$$\beta \Delta t \boldsymbol{v}_{n+1} = \gamma(\boldsymbol{x}_{n+1} - \boldsymbol{x}_n) + (\beta - \gamma)\Delta t \boldsymbol{v}_n + \left(\beta - \frac{\gamma}{2}\right)\Delta t^2 \boldsymbol{a}_n. \qquad (7.6.20)$$

Substituting for $\boldsymbol{a}_{n+1}$ and $\boldsymbol{v}_{n+1}$ into eqn (7.6.13) we have the following:

$$\begin{aligned}&\boldsymbol{M}\{\boldsymbol{x}_{n+1} - \boldsymbol{x}_n - \Delta t \boldsymbol{v}_n - (\tfrac{1}{2} - \beta)\Delta t^2 \boldsymbol{a}_n\} \\ &\quad + \Delta t \boldsymbol{C}\left\{\gamma(\boldsymbol{x}_{n+1} - \boldsymbol{x}_n) + (\beta - \gamma)\Delta t \boldsymbol{v}_n + \left(\beta - \frac{\gamma}{2}\right)\Delta t^2 \boldsymbol{a}_n\right\} \\ &\quad\quad + \beta \Delta t^2 \boldsymbol{P}(\boldsymbol{x}_{n+1}) - \beta \Delta t^2 \boldsymbol{f}_{n+1} = \boldsymbol{0}. \qquad (7.6.21)\end{aligned}$$

We can rearrange eqn (7.6.21) as

$$(\boldsymbol{M} + \gamma \Delta t \boldsymbol{C})\boldsymbol{x}_{n+1} + \beta \Delta t^2 \boldsymbol{P}(\boldsymbol{x}_{n+1}) - \hat{\boldsymbol{F}}_n = \boldsymbol{0}, \qquad (7.6.22)$$

which gives the non-linear equation to be solved for $\boldsymbol{x}_{n+1}$ with the term $\hat{\boldsymbol{F}}_n$ including all the known values from the beginning of the time step and the forcing function. The Jacobian matrix at the ith iteration is now

$$\boldsymbol{J}_b^i = \boldsymbol{M} + \gamma \Delta t \boldsymbol{C} + \beta \Delta t^2 \boldsymbol{K}(\boldsymbol{x}_{n-1}^i) \qquad (7.6.23)$$

and the iteration consists of solving

$$\boldsymbol{J}_b^i \boldsymbol{\epsilon}^i = \hat{\boldsymbol{F}}_n - (\boldsymbol{M} + \gamma \Delta t \boldsymbol{C})\boldsymbol{x}_{n+1}^i - \beta \Delta t^2 \boldsymbol{P}(\boldsymbol{x}_{n+1}^i) \qquad (7.6.24)$$

and updating by

$$\boldsymbol{x}_{n+1}^{i+1} = \boldsymbol{x}_{n+1}^{i} + \boldsymbol{\epsilon}^{i},$$

until convergence.

In the application of the Newton method to the numerical solution of eqn (7.6.1) via the approach represented in eqn (7.6.18) or eqn (7.6.24) the matrix $\boldsymbol{K}$ is called the tangent stiffness matrix. When both $\boldsymbol{K}$ and the right-hand side of eqn (7.6.24) are updated at each iteration (which corresponds to variable tangent in the single variable case, that is, the full application of the Newton method) this is called the tangent stiffness method. When the stiffness matrix $\boldsymbol{K}$ is kept constant at the value $\boldsymbol{K}(\boldsymbol{x}_n)$ during the iteration while the right-hand side of eqn (7.6.24) is updated (which corresponds to constant tangent in the single variable case), the method is called a pseudo-force formulation (Owen 1978).

We next look at the application of the single-step algorithm SS22 to the numerical solution of eqn (7.6.1), illustrating this on the equation with zero damping and forcing function because the main interest is in the treatment of the non-linear $\boldsymbol{P}(\boldsymbol{x})$ term. From the basic theory in §4.5 we now have the weighted residual equation

$$\int_0^{\Delta t} W(\tau)\left\{\boldsymbol{M}\boldsymbol{\alpha}_n + \boldsymbol{P}\left(\boldsymbol{x}_n + \tau\dot{\boldsymbol{x}}_n + \frac{\tau^2}{2}\boldsymbol{\alpha}_n\right)\right\} \mathrm{d}\tau = \boldsymbol{0}, \quad (7.6.25)$$

which leads to

$$\boldsymbol{M}\boldsymbol{\alpha}_n + \frac{1}{I}\int_0^{\Delta t} W(\tau)\boldsymbol{P}\left(\boldsymbol{x}_n + \tau\dot{\boldsymbol{x}}_n + \frac{\tau^2}{2}\boldsymbol{\alpha}_n\right) \mathrm{d}\tau = \boldsymbol{0}, \quad (7.6.26)$$

where $I = \int_0^{\Delta t} W(\tau)\,\mathrm{d}\tau$. We now have to decide how to approximate the integral term in eqn (7.6.26). (The difficulty in deciding how to do this with SS22 tends to make Newmark more popular.) We can simply replace it by $\boldsymbol{P}(\boldsymbol{x}_n + \theta_1\Delta t\dot{\boldsymbol{x}}_n + (\theta_2/2)\Delta t^2\boldsymbol{\alpha}_n)$ which is the method adopted by Chan (1988).

Alternatively we can expand the integrand to give:

SS22a, where we take

$$\frac{1}{I}\int_0^{\Delta t} W(\tau)\left\{\boldsymbol{P}(\boldsymbol{x}_n) + \boldsymbol{K}(\hat{\boldsymbol{x}}_n)\left(\tau\dot{\boldsymbol{x}}_n + \frac{\tau^2}{2}\boldsymbol{\alpha}_n\right)\right\}\mathrm{d}\tau$$

$$\approx \boldsymbol{P}(\boldsymbol{x}_n) + \boldsymbol{K}(\hat{\boldsymbol{x}}_n)\left(\theta_1\Delta t\dot{\boldsymbol{x}}_n + \theta_2\frac{\Delta t^2}{2}\boldsymbol{\alpha}_n\right)$$

and we can choose what to use for $\hat{\boldsymbol{x}}_n$. The equation to be solved is then

$$\boldsymbol{M\alpha}_n + \boldsymbol{P}(\boldsymbol{x}_n) + \boldsymbol{K}(\hat{\boldsymbol{x}}_n)(\theta_1\Delta t\dot{\boldsymbol{x}}_n + \theta_2\Delta t^2\boldsymbol{\alpha}_n) = \boldsymbol{0}. \quad (7.6.27)$$

Wood and Oduor (1988) and Carroll (1988) experiment with

$$\theta_1 = \theta_2 = 0.5, \quad \text{and} \quad \hat{\boldsymbol{x}}_n = (1-\Phi)\boldsymbol{x}_n + \Phi\boldsymbol{x}_{n+1} \quad (7.6.28)$$

with three values of Φ given by

(1) $\Phi = 0$ which has the effect of linearizing;

(2) $\Phi = 0.5$, which means that the tangent matrix is evaluated at an average value of the displacement; and

(3) $\Phi = 1$, which means that the tangent matrix is evaluated at the end point.

The iteration for the increment

$$\Delta\boldsymbol{\alpha}_n^i = \boldsymbol{\alpha}_n^{i+1} - \boldsymbol{\alpha}_n^i \quad (7.6.29)$$

in SS22a is given by

$$\Delta\boldsymbol{\alpha}_n^i = -\{\boldsymbol{M} + \theta_2\Delta t^2\boldsymbol{K}(\boldsymbol{x}_n^*)\}^{-1}$$
$$\times\left\{\boldsymbol{M\alpha}_n^i + \boldsymbol{P}(\boldsymbol{x}_n) + \boldsymbol{K}(\boldsymbol{x}_n^*)\left(\theta_1\Delta t\dot{\boldsymbol{x}}_n + \theta_2\frac{\Delta t^2}{2}\boldsymbol{\alpha}_n^i\right)\right\} \quad (7.6.30)$$

where

$$\boldsymbol{x}_n^* = \boldsymbol{x}_n + \Phi\left(\Delta t\dot{\boldsymbol{x}}_n + \frac{\Delta t^2}{2}\boldsymbol{\alpha}_n^i\right).$$

SS22b. Another way of writing the weighted residual equation

(7.6.26) is given by

$$\int_0^{\Delta t} W(\tau)\left\{\boldsymbol{M}(\boldsymbol{\alpha}_n^i + \Delta\boldsymbol{\alpha}_n^i) + \boldsymbol{P}(\tilde{\boldsymbol{x}}_n) + \frac{\tau^2}{2}\boldsymbol{K}(\tilde{\boldsymbol{x}}_n)\Delta\boldsymbol{\alpha}_n^i\right\}\mathrm{d}\tau = \boldsymbol{0}, \tag{7.6.31}$$

where $\tilde{\boldsymbol{x}}_n$ is some average value. Equation (7.6.31) can then be approximated by

$$\boldsymbol{M}(\boldsymbol{\alpha}_n^i + \Delta\boldsymbol{\alpha}_n^i) + \boldsymbol{P}(\bar{\boldsymbol{x}}_n^i) + \theta_2\frac{\Delta t^2}{2}\boldsymbol{K}(\bar{\boldsymbol{x}}_n^i)\Delta\boldsymbol{\alpha}_n^i = \boldsymbol{0}, \tag{7.6.32}$$

where

$$\bar{\boldsymbol{x}}_n^i = \boldsymbol{x}_n + \theta_1\Delta t\dot{\boldsymbol{x}}_n + \theta_2\frac{\Delta t^2}{2}\boldsymbol{\alpha}_n^i. \tag{7.6.33}$$

The iteration is now given by

$$\Delta\boldsymbol{\alpha}_n^i = -\left\{\boldsymbol{M} + \theta_2\frac{\Delta t^2}{2}\boldsymbol{K}(\bar{\boldsymbol{x}}_n^i)\right\}^{-1}\{\boldsymbol{M}\boldsymbol{\alpha}_n^i + \boldsymbol{P}(\bar{\boldsymbol{x}}_n^i)\}. \tag{7.6.34}$$

The convergence criteria can be of the form suggested by Hughes (1976), that is

for Newmark: $\quad \|\Delta\boldsymbol{a}_{n+1}^i\| \le \sigma_N \|\boldsymbol{a}_{n+1}^{i+1} - \boldsymbol{a}_n\| \qquad (7.6.35)$

or for SS22: $\quad \|\Delta\boldsymbol{\alpha}_n^i\| \le \sigma_S \|\boldsymbol{\alpha}_n^{i+1} - \boldsymbol{\alpha}_{n-1}\|, \qquad (7.6.36)$

with some convenient norm, where σ_N, σ_S are given tolerances. The convergence criterion (7.6.35) or (7.6.36) can be combined with a criterion on the residual or 'out of balance force' in the equation which is being approximately satisfied, which is often used with Newton iteration (Dennis and Moré 1977). With Newmark, from eqn (7.6.13), with zero damping and forcing function, the residual at the ith iteration is given by

$$\boldsymbol{f}_{n+1,i}^{\text{err}} = \boldsymbol{M}\boldsymbol{a}_{n+1}^i + \boldsymbol{P}(\boldsymbol{x}_n^* + \beta\Delta t^2\boldsymbol{a}_{n+1}^i), \tag{7.6.37}$$

where $\boldsymbol{x}_n^*$ includes the known values from the beginning of the time step. With SS22 we can calculate the residual from the equation we are iterating to satisfy. For SS22a, from eqn (7.6.27), the residual at the ith iteration is given by

$$\boldsymbol{f}_{n+1,i}^{\text{err}} = \boldsymbol{M}\boldsymbol{\alpha}_n^i + \boldsymbol{P}(\boldsymbol{x}_n) + \boldsymbol{K}(\hat{\boldsymbol{x}}_n^i)\left(\theta_1\Delta t\boldsymbol{v}_n + \theta_2\frac{\Delta t^2}{2}\boldsymbol{\alpha}_n^i\right), \tag{7.6.38}$$

where $\hat{\boldsymbol{x}}_n^i = \boldsymbol{x}_n + \Phi(\Delta t \boldsymbol{v}_n + (\Delta t^2/2)\boldsymbol{\alpha}_n^i)$. For SS22b, from eqn (7.6.34), the residual at the ith iteration is given by

$$f_{n+1,i}^{\text{err}} = \boldsymbol{M}\boldsymbol{\alpha}_n^i + \boldsymbol{P}(\bar{\boldsymbol{x}}_n^i) \qquad (7.6.39)$$

where $\bar{\boldsymbol{x}}_n^i = \boldsymbol{x}_n + \theta_1 \Delta t \dot{\boldsymbol{x}}_n + \theta_2 \Delta t^2 \boldsymbol{\alpha}_n^i$.

It is well known that the convergence of the Newton iteration to solve the non-linear equation at each time step may depend crucially on a good starting value. Starting values considered for the iteration include:

(1) Newmark, $\boldsymbol{a}_{n+1}^0 = 0$, SS22, $\boldsymbol{\alpha}_n^0 = 0$ (Zienkiewicz *et al* 1984, Chan 1988);

(2) using the value from the end of the last time step: $\boldsymbol{a}_{n+1}^0 = \boldsymbol{a}_n$, $\boldsymbol{\alpha}_n^0 = \boldsymbol{\alpha}_{n-1}$, which seems the most obvious; and

(3) predictor values obtained from the information available without incurring unnecessary storage. These can be obtained by using the algorithm in explicit form. For example, for Newmark, $\beta = 0$ in eqn (7.6.16) with zero damping and forcing function gives

$$\boldsymbol{a}_{n+1}^0 = -\boldsymbol{M}^{-1}\boldsymbol{P}\left(\boldsymbol{x}_n + \Delta t \boldsymbol{v}_n + \frac{\Delta t^2}{2}\boldsymbol{a}_n\right), \qquad (7.6.40)$$

SS22a with $\theta_2 = 0$ and $\Phi = 0$ in eqn (7.6.27) gives

$$\boldsymbol{\alpha}_n^0 = -\boldsymbol{M}^{-1}\{(\boldsymbol{P}(\boldsymbol{x}_n) + \theta_1 \Delta t \boldsymbol{K}(\boldsymbol{x}_n)\dot{\boldsymbol{x}}_n\}, \qquad (7.6.41)$$

and SS22b with $\theta_2 = 0$ in eqn (7.6.32) gives

$$\boldsymbol{\alpha}_n^0 = -\boldsymbol{M}^{-1}\boldsymbol{P}(\boldsymbol{x}_n + \theta_1 \Delta t \dot{\boldsymbol{x}}_n). \qquad (7.6.42)$$

There is also the question of how often the Jacobian iteration matrix in eqns (7.6.18) and (7.6.24) for Newmark or eqns (7.6.30) and (7.6.34) for SS22 should be updated in the time step; with reasonably small Δt it can be kept constant.

Figure 7.6.1 (from Wood and Oduor 1988) shows for the hardening spring whose equation is given in (7.6.9), an example where Newmark (with $2\beta = \gamma = 0.5$), remains stable and SS22a (with $\theta_1 = \theta_2 = 0.5$, $\Phi = 0.5$), is unstable. This shows the same effect as for the energy ratio curves in Figs. 6.5.1 and 6.5.2. Other results in Wood and Oduor (1988) confirm the effects shown with the energy ratio curves in Figs. 6.5.3 and 6.5.4 for

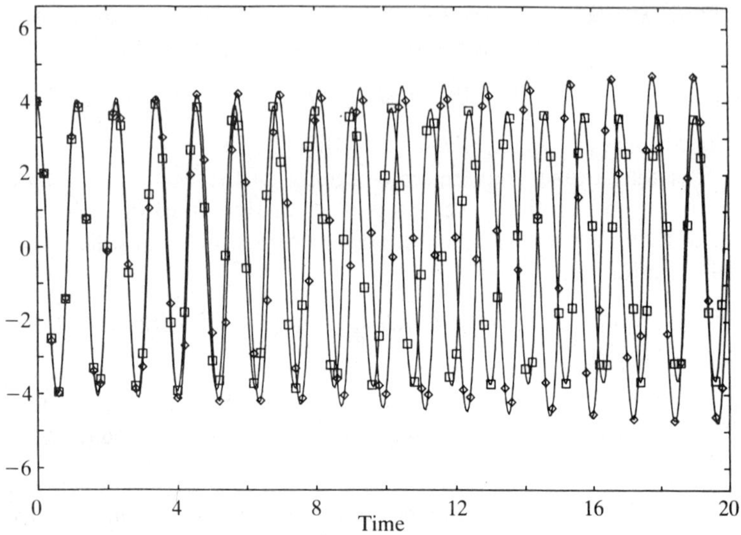

Fig. 7.6.2. Softening spring with SS22a, $\Phi = 0$, $\Phi = 0.5$. $\ddot{u} + 100 \tanh u = 0$, $u_0 = 4.0$, $v_0 = 0$, $\Delta t = 0.04$. SS22 (i) $\hat{u}_n = u_n$, (ii) $\hat{u}_n = \frac{1}{2}(u_n + u_{n+1})$, $\theta_1 = \theta_2 = 0.5$.

the softening spring problem (eqn 7.6.11), that Newmark (with $2\beta = \gamma = 0.5$) becomes unstable and SS22a (with $\theta_1 = \theta_2 = 0.5$, $\Phi = 0.5$), is damped. Figure 7.6.2 (from Wood and Oduor 1988) shows an example where for the softening spring problem with SS22a (with $\theta_1 = \theta_2 = 0.5$), $\Phi = 0$ gives an unstable solution and $\Phi = 0.5$ gives a damped solution.

Carroll (1988) reports numerical experiments with a non-linear problem of two masses m_1 and m_2 and connecting springs as shown in Fig. 7.6.3. The restoring forces in springs A and B when extended by a distance x are $\lambda_A(x)$ and $\lambda_B(x)$, respectively.

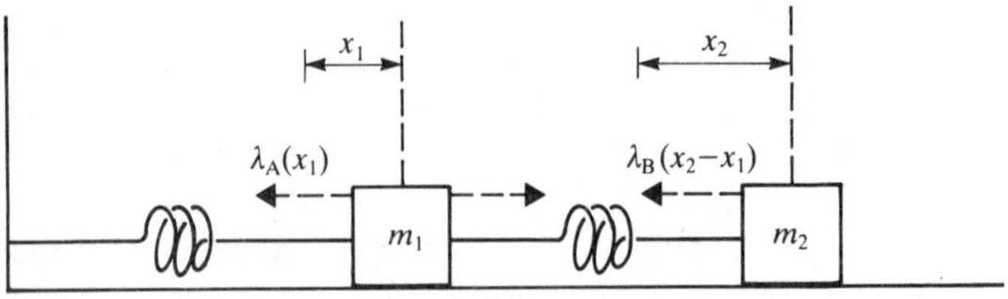

Fig. 7.6.3. Two-mass-spring system.

The motion of the masses is given by the two equations:

$$m_1\ddot{x}_1 + \lambda_A(x_1) - \lambda_B(x_2 - x_1) = 0 \tag{7.6.43}$$

and

$$m_2\ddot{x}_2 + \lambda_B(x_2 - x_1) = 0. \tag{7.6.44}$$

The restoring forces in the springs are taken as combinations of

(1) hardening, given by $\lambda(x) = k_1x + k_2x^3$, $k_1, k_2 > 0$, (7.6.45)

and

(2) softening, given by $\lambda(x) = k \tanh x$, $k > 0$. (7.6.46)

The combinations considered are hard–hard, soft–soft, hard–soft, and soft–hard.

The experiments are made with the trapezium rule equivalents of Newmark ($2\beta = \gamma = 0.5$) and SS22 ($\theta_1 = \theta_2 = 0.5$), and in the form of iterations for the increment in the acceleration. The convergence criteria are of the form shown in (7.6.35) or (7.6.36).

Since there is no damping and no external force, we have constant energy E, the sum of the kinetic and potential energies of the two masses given by the equation

$$E = \tfrac{1}{2}(m_1v_1^2 + m_2v_2^2) + \int \lambda(x_1)\,dx_1 + \int \lambda(x_2 - x_1)\,d(x_2 - x_1). \tag{7.6.47}$$

We can integrate the non-linear expressions for the spring tensions, that is, for the hardening spring,

$$\int (k_1x + k_2x^3)\,dx = \tfrac{1}{2}k_1x^2 + \tfrac{1}{4}k_2x^4. \tag{7.6.48}$$

and for the softening spring,

$$\int k \tanh x\,dx = k \ln(\cosh x), \tag{7.6.49}$$

and obtain an expression for the approximate energy in terms of the displacements and velocities from a numerical solution at any time.

Carroll (1988) reports that using the predictor values from eqns (7.6.40), (7.6.41), or (7.6.42) gives the same solution as that using the value from the previous time step in marginally fewer iterations but more operations are required. Using the zero starting value is only slightly less efficient than using the value from the previous time step. Taking all possible combinations of the spring forces and iteration starting conditions, the solution obtained and the number of iterations required are found to be independent of the frequency of update of the Jacobian. Hence for the results in Carroll (1988), the Jacobian matrix is calculated only once per time step using the acceleration from the previous time step. This acceleration is also used for the iteration starting value. Results are given for the combinations hard–hard, soft–soft, hard–soft and soft–hard with two kinds of starting conditions which make the problem less or more non-linear. The tolerance parameters in (7.6.35), (7.6.36) are taken as $\sigma_N = \sigma_S = 10^{-4}$. The number of iterations per time step is limited to four; if this is not sufficient an error message is returned. The program also checks the energy E_n calculated by substituting the displacement and velocity values given by the algorithms at time $t = n\Delta t$ into the corresponding form of the expression in eqn (7.6.47). If the condition

$$E_{n+1} \leq (1+\epsilon)E_0 \tag{7.6.50}$$

with $\epsilon = 10^{-4}$ is violated, the program returns an energy error message. Thus an algorithm can fail either because the Newton method does not converge in four iterations or because the bounded energy growth condition (7.6.50) fails.

In the numerical experiments the masses m_1, m_2 are taken equal to unity. The initial conditions are

$$(1)\quad x(0) = [0.2, 0]^T, \qquad v(0) = [0, 0]^T \tag{7.6.51}$$

$$(2)\quad x(0) = [1.5, 1.5]^T, \qquad v(0) = [0, 0]^T \tag{7.6.52}$$

In the use of SS22, α_0^0 is obtained from the differential equation: $\alpha_0^0 = -M^{-1}P(x(0))$.

For the hardening spring, in each case, $k_1 = 100$ and $k_2 = 1000$ in eqn (7.6.45). For the softening spring, in each case, $k = 100$ in eqn (7.6.46). Tables 7.6.1 to 4 show the results with the various combinations of springs and starting conditions. The time step is gradually increased until the program registers failure either due to the non-convergence in the specified four iterations or to the non-fulfilment of the energy condition (7.6.50). With an adjustable time-step scheme, either of these criteria could be used to signal an adjustment in Δt. It is interesting that the SS22b scheme compares well with Newmark except in the soft-soft combination.

Table 7.6.1. Summary of results from Carroll (1988).

Hard–hard;	$x_1(0)=0.2, x_2(0)=0$		$x_1(0)=x_2(0)=1.5$	
Method	Δt at failure	Reason	Δt at failure	Reason
Newmark	0.0052	Energy	0.0015	Energy
SS22a $\Phi=0$	Stable because overdamped			
$\Phi=0.5$	0.0052	Energy	0.0015	Energy
$\Phi=1$	0.0032	Energy	0.0010	Energy
SS2b	0.0068	Energy	0.0136	Newton

Table 7.6.2. Summary of results from Carroll (1988).

Soft–soft;	$x_1(0)=0.2, x_2(0)=0$		$x_1(0)=x_2(0)=1.5$	
Method	Δt at failure	Reason	Δt at failure	Reason
Newmark	0.017	Energy	0.045	Energy
SS22a $\Phi=0$	0.013	Energy	0.008	Energy
$\Phi=0.5$	0.078	Newton	0.043	Newton
$\Phi=1$	0.011	Energy	0.008	Energy
SS22b	0.021	Energy	0.012	Energy

Table 7.6.3. Summary of results from Carroll (1988).

Hard–soft;	$x_1(0)=0.2, x_2(0)=0$		$x_1(0)=x_2(0)=1.5$	
Method	Δt at failure	Reason	Δt at failure	Reason
Newmark	0.0090	Energy	0.0014	Energy
SS22a $\Phi=0$	Stable because overdamped			
$\Phi=0.5$	0.0090	Energy	0.0014	Energy
$\Phi=1$	0.0060	Energy	0.0009	Energy
SS2b	0.0400	Energy	0.0095	Energy

Table 7.6.4. Summary of results from Carroll (1988).
Soft–hard; $x_1(0)=0.2, x_2(0)=0$ $x_1(0)=x_2(0)=1.5$

Method	Δt at failure	Reason	Δt at failure	Reason
Newmark	0.006	Energy	0.009	Energy
SS22a $\Phi=0$	0.005	Energy	0.006	Energy
$\Phi=0.5$	0.006	Energy	0.008	Energy
$\Phi=1$	0.004	Energy	0.007	Energy
SS22b	0.007	Energy	0.010	Energy

7.7. Starting conditions

The following comments on starting conditions apply to both linear and non-linear problems. A problem which is first order in time such as the subsurface flow problem has to have given starting values of the function, in this case, the hydraulic potential. A problem which is second order in time such as the dynamic vibration equation has to have given starting values of displacement and velocity. With multistep methods or the single-step SS*pj*, $p \geq j$, it is usually assumed that the further starting values needed can be generated from the differential equation. Often the main interest is on the response of the system to some external forcing function. In the subsurface flow problem, this may be the effect of rainfall and/or extraction wells. In the dynamic vibration problem, the external force may be due to an applied load or earthquake tremors. There is then a choice as to how the system is to be started and allowance can be made for a 'running in' period.

As a general rule, it is bad to have a jerky start because it puts into the system high frequency components which may take a long time to die away. This effect is illustrated in Wood and Lewis (1975) which describes the problem of heat conduction in a bar represented by the non-dimensionalized equation

$$\frac{\partial \Phi}{\partial t}=\frac{\partial^2 \Phi}{\partial x^2}, \qquad 0 \leq x \leq 4 \tag{7.7.1}$$

where Φ is the temperature, with

(1) the temperature at $x=0$ suddenly raised from zero to unity at $t=0$; or

(2) the temperature at $x = 0$ raised from zero to unity in two halves over the first two time-steps.

The oscillation produced by using the Crank–Nicolson time stepping scheme (§2.6) is much reduced by (2). Zienkiewicz and Wood (1987) advocate always starting from everything at rest, for both first and second order problems, and bringing in the forcing function smoothly over two time-steps. The swing up from one step function compensates for the swing down from that of the previous time step. Hine (1985) shows results for the numerical solution of the motion of a cantilever with a force applied at the free end for both Newmark, and SS22 with and without smoothing. The smoothed forcing function is represented by a linear increase over the first time-step. The results for the displacement are always fairly good; the effect of smoothing on the velocity and acceleration is variable. There may be a sudden change in the forcing function later in the run which introduces oscillations into the numerical solution as shown in Penry and Wood (1985). These are reduced by decreasing the time step which suggests a time-step control should be used.

Rushton and Wedderburn (1973) compare the results of different starting conditions for the groundwater flow problem, represented by the vertically averaged, linearized form using transmissivity, which has the one-dimensional form:

$$S\frac{\partial h}{\partial t} = T\frac{\partial^2 h}{\partial x^2}, \tag{7.7.2}$$

where h is the piezometric head, S is the storage, and T the transmissivity. In particular Rushton *et al.* conclude that starting with the heads zero everywhere and a sinusoidally varying input representing annual rainfall applied at the same instant as a discharge from a drain is switched on gives a head always lower than the correct solution; it is better to attain a 'dynamic balance' with the annually varying input first and then switch on the drain. The present author's experience with the Griffiths (1978) program applied to a large aquifer is that the running in to a repeat response to an annual cycle itself, can give no trouble. Starting either with some known water levels with interpolation, or from a steady state, small changes introduced with small time steps avoid the kick in the solution. Here a repeated annual cycle input of rainfall and pumping gives acceptable results after only two years.

There is a different situation with the subsurface flow problem discussed in Calver and Wood (1989). This is flow in a thin layer of soil on a hillside and the equation to be solved is Richards' eqn (7.3.1) for the total hydraulic potential. It is not convenient to have the total hydraulic potential everywhere equal to zero and it is difficult to achieve a 'dynamic balance' here. This is not a situation where the response to an annual cycle of input is appropriate. The interest is in the response to an isolated incident such as a shower of rain. Hence we expect to have to 'run in' with zero forcing function to allow for time for any spurious oscillations to die down. Figure 7.7.1 from Calver and Wood (1989) shows how the run-in time is reduced by finer grids in the space discretization the key to which is given in Table 7.4.1. This figure shows the hillslope discharge obtained by the method of Lynch (1984) described in §7.4 from the solutions for the potential calculated at half-hour intervals. Run 1a from an overall fine grid is taken as the baseline solution. This run gives a very smooth run-in and a peak of 9.4×10^{-3} cubic metres per

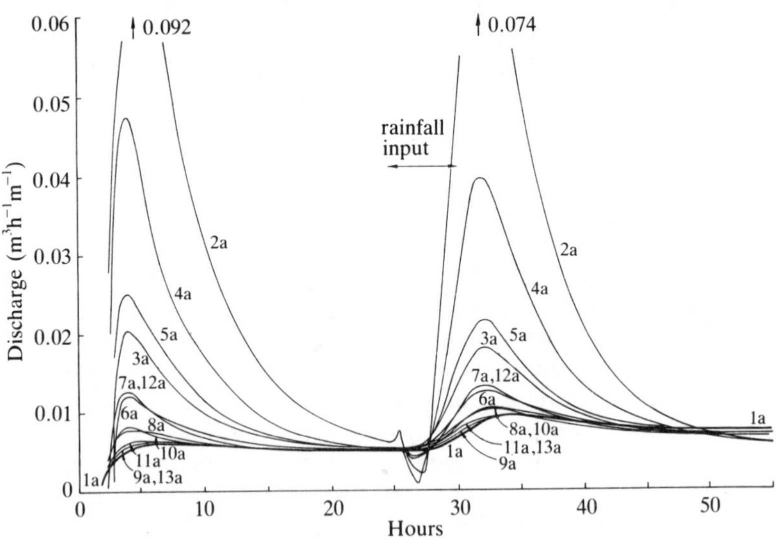

Fig. 7.7.1. Hydrographs from runs specified in Table 7.4.1. (From Calver and Wood 1989.)

hour about four hours after the end of the shower. The grids with large elements near the base of the slope in Fig. 7.4.1 require longer run-in time and produce spurious oscillations at the onset of the discharge as well as distorted peaks. The obviously spurious oscillations are an indication of inappropriate discretization.

7.8. Time-step adjustment

It is frequently cost-effective in the numerical solution of time-dependent problems to use some form of time-step adjustment. this may be

(1) because the external forcing function changes at different rates, sometimes fast, sometimes slow; or

(2) because the adequate representation of the non-linearity of the problem requires different time steps at different times.

The usual criterion for adjusting the time-step size depends on putting a tolerance limit on an estimate of the local truncation error. Park and Underwood (1980) and Underwood and Park (1980) discuss the solution of the dynamic vibration equation by the central difference method which corresponds to Newmark with $\beta = 0$ and $\gamma = 0.5$. As shown in §2.3 this method is explicit and conditionally stable. Park and Underwood argue that the truncation error criterion is not suitable when only low accuracy is required and propose instead using an 'apparent frequency' (§7.6). Their method has the disadvantage that the time-stepping formula used is over more than one time step and this complicates changing the size of time step. It is much simpler to use time-step adjustment with a one-step scheme.

Zienkiewicz *et al.* (1984) introduce the SS*pj* family of single step algorithms described in §4.5. These algorithms use the function and its derivatives up to and including the pth derivative at each time level and are based on the expansion within the time step:

$$\boldsymbol{x} = \boldsymbol{x}_n + \tau\dot{\boldsymbol{x}}_n + \frac{\tau^2}{2}\ddot{\boldsymbol{x}}_n + \cdots + \frac{\tau^p}{p!}\boldsymbol{\alpha}_n^p, \tag{7.8.1}$$

where $0 \leq \tau \leq \Delta t$. It hence seems reasonable to take the local truncation error as an estimate of the next term in the Taylor series expansion:

$$\boldsymbol{E} = \frac{\Delta t^p}{(p+1)!}(\boldsymbol{\alpha}_n^p - \boldsymbol{\alpha}_{n-1}^p). \tag{7.8.2}$$

The norm of this error is then proposed for use in the criterion for changing the size of the time step. For example, the time step is halved if $\|\boldsymbol{E}\| > E_{\max}$, and the time step is doubled if $\|\boldsymbol{E}\| < cE_{\max}$, where $E_{\max}$ is the tolerance and c is typically equal to 0.5. The tolerance $E_{\max}$ depends on the problem; it may be related to the maximum value of the function x during the course of the calculation. It is also usual to specify a minimum value of the time step.

Thomas and Gladwell (1988) in their generalization of SSp2 algorithms in the form of p-stage algorithms (§4.9) propose that the local error in the solution can be estimated as the difference between the results with a p-stage and a $(p+1)$-stage algorithm used on the same time step. The time step is then adjusted to keep this error less than the tolerance.

With implicit schemes, a matrix has to be inverted at each step and if the equation is integrated over a fixed time interval, larger time steps mean less cost from this. For non-linear problems there is also the cost of evaluating the non-linear terms; with smaller time steps it may be possible to hold the iteration matrix constant within the time step without updating.

Chan (1988) takes into account the relative effect of the non-linearity. Chan is using the non-linear equation

$$\boldsymbol{M}\ddot{\boldsymbol{x}} + \boldsymbol{P}(\boldsymbol{x}) = \boldsymbol{f} \tag{7.8.3}$$

and the SS22 algorithm but the criteria used here could be adapted to apply to the solution of any non-linear equation using the SSpj algorithm. If SS22 is used iteratively to solve for the increment $\Delta\boldsymbol{\alpha}_n^{m-1}$ at the mth iteration on the time step from t_{n-1} to $t_n = t_{n-1} + \Delta t$, then $\boldsymbol{\alpha}_n^m = \boldsymbol{\alpha}_n^{m-1} + \Delta\boldsymbol{\alpha}_n^{m-1}$. The following criteria are then examined:

1. If

$$\left\| \frac{\Delta t^2}{6}(\boldsymbol{\alpha}_n^m - \boldsymbol{\alpha}_{n-1}) \right\| > \text{TOL},$$

where TOL is the tolerance, the time step should be halved and the step repeated.

2. If

$$\left\| \frac{\Delta t^2}{2} \Delta \boldsymbol{\alpha}_n^{m-1} \right\| < c \left\| \frac{\Delta t^2}{6} (\boldsymbol{\alpha}_n^m - \boldsymbol{\alpha}_{n-1}) \right\|,$$

   where c is a preset parameter $0 < c \leq 1$, the iteration increment in the displacement equation (7.8.1) is less than the current assessment of the local error. If also

$$\left\| \frac{\Delta t^2}{6} (\boldsymbol{\alpha}_n^m - \boldsymbol{\alpha}_{n-1}) \right\| < \text{TOL},$$

   this means that the iteration stops and the time step may be doubled. Chan suggests that this is only done after five consecutive such signals.

3. If neither (1) nor (2) is satisfied the iteration is repeated until m exceeds some set number when the time step should be halved and the process repeated.

Enright (1986) proposes an error control based on residuals from an interpolation formula in the time step which is precisely what is being used in the SS*pj* methods and their generalization by Thomas and Gladwell (1988).

A different kind of approach to time-step adjustment can be used when the problem being solved has some quantity associated with it which ought in theory to remain constant. Where it is possible to construct an expression for the energy, for example (Chapter 6), and this should remain constant, the time step can be adjusted to keep any change in this less than some tolerance. In the groundwater flow problem there should be a mass balance of water, that is, the amount of water which has entered the aquifer should equal the amount that has run out plus the extra water in storage, to within some acceptable tolerance. This requires a calculation of the mass flux at the 'known head' (Dirichlet) boundary nodes. The finite difference water engineers have a low opinion of the finite element solution using four-node quadrilaterals for the solution of groundwater flow particularly because these give poor results for the calculation of boundary fluxes from the gradients across the boundary elements. This can be remedied by calculating the boundary fluxes using the Lynch (1984) method (described in §7.4) which gives an 'exact' balance on the steady state problem and takes account of the change in storage in the boundary elements in the time-dependent problem. The Lynch method can be used for calculating the discharge in a time step but, as shown in Table 7.4.1 from Calver and Wood (1989), it can give a specious result for the cumulative mass balance error in the time dependent problem. The cumulative mass balance error in run 2a looks very good but the hydrograph in Fig. 7.7.1 shows that this is a very bad result. The IHDM4 program used by Calver and Wood, which uses four-node quadrilaterals, controls the time step by checking the number of iterations required to attain convergence. Matthews and Wood (1987) uses eight-node quadrilaterals for the flow in a phreatic aquifer, vertically integrated to two dimensions in plan. The entries in the stiffness matrix require calculation of the gradients of the basis functions at the two-by-two Gauss points in the element, hence it is easy to pick up the gradients of the solution here. These points are the Barlow stress points (Barlow 1976), where the gradient is superconvergent. A parabola through three of the points, conveniently situated, is extrapolated to the boundary to give the boundary flux. With a steady-state aquifer problem, this method can give a mass balance error within 0.1%. This method can be combined with a simple time-stepping scheme with allowance made for the change of storage to control the size of time step in order to keep the mass balance error within an acceptable limit.

7.9. Conjugate gradient methods

Suppose we have a first order non-linear problem which after space discretization gives a system of equations of the form

$$\boldsymbol{M}(\boldsymbol{\Phi})\dot{\boldsymbol{\Phi}} + \boldsymbol{K}(\boldsymbol{\Phi}) = \boldsymbol{f}(t). \tag{7.9.1}$$

With the use of a time-stepping scheme we have a non-linear equation like (7.3.21) or (7.3.24) to solve at each time step, i.e. a system of equations of the form

$$\boldsymbol{A}\boldsymbol{x} = \boldsymbol{b}, \tag{7.9.2}$$

where the matrix $\boldsymbol{A}$ is symmetric. In §7.3, methods have been described which solve this non-linear equation by some kind of iteration. It is usually understood that the linearized system is solved by some convenient standard package Gauss or Cholesky at each iteration. When it is convenient to form the Jacobian matrix or some approximation to it then some Newton or Newton-like method is frequently used as described in §7.5. Another method for solving the non-linear system of equations (7.9.2) which has become popular with practical scientists and engineers is the Preconditioned Conjugate Gradient (PCG) algorithm. We start by describing the Conjugate Gradient (CG) algorithm (Hestenes and Stiefel 1952). This is one of the gradient algorithms which use the fact that the matrix $\boldsymbol{A}$ is symmetric to associate eqn (7.9.2) with the minimization of the quadratic form

$$F(x) = \tfrac{1}{2}\boldsymbol{x}^{\mathrm{T}}\boldsymbol{A}\boldsymbol{x} - \boldsymbol{b}^{\mathrm{T}}\boldsymbol{x}. \qquad (7.9.3)$$

The solution of eqn (7.9.2) minimizes $F(\boldsymbol{x})$ with respect to $\boldsymbol{x}$. We can visualize $F(\boldsymbol{x})$ as representing a kind of bowl in N-dimensional space, where N is the order of the matrix $\boldsymbol{A}$, the number of equations in the system (7.9.2). We start from some initial approximation and want to travel down the bowl to reach the lowest point. The CG method is a particular strategy for this. Suppose $\boldsymbol{x}_k$ is the kth iterate in the process of solution and we form the next iterate as

$$\boldsymbol{x}_{k+1} = \boldsymbol{x}_k + \alpha_k \boldsymbol{w}_k. \qquad (7.9.4)$$

This produces a change in $F(\boldsymbol{x})$ equal to

$$F(\boldsymbol{x}_{k+1}) - F(\boldsymbol{x}_k) = \alpha_k^2 \boldsymbol{w}_k^{\mathrm{T}}\boldsymbol{A}\boldsymbol{w}_k - 2\alpha_k \boldsymbol{w}_k^{\mathrm{T}}\boldsymbol{r}_k, \qquad (7.9.5)$$

where $\boldsymbol{r}_k$ is the residual, $\boldsymbol{r}_k = \boldsymbol{b} - \boldsymbol{A}\boldsymbol{x}_k$. Then $F(\boldsymbol{x}_{k+1})$ is a minimum, i.e. we are going as far as possible down the bowl in the direction $\boldsymbol{w}_k$, if

$$\alpha_k = \boldsymbol{w}_k^{\mathrm{T}}\boldsymbol{r}_k / \mathbf{w}_k^{\mathrm{T}}\boldsymbol{A}\boldsymbol{w}_k. \qquad (7.9.6)$$

Now we choose the direction $\boldsymbol{w}_k$ as a combination of the residual $\boldsymbol{r}_k$ and the previous direction $\boldsymbol{w}_{k-1}$ in the form

$$\boldsymbol{w}_k = \boldsymbol{r}_k + \beta_k \boldsymbol{w}_{k-1} \qquad (7.9.7)$$

and we choose the parameters β_k so that (in theory) the process should terminate in N steps where N is the number of equations in the system (7.9.2). This is done by making $\boldsymbol{w}_k$ 'conjugate', which signifies orthogonal with respect to the matrix $\boldsymbol{A}$, to the previous direction $\boldsymbol{w}_{k-1}$, hence we take

$$\boldsymbol{w}_k^{\mathrm{T}}\boldsymbol{A}\boldsymbol{w}_{k-1} = 0. \qquad (7.9.8)$$

Then the corresponding value of the parameter β is given by

$$\beta_k = -\boldsymbol{r}_k^{\mathrm{T}}\boldsymbol{A}\boldsymbol{w}_{k-1}/\boldsymbol{w}_{k-1}^{\mathrm{T}}\boldsymbol{A}\boldsymbol{w}_{k-1}. \tag{7.9.9}$$

The CG algorithm is given by the following. Given the starting value $\boldsymbol{x}_0$ (in the context of time-stepping schemes this is obtained from the value at the end of the previous time step or by some extrapolation), take $\boldsymbol{r}_0 = \boldsymbol{b} - \boldsymbol{A}\boldsymbol{x}_0$ and $\boldsymbol{w}_0 = \boldsymbol{r}_0$, then iterate on

$$\begin{aligned}
&\alpha_k = \boldsymbol{r}_k^{\mathrm{T}}\boldsymbol{w}_k/\boldsymbol{w}_k^{\mathrm{T}}\boldsymbol{A}\boldsymbol{w}_k, \quad \boldsymbol{x}_{k+1} = \boldsymbol{x}_k + \alpha_k\boldsymbol{w}_k,\\
&\boldsymbol{r}_{k+1} = \boldsymbol{b} - \boldsymbol{A}\boldsymbol{x}_{k+1} = \boldsymbol{r}_k - \alpha_k\boldsymbol{A}\boldsymbol{w}_k, \quad \beta_{k+1} = -\boldsymbol{r}_{k+1}^{\mathrm{T}}\boldsymbol{A}\boldsymbol{w}_k/\boldsymbol{w}_k^{\mathrm{T}}\boldsymbol{A}\boldsymbol{w}_k,\\
&\boldsymbol{w}_{k+1} = \boldsymbol{r}_{k+1} + \beta_{k+1}\boldsymbol{w}_k,
\end{aligned} \tag{7.9.10}$$

$k \to k+1$, until convergence.

By induction it can be shown that if there are N equations in the system in theory the residual $r_N \equiv 0$ and the procedure terminates. In practice, rounding errors in the process of computation may leave a non-zero residual r_N. Fox (1964) reports that with well-conditioned problems one more step is usually sufficient but it is clearly desirable to try to make it converge faster.

Evidence of the practical use of the conjugate gradient method is provided by a paper in the Proceedings of the Institution of Civil Engineers (1986) by Buchholdt *et al.*, on the non-linear dynamic analysis of guyed masts. These authors remark that using Conjugate Gradient rather than Newton makes a compact algorithm which is very suitable for use with micro-computers. Their method uses the Newmark formulae to express the displacement, velocity and acceleration at time t_{n+1} in terms of the displacement increment and the values at t_n, given by the formulae:

$$\boldsymbol{x}_{n+1} = \boldsymbol{x}_n + \Delta\boldsymbol{x}_n, \quad \dot{\boldsymbol{x}}_{n+1} = \frac{2}{\Delta t}\Delta\boldsymbol{x}_n - \dot{\boldsymbol{x}}_n,$$

$$\ddot{\boldsymbol{x}}_{n+1} = \frac{4}{\Delta t^2}\Delta\boldsymbol{x}_n - \frac{4}{\Delta t}\dot{\boldsymbol{x}}_n - \ddot{\boldsymbol{x}}_n.$$

The total potential work, W, at time t_{n+1} is expressed in terms of the Δx_s, $s = 1, 2, \ldots, N$, the incremental displacements at the nodes, the known values at the start of the time step plus the effect of the supports, wind loading, etc. Then W is minimized by the conjugate gradient method with respect to the nodal incremental displacements Δx_s. Buchholdt *et al.* mention the use of diagonal scaling of the matrix to help with convergence.

The Preconditioned Conjugate Gradient (PCG) method speeds up the conjugate gradient process by finding a matrix $\boldsymbol{B}$ which is an approximation to the matrix $\boldsymbol{A}$ and solving the equivalent system

$$\boldsymbol{B}^{-1}\boldsymbol{A}\boldsymbol{x} = \boldsymbol{B}^{-1}\boldsymbol{b} \tag{7.9.11}$$

with the idea that this will converge much faster. (If $\boldsymbol{B} \equiv \boldsymbol{A}$ the objective is attained immediately.) Since the matrix of the system being solved has to be symmetric, we can suppose that there exists a non-singular matrix $\boldsymbol{P}$ such that $\hat{\boldsymbol{A}} = \boldsymbol{P}\boldsymbol{B}^{-1}\boldsymbol{A}\boldsymbol{P}^{-1}$ is symmetric, which is possible if the matrices $\boldsymbol{A}$ and $\boldsymbol{B}$ are each symmetric and positive definite. (This is like the situation with the mass and stiffness matrices in §1.2.)

The cost-efficiency of the Preconditioned Conjugate Gradient method depends crucially on obtaining a good approximation matrix $\boldsymbol{B}$. Many people simply use $\boldsymbol{B} = \text{diag}(\boldsymbol{A})$. Winget and Hughes (1985) recommend combining PCG with an element by element (EBE) approximate factorization (§5.6). The matrix $\boldsymbol{B}$ which is to approximate the matrix $\boldsymbol{A}$ is defined as

$$\boldsymbol{B} = \boldsymbol{W}^{\frac{1}{2}}(\boldsymbol{I} + \epsilon\bar{\boldsymbol{A}})\boldsymbol{W}^{\frac{1}{2}}, \tag{7.9.12}$$

where $\boldsymbol{W}$ is a positive definite diagonal scaling matrix which reduces the entries of the matrix $\boldsymbol{A}$ to O(1). The parameter ϵ is a small positive number and $\bar{\boldsymbol{A}}$ is a matrix with the same sparsity structure as $\boldsymbol{A}$.

Further results on PCG methods may be found in Crisfield (1979), (1985); Carey and Jiang (1986); Lai and Liddell (1988); Wathen (1988); Dupont and Marchal (1988); Prakhya (1988) and the references therein.

Crisfield (1979) points out the connection between Dynamic Relaxation and the conjugate gradient methods. Dynamic Relaxation is an iterative method for the solution of static problems originally proposed by Otter (1965) with the idea that the static problem can be regarded as the end result of a damped oscillation. This visualizable concept has made the method popular with engineers although it is not the most efficient way to solve the problem. It is compared with some other iterative methods in Wood (1967). There is a very comprehensive survey of the Dynamic Relaxation method by Underwood in Belytschko and Hughes (1983).

References

Anderson, C. A. and Zienkiewicz, O. C. (1974). Spontaneous ignition: finite element solutions for steady and transient conditions. *Journal of Heat Transfer, Trans. ASME,* 398–404.

Barlow, J. (1976). Optimal stress location in finite element models. *International Journal of Numerical Methods in Engineering,* **10,** 243–251.

Bear, J. (1979). *Hydraulics of groundwater.* McGraw-Hill.

Belytschko, T. and Hughes, T. J. R. (eds.) (1983). *Computational methods for transient analysis.* North-Holland.

Belytschko, T., Chiapetta, R. L., and Bartel, H. D. (1976). Efficient large-scale non-linear transient analysis by finite elements. *International Journal of Numerical Methods in Engineering,* **10,** 579–596.

Belytschko, T., Holmes, N., and Mullen, R. (1975). Explicit integration-stability, solution properties, cost. In *ASME applied mechanics symposia series. Finite element analysis of transient non-linear structural behaviour* (eds. T. Belytschko *et al.*), Vol. AMD14, pp. 1–21.

Beven, K., Calver, A., and Morris, E. M. (1987). *The Institute of Hydrology Distributed Model, Report no. 98*. Institute of Hydrology, Wallingford, UK OX10 8BB.

Buchholdt, H. A., Moosavinejad, S., and Iannuzzi, A. (1986). Non-linear dynamic analysis of guyed masts subjected to wind and guy ruptures, Paper 9015. *Proceedings of the Institute of Civil Engineering,* **81,** 353–395.

Calver, A. and Wood, W. L. (1989). On the discretization and cost-effectiveness of a finite element solution for hillslope subsurface flow. *Journal of Hydrology,* **110,** 165–179.

Carey, G. F. and Jiang, B.-N. (1986). Element by element linear and non-linear solution schemes. *Communications in Applied Numerical Methods,* **2,** 145–153.

Carroll, J. (1988). Investigation of the properties of Newmark and the SS22 methods for the numerical solution of non-linear dynamic vibration equations. *M.Sc. Dissertation.* University of Reading.

Chan, A. C. H. (1988). A unified finite element solution to state and dynamic problems of geomechanics. *Ph. D. Thesis*. University College of Swansea.

Connorton, B. J. and Wood, W. L. (1983). Noise problems with ephemeral streams in aquifer models. *International Journal for Numerical Methods in Fluids*, **3**, 201–208.

Conte, S. D. and de Boor, C. (1980). *Elementary numerical analysis* (3rd edn). McGraw-Hill.

Cooley, R. L. (1983). Some new procedures for numerical solution of variably saturated flow problems. *Water Resources Research*, **19**(5) 1271–1285.

Crisfield, M. A. (1979). Iterative solution procedures for linear and non-linear structural analysis. *TRRL Report 900.* Transport and Road Research Laboratory, Crowthorne, Berks.

Crisfield, M. A. (1985). New solution procedures for linear and non-linear finite element analysis. In *Proceedings of the conference on the mathematics of finite elements and applications V* (ed. J. R. Whiteman). Academic Press.

Culham, W. E. and Varga, R. S. (1971). Numerical methods for time-dependent non–linear boundary value problems. *Journal of the Society Petroleum Engineers*, Dec., 374–388.

Dahlquist, G. (1975). Error analysis for a class of methods for stiff non-linear initial value problems. *Numerical analysis, Dundee, Springer Lecture Notes in Mathematics*, **506**, 60–74.

Dennis, J. E. and Moré, J. J. (1977). Quasi–Newton methods, motivation and theory. *SIAM Review*, **19**, 46–89.

Dupont, S. and Marchal, J. M. (1988). Preconditioned conjugate gradients for solving the transient Boussinesq equations in three-dimensional geometries. *International Journal of Numerical Methods in Fluids*, **8**, 283–303.

Enright, W. H. (1986). A new error-control for initial value solvers. *Numerical analysis report no. 122*. Department of Mathematics, University of Manchester.

Fox, L. (1964). *Introduction to numerical linear algebra.* Clarendon Press, Oxford.

Fox, L. and Mayers, D. F. (1968) *Computing methods for engineers and Scientists.* Clarendon Press, Oxford.

Frind, E. O. and Verge, M. J. (1978). Three-dimensional modelling of groundwater flow systems. *Water Resources Research*, **14**(5) 844–856.

Fujii, H. (1973). Some remarks on finite element analysis of time-dependent field problems. In *Theory and practice in finite element structural analysis.* University of Tokyo Press.

Gellert, M. (1985). Discrete numerical solution of the generalized Boussinesq problem. *International Journal of Numerical Methods in Engineering*, **21**, 2131–2144.

Gourlay, A. R. (1970). A note on trapezoidal methods for the solution of initial value problems. *Mathematics of Computation*, **24**, 629–633.

Gresho, P. M., Lee, R. L., and Sani, R. L. (1978). Advection dominated flows, with emphasis on the consequences of mass lumping. In *Finite elements in fluids* (eds. R. H. Gallagher, O. C. Zienkiewicz, J. T. Oden, M. Morandi Cecchi, and C. Taylor), Vol. 3. Wiley

Griffiths, A. L. (1978). A finite element model of groundwater flow in unconfined aquifers. *M.Sc. Dissertation.* University of Reading.

Henrici, P. (1962). *Discrete variable methods in ordinary differential equations.* Wiley.

Hestenes, M. R. and Stiefel, E. (1952). Methods of conjugate gradients for solving linear systems. *NBS Journal of Research,* **49,** 409–436.

Hine, N. W., (1985). Starting conditions and procedures for the solution of a 1-D dynamic elastic bar problem usng the SS22 and Newmark single step time integration methods. *Report C/R/530/85.* Institute for Numerical Methods in Engineering, Swansea.

Hromadka, T. V., II and Guymon, G. L. (1980). Some effects of linearizing the unsaturated soil moisture transfer diffusivity model. *Water Resources Research,* **16,** 643–650.

Hughes, T. J. R. (1977). Unconditionally stable algorithms for non-linear heat conduction. *Computational Methods in Applied Mechanics and Engineering,* **10,** 135–139.

Huyakorn, P. S. and Pinder, G. F. (1983). *Computational methods in subsurface flow.* Academic Press.

Kukawski, J. and Desai, C. S. (1984). Generalized time finite element algorithm for non-linear dynamic problems. *Engineering Computation,* **1,** 247–251.

Lai, C. H. and Liddell, H. M. (1988). Preconditioned Conjugate Gradient methods on the DAP. *Proceedings of mathematics of finite elements and applications VI* (ed. J. R. Whiteman). Academic Press.

Lees, M. (1966). A linear three-level difference scheme for quasi-linear parabolic equations. *Mathematics of Computation,* **20,** 516–622.

Lynch, D. R., (1984) Mass conservation in finite element groundwater models. *Advances in Water Resources,* **7,** 67–75.

Mariño, M. A. (1981). Analysis of the transient movement of water and solutes in stream-aquifer systems. *Journal of Hydrology,* **49,** 1–17.

Matthews, R. S. and Wood, W. L. (1987). A cost-efficiency analysis of boundary flow calculations from a finite element solution to groundwater flow problems. *Communications in Applied Numerical Methods,* **3,** 145–154.

McNamara, J. F. (1974). Solution schemes for problems of nonlinear structural dynamics. *Transactions of ASME, Journal of Pressure Vessel Technology,* **96,** 96–102.

Narasimhan, T. N. (1978). A perspective on numerical analysis of the diffusion equation. *Advances in Water Resources,* **1,** 147–155.

Neuman, S. P. (1973). Saturated–Unsaturated seepage by finite elements. *Journal of Hydraulics Division, Proceedings of ASCE,* **99**(HY12) 2233–2250.

Oden, J. T. and Fost, R. B. (1973). Convergence, accuracy and stability of finite element approximations of a class of nonlinear hyperbolic equations. *International Journal of Numerical Methods in Engineering,* **6,** 357–365.

Otter, J. R. H. (1965). Computations for prestressed concrete reactor pressure vessels using dynamic relaxation. *Nuclear Structural Engineering,* **1,** 61–75.

Owen, D. R. J., (1978). Implicit finite element methods for the dynamic transient analysis of solids with particular reference to non-linear situations. *Advanced Course on Structural Dynamics.* EURATOM, Ispra, Italy.

Park, K. C. (1975). An improved stiffly stable method for direct integration of nonlinear structural dynamic equations. *Transactions of American Society of Mechanical Engineers,* **42,** 464–470.

Park, K. C. (1977). Practical aspects of numerical time integration. *Journal of Computers and Structures,* **7,** 343–353.

Park, K. C. and Underwood, P. G. (1980). A variable-step central difference method for structural dynamics analysis. Part 1. Theoretical aspects. *Computer Methods in Applied Mechanics and Engineering,* **22,** 241–258.

Penry, S. N. and Wood, W. L. (1985). Comparison of some single step methods for the numerical solution of the structural dynamic equation. *International Journal of Numerical Methods in Engineering,* **21,** 1941–1955.

Phillips, G. M. and Taylor, P. J. (1973). *Theory and applications of numerical analysis.* Academic Press.

Prakhya, K. V. G. (1988). Some conjugate gradient methods for symmetric and nonsymmetric systems. *Communications in Applied Numerical Methods,* **4,** 531–539.

Richards, L. A. (1931). Capillary conduction of liquids through porous mediums. *Physics,* **1,** 318–333.

Rushton, K. R. and Wedderburn, L. A. (1973). Starting conditions for aquifer simulations. *Ground Water,* **11,** 37–42.

Stakgold, I. (1967). *Boundary value problems of mathematical physics.* Macmillan.

Strang, G. and Fix, G. J. (1973). *An analysis of the finite element method.* Prentice-Hall.

Thomas, R. M. and Gladwell, I. (1988). Variable-order variable-step algorithms for second order problems. *International Journal of Numerical Methods in Engineering,* **26,** 39–53 and 55–80.

Underwood, P. (1983). Dynamic relaxation. In *Computational methods for transient analysis* (eds. T. Belytschko and T. R. J. Hughes), Ch. 5. North-Holland.

Underwood, P. G. and Park, K. C. (1980). A variable step central difference method for structural dynamics analysis. Part 2: Implementation and performance evaluation. *Computer Methods in Applied Mechanics and Engineering,* **23,** 259–279.

Varga, R. S., (1962). *Matrix iterative analysis.* Prentice-Hall.

Wathen, A. J., (1988). Spectral bounds and preconditioning methods using element-by-element analysis for Galerkin finite element equations. *Proceedings of Mathematics of Finite Elements and Applications VI* (ed. J. R. Whiteman). Academic Press.

Winget, J. M. and Hughes, T. J. R. (1985). Solution algorithms for nonlinear transient heat conduction analysis employing element-by-element iterative strategies. *Computer Methods in Applied Mechanics and Engineering,* **52,** 711–815.

Wood, W. L. (1967). Comparison of dynamic relaxation with three other iterative methods. *Engineer,* **Nov. 1967,** 683–687.

Wood, W. L. and Calver, A. (1990). Lumped versus distributed mass matrices in subsurface flow. To appear in *Water Resources Research.*

Wood, W. L. and Lewis, R. W. (1975). A comparison of time marching schemes for the transient heat conduction equation. *International Journal of Numerical Methods in Engineering,* **9,** 679–689.

Wood, W. L. and Oduor, M. E. (1988). Stability properties of some algorithms for the solution of non-linear dynamic vibration equations. *Communications in Applied Numerical Methods*, **4,** 205–212.

Yeh, G. T. (1985). Comparisons of successive iteration and direct methods to solve finite element equations of aquifer contaminant transport. *Water Resources Research,* **21,** 272–280.

Yoon, Y. S. and Yeh, W. W.-G. (1975). The Galerkin method for nonlinear parabolic equations of unsteady groundwater flow. *Water Resources Research,* **11,** 751–754.

Zienkiewicz, O. C. (1977). *The finite element method.* McGraw-Hill.

Zienkiewicz, O. C. and Wood, W. L. (1987). Transient response analysis. In *Finite element handbook* (ed. H. Hardestuncer), Ch. 8. McGraw-Hill.

Zienkiewicz, O. C., Li Xi-Kui, and Nakazawa, S. (1986). Dynamic transient analysis. *International Journal of Numerical Methods in Engineering,* **23,** 1343–1353.

Zienkiewicz, O. C., Taylor, R. L. and Too, J. M. (1971). Reduced integration technique in general analysis of plates and shells. *International Journal of Numerical Methods in Engineering,* **3,** 275–290.

Zienkiewicz, O. C., Wood, W. L., Hine, N. W., and Taylor, R. L. (1984). A unified set of single step algorithms. Part 1: general formulation and applications. *International Journal of Numerical Methods in Engineering,* **20,** 1529–1552.

8 Coupled systems

8.1. Introduction

In this chapter we consider simple practical methods for the numerical solution of time-dependent problems where there is more than one state variable to be obtained. The following is a list of problems included.

1. Shallow water problems, where the Navier–Stokes and continuity equations are depth-integrated and the resulting system is to be solved for the depth of the water and the horizontal components of the depth-averaged velocity. This problem can be extended to include the transport of effluent.
2. Problems of fluid flow in a rigid porous medium. These can also be extended to include the transport of effluent.
3. Problems of heat and fluid flow in a deforming porous medium. These involve solution for the displacement in the medium, the pore pressure and the temperature.
4. Problems of structure–fluid interaction.
5. Problems of structure–soil interaction

These are all treated in sets of either first order or second order equations. Generally it is not considered a good idea to eliminate one of the unknowns to produce an equation of higher order. Table 8.1 shows a list of the types of problem used in this chapter to illustrate various approaches to the choice of time-stepping scheme with the orders of the differential equations involved.

The methods for the numerical solution of these problems can be divided into:

(1) monolithic schemes where the differential equations for the different variables are all solved together;

(2) staggered schemes where the different variables are solved for separately and there may or may not be iteration between them.

Table 8.1. Examples of coupled systems.

Problem	Function	Order of equation	References
(1) Shallow water flow	Depth	1	Taylor *et al.* (1972)
	Velocity	1	Taylor *et al.* (1974a)
(1a) Plus effluent same density	Concentration	1	Taylor *et al.* (1974b)
(2) Two-phase flow in rigid porous medium	Phase-potentials	1	White (1978) White *et al.* (1981)
(3) Fluid in deformable medium	Pore pressure	1	
	Medium-displacement	1	Lewis and Schrefler (1987)
		2	Zienkiewicz *et al.* (1985, 1988)
(3a) Plus heat flow	Temperature	1	Lewis and Schrefler (1987)
(4) Structure–fluid	Displacement	2	
	Pressure-integral	1	Park *et al.* (1977)
	Pressure	2	Zienkiewicz *et al.* (1978)
(5) Soil–structure	Displacement	2	
	Displacement	2	Smith (1975)

8.2. Monolithic schemes

We illustrate this approach with examples.

The equations for shallow water flow are the equation of continuity and the Navier–Stokes equations. As shown, for instance, in Kawahara *et al.* (1982), when depth-integrated and space-discretized by finite elements these produce equations of the form

$$\boldsymbol{M}_1\dot{\boldsymbol{h}} = \boldsymbol{F}_1(\boldsymbol{h}, \boldsymbol{U}) \tag{8.2.1}$$

$$\boldsymbol{M}_2\dot{\boldsymbol{U}} = \boldsymbol{F}_2(\boldsymbol{h}, \boldsymbol{U}), \tag{8.2.2}$$

where $\boldsymbol{h}$ = the vector of the depths of water at the nodes, $\boldsymbol{U} = [\boldsymbol{u}, \boldsymbol{v}]^{\mathrm{T}}$, where $\boldsymbol{u}, \boldsymbol{v}$ are the vectors of the horizontal components of the depth-averaged velocity at the nodes, $\boldsymbol{M}_1$, $\boldsymbol{M}_2$ are the mass matrices, the same basis functions being used for the unknowns $h, u,$ and v, and $\boldsymbol{F}_1(\boldsymbol{h}, \boldsymbol{U})$, $\boldsymbol{F}_2(\boldsymbol{h}, \boldsymbol{U})$ include the non-linear terms and the extras such as rainfall, infiltration, and bottom friction effects. In the monolithic approach, the three equations are assembled into the single set

$$\boldsymbol{M}\dot{\boldsymbol{q}} = \boldsymbol{F}, \tag{8.2.3}$$

where $\boldsymbol{q}^{\mathrm{T}} = [\boldsymbol{h}, \boldsymbol{u}, \boldsymbol{v}]$. Taylor *et al.* (1972) use equations of the form of (8.2.1) and (8.2.2) to represent watershed run-off and solve the monolithic system (8.2.3). They say that the boundary conditions are more easily applied in the monolithic approach because they can then be simultaneously included. The trapezium rule Crank–Nicolson time-stepping algorithm is used in the form

$$\boldsymbol{M}\boldsymbol{q}_{n+1} = \boldsymbol{M}\boldsymbol{q}_n + \frac{\Delta t}{2}(\boldsymbol{F}_{n+1} + \boldsymbol{F}_n). \tag{8.2.4}$$

The non-linearity is dealt with by iteration starting from the explicit predictor value:

$$\boldsymbol{q}_{n+1}^0 = \boldsymbol{q}_n + \Delta t\boldsymbol{M}^{-1}\boldsymbol{F}_n.$$

This predictor value is used to form $\boldsymbol{F}_{n+1}^0$ and then eqn (8.2.4) is used to give the iteration scheme

$$\boldsymbol{M}\boldsymbol{q}_{n+1}^i = \boldsymbol{M}\boldsymbol{q}_n + \frac{\Delta t}{2}(\boldsymbol{F}_{n+1}^{i-1} + \boldsymbol{F}_n) \tag{8.2.5}$$

and so on. Taylor *et al.* state that the convergence can be accelerated by the relaxation process

$$\boldsymbol{q}^{i+1} = \boldsymbol{q}^{i-1} + \omega(\boldsymbol{q}^i - \boldsymbol{q}^{i-1}), \tag{8.2.6}$$

where the factor ω usually lies in the range $0.5 < \omega < 1.5$ (so it is not clear whether they are recommending under- or over-relaxation here).

A similar monolithic approach is used in Taylor and Davis (1974a) for a finite element model of tides in estuaries again producing a system of equations of the form (8.2.3) which is solved by the trapezium rule scheme (8.2.4) with iteration and found to be unconditionally stable. The main disadvantages are stated to be the lack of prototype data and adequate measurement of parameters such as the bottom surface roughness (this is always a difficulty).

With the iteration arranged as in eqn (8.2.5) it does not matter particularly whether the stiffness type matrices are symmetric or not as they are on the right-hand side. The mass matrix is constant so the steps of the interation can be arranged efficiently but it is not stated how many iterations are usually required. The convergence criterion is on the change in the solution. This method comes in the category of *ad hoc* iterative methods described in §7.3.

Kawahara *et al.* (1978), Kawahara *et al.* (1982a), and Kawahara and Hirano (1983) use a two-step time-stepping scheme for the single monolithic set of shallow water equations as given in eqn (8.2.3) with selective mass lumping introduced in the later papers. Applied to eqn (8.2.3) for the shallow water equations with no viscosity and no bottom friction term this corresponds to using the algorithm

$$\begin{aligned} \bar{\boldsymbol{M}}\boldsymbol{q}_{n+\frac{1}{2}} &= \tilde{\boldsymbol{M}}\boldsymbol{q}_n + \frac{\Delta t}{2}\boldsymbol{F}_n, \\ \bar{\boldsymbol{M}}\boldsymbol{q}_{n+1} &= \tilde{\boldsymbol{M}}\boldsymbol{q}_n + \Delta t\boldsymbol{F}_{n+\frac{1}{2}} \end{aligned} \tag{8.2.7}$$

to advance from t_n to $t_n + \Delta t$, where

$$\tilde{\boldsymbol{M}} = e\bar{\boldsymbol{M}} + (1-e)\boldsymbol{M}, \tag{8.2.8}$$

and $\bar{\boldsymbol{M}}$ is the lumped mass matrix. The parameter e is called the selective lumping parameter and it is there to control the numerical damping and stability. This is an explicit scheme and the limitation on the time step has the familiar association with the speed $\sqrt{(gH)}$ with which waves travel in shallow water, average depth H, and is given in Kawahara *et al.* (1982a) as

$$\Delta t < \frac{(2-e)}{3\sqrt{2}}\frac{\Delta x}{\sqrt{(gH)}}, \tag{8.2.9}$$

where Δx is the minimum size of element. Kawahara *et al.* (1982a) also advise that $\Delta x/\sqrt{(gH)}$ should be kept constant over the domain. The value for the parameter e is chosen from numerical experiments with the one-dimensional shallow water equation, and recommended to be chosen in the interval (0.8, 0.95). The authors say that their experiments show that then excessive numerical damping is not produced. The method of Kawahara *et al.* (1982a) is applied to various problems including tsunami wave propagation and tidal flow in Osaka and Yatsuhiro Bays.

The system of equations (8.2.3) can be augmented to include also the set of equations for the convection and diffusion of effluent. If the effluent is assumed to be of the same density as the water in the estuary and to be instantaneously well mixed over the depth, then the depth-averaging again applies. As shown for example, in Taylor and Davis (1974b), the result is that the equation for the concentration c of effluent, when space-discretized, produces another equation of the same form as (8.2.1) and (8.2.2):

$$\boldsymbol{M}\dot{\boldsymbol{c}} = \boldsymbol{F}_3, \tag{8.2.10}$$

where $\boldsymbol{c}$ is the vector of nodal values of the concentration, and $\boldsymbol{F}_3$ depends on $\boldsymbol{h}$, $\boldsymbol{u}$, $\boldsymbol{v}$, and $\boldsymbol{c}$. The eqns (8.2.3) and (8.2.10) can then be put together in one monolithic system. Taylor and Davis use this method for a numerical model of dispersion in estuaries and solve the monolithic system by the algorithm given by eqns (8.2.4) to (8.2.6).

Lewis and Schrefler (1987) (Chapter 7) give a detailed discussion of the mathematical modelling of the problem of heat and fluid flow in a porous medium. The reader is referred to Lewis and Schrefler for the background physics and the finite element space discretization details. With the same basis functions for the displacement u of the medium, the pore pressure p and the temperature T, the result is a system of first order equations of the form

$$\boldsymbol{K}\dot{\boldsymbol{u}} + \boldsymbol{L}\dot{\boldsymbol{p}} + \boldsymbol{T}_{\mathrm{U}}\dot{\boldsymbol{T}} = \boldsymbol{f}_1 \tag{8.2.11}$$

$$\boldsymbol{L}^{\mathrm{T}}\dot{\boldsymbol{U}} + \boldsymbol{S}\dot{\boldsymbol{p}} + \boldsymbol{T}_{\mathrm{P}}\dot{\boldsymbol{T}} + \boldsymbol{H}\boldsymbol{p} = \boldsymbol{f}_2 \tag{8.2.12}$$

$$\boldsymbol{T}_{\mathrm{L}}\dot{\boldsymbol{p}} + \boldsymbol{T}_{\mathrm{S}}\dot{\boldsymbol{T}} + \boldsymbol{T}_{\mathrm{R}}\boldsymbol{T} = \boldsymbol{f}_3, \tag{8.2.13}$$

where the matrices $\boldsymbol{T}_{\mathrm{L}}$ and $\boldsymbol{T}_{\mathrm{S}}$ depend on the temperature $\boldsymbol{T}$ and the other terms are linear if a linear constitutive relationship and constant material parameters are assumed. For the monolithic approach, the equations can be written in the concise form

$$\boldsymbol{B}\dot{\boldsymbol{X}} + \boldsymbol{C}\boldsymbol{X} = \boldsymbol{F}, \tag{8.2.14}$$

where

$$\boldsymbol{B} = \begin{bmatrix} \boldsymbol{K} & \boldsymbol{L} & \boldsymbol{T}_{\mathrm{U}} \\ \boldsymbol{L}^{\mathrm{T}} & \boldsymbol{S} & \boldsymbol{T}_{\mathrm{P}} \\ \boldsymbol{0} & \boldsymbol{T}_{\mathrm{L}} & \boldsymbol{T}_{\mathrm{S}} \end{bmatrix} \tag{8.2.15}$$

$$\boldsymbol{C} = \begin{bmatrix} \boldsymbol{0} & \boldsymbol{0} & \boldsymbol{0} \\ \boldsymbol{0} & \boldsymbol{H} & \boldsymbol{0} \\ \boldsymbol{0} & \boldsymbol{0} & \boldsymbol{T}_{\mathrm{R}} \end{bmatrix}. \tag{8.2.16}$$

The matrix $\boldsymbol{B}$ is not symmetric but Lewis and Schrefler (1987) give this monolithic approach as an option with the time-stepping formula

$$\frac{1}{\Delta t}\boldsymbol{B}(\boldsymbol{X}_{n+1}-\boldsymbol{X}_n)+\boldsymbol{C}\{(1-\alpha)\boldsymbol{X}_n+\alpha\boldsymbol{X}_{n+1}\}=\boldsymbol{F}_n, \quad (8.2.17)$$

where they say the matrices need to be evaluated only once per time step. The parameter α in eqn (8.2.17) is like the θ parameter in the θ-method, but Lewis and Schrefler mention other possibilities than the usual 0, $\frac{1}{2}$, $\frac{2}{3}$, 1 values with references (in Chapter 3), to values including logarithmic functions used by Sandhu (1968) and Hwang *et al.* (1971). For the linear equation, the method given by eqn (8.2.17) has unconditional stability with the parameter $\alpha \geq \frac{1}{2}$ as shown for the corresponding θ-method in §2.6.

Another approach described by Lewis and Schrefler here is to partition the matrix $\boldsymbol{B}$ into a symmetric part to be used in an implicit part of the scheme and the rest to be used in the explicit part. Thus $\boldsymbol{B}$ is written as

$$\boldsymbol{B}=\boldsymbol{B}^{\mathrm{I}}+\boldsymbol{B}^{\mathrm{E}}, \quad (8.2.18)$$

where

$$\boldsymbol{B}^{\mathrm{I}}=\begin{bmatrix}\boldsymbol{K} & \boldsymbol{L} & \boldsymbol{0}\\ \boldsymbol{L}^{\mathrm{T}} & \boldsymbol{S} & \boldsymbol{0}\\ \boldsymbol{0} & \boldsymbol{0} & \boldsymbol{T}_{\mathrm{S}}\end{bmatrix} \qquad \boldsymbol{B}^{\mathrm{E}}=\begin{bmatrix}\boldsymbol{0} & \boldsymbol{0} & \boldsymbol{T}_{\mathrm{U}}\\ \boldsymbol{0} & \boldsymbol{0} & \boldsymbol{T}_{\mathrm{P}}\\ \boldsymbol{0} & \boldsymbol{T}_{\mathrm{L}} & \boldsymbol{0}\end{bmatrix}.$$

Then the first two sets of equations (8.2.17) are solved with a predicted value for the temperature, that is, $\boldsymbol{T}_{n+1}$ is replaced by $\boldsymbol{T}^p_{n+1}$ to give the following equation

$$\begin{bmatrix}\boldsymbol{K} & \boldsymbol{L}\\ \boldsymbol{L}^{\mathrm{T}} & \boldsymbol{S}+\alpha\Delta t\boldsymbol{H}\end{bmatrix}\begin{bmatrix}\boldsymbol{u}_{n+1}\\ \boldsymbol{p}_{n+1}\end{bmatrix}=\begin{bmatrix}\boldsymbol{K} & \boldsymbol{L} & \boldsymbol{T}_{\mathrm{U}}\\ \boldsymbol{L}^{\mathrm{T}} & \boldsymbol{S}-(1-\alpha)\Delta t\boldsymbol{H} & \boldsymbol{T}_{\mathrm{P}}\end{bmatrix}\begin{bmatrix}\boldsymbol{u}_n\\ \boldsymbol{p}_n\\ \boldsymbol{T}_n\end{bmatrix}$$
$$+\Delta t\begin{bmatrix}\boldsymbol{f}_1\\ \boldsymbol{f}_2\end{bmatrix}-\begin{bmatrix}\boldsymbol{T}_{\mathrm{U}}\boldsymbol{T}^p_{n+1}\\ \boldsymbol{T}_{\mathrm{P}}\boldsymbol{T}^p_{n+1}\end{bmatrix}. \quad (8.2.19)$$

Then $\boldsymbol{T}_{n+1}$ is obtained from the third set of equations:

$$(\boldsymbol{T}_{\mathrm{S}}+\alpha\Delta t\boldsymbol{T}_{\mathrm{R}})\boldsymbol{T}_{n+1}=\{\boldsymbol{T}_{\mathrm{S}}-(1-\alpha)\Delta t\boldsymbol{T}_{\mathrm{R}}\}\boldsymbol{T}_n$$
$$-\boldsymbol{T}_{\mathrm{L}}(\boldsymbol{p}_{n+1}-\boldsymbol{p}_n)+\Delta t\boldsymbol{f}_3 \quad (8.2.20)$$

apparently with no iteration.

This partitioning resembles the partitioning methods proposed by Park and Felippa (1983) for the solution of problems of the interaction of a structure with its environment.

Another type of coupled problem producing a first order system of differential equations is that of two-phase flow through a rigid porous medium. The problem can be expressed in terms of the potentials Ψ_n and Ψ_w of the non-wetting and wetting phases with the equations

$$\delta_\pi \boldsymbol{C} \frac{\mathrm{d}}{\mathrm{d}t}(\boldsymbol{\Psi}_{\mathrm{n}} - \boldsymbol{\Psi}_{\mathrm{w}}) = \boldsymbol{K}_\pi \boldsymbol{\Psi}_\pi + \boldsymbol{Q}_\pi, \tag{8.2.21}$$

$\pi = \mathrm{n, w}$, where $\boldsymbol{C}$ is the capacitance (mass) matrix, δ_π is the Kronecker delta in the sense that $\delta_{\mathrm{n}}\boldsymbol{\Psi}_{\mathrm{n}} = \boldsymbol{\Psi}_{\mathrm{n}}$, $\delta_{\mathrm{n}}\boldsymbol{\Psi}_{\mathrm{w}} = \boldsymbol{0}, \ldots, \boldsymbol{K}_\pi$ is a stiffness type matrix and $\boldsymbol{Q}_\pi$ contains the boundary conditions. It is convenient to make the substitution

$$\boldsymbol{P} = (\boldsymbol{\Psi}_{\mathrm{n}} + \boldsymbol{\Psi}_{\mathrm{w}})/2, \qquad \boldsymbol{R} = (\boldsymbol{\Psi}_{\mathrm{n}} - \boldsymbol{\Psi}_{\mathrm{w}})/2.$$

Then the equations for the flow become

$$4\boldsymbol{C}\dot{\boldsymbol{R}} = \boldsymbol{M}\boldsymbol{R} + \boldsymbol{N}\boldsymbol{P} + \boldsymbol{Q}_{\mathrm{d}} = \boldsymbol{F}, \text{ say}, \tag{8.2.22}$$

and

$$\boldsymbol{0} = \boldsymbol{M}\boldsymbol{P} + \boldsymbol{N}\boldsymbol{R} + \boldsymbol{Q}_{\mathrm{s}}, \tag{8.2.23}$$

where $\boldsymbol{M} = \boldsymbol{K}_{\mathrm{n}} + \boldsymbol{K}_{\mathrm{w}}$, and $\boldsymbol{N} = \boldsymbol{K}_{\mathrm{n}} - \boldsymbol{K}_{\mathrm{w}}$ are stiffness type matrices. The matrices $\boldsymbol{M}, \boldsymbol{N}$ and $\boldsymbol{C}$ depend on $\boldsymbol{R}$ not on $\boldsymbol{P}$. The terms $\boldsymbol{Q}_{\mathrm{d}}$ and $\boldsymbol{Q}_{\mathrm{s}}$ include the boundary conditions. Verner *et al.* (1974) solve equations (8.2.22) and (8.2.23) with the trapezium rule Crank–Nicolson used on (8.2.22):

$$\boldsymbol{R}_{n+1} - \boldsymbol{R}_n = \frac{\Delta t}{8}(\boldsymbol{C}_{n+1}^{-1}\boldsymbol{F}_{n+1} + \boldsymbol{C}_n^{-1}\boldsymbol{F}_n) \tag{8.2.24}$$

and iteration to satisfy this equation and eqn (8.2.23) at each time step for the new values $\boldsymbol{R}_{n+1}$ and $\boldsymbol{P}_{n+1}$. Time steps are chosen to limit the change in saturation at any time step. White (1978) and White *et al.* (1981) eliminate the variable $\boldsymbol{P}$ from eqns (8.2.22), (8.2.23) and solve the equation

$$\dot{\boldsymbol{R}} = \boldsymbol{A}\boldsymbol{R} + \boldsymbol{V}, \tag{8.2.25}$$

where $\boldsymbol{A} = \frac{1}{4}\boldsymbol{C}^{-1}(\boldsymbol{M} - \boldsymbol{N}\boldsymbol{M}^{-1}\boldsymbol{N})$, and $\boldsymbol{V} = \frac{1}{4}\boldsymbol{C}^{-1}(-\boldsymbol{N}\boldsymbol{M}^{-1}\boldsymbol{Q}_{\mathrm{s}} + \boldsymbol{Q}_{\mathrm{d}})$, which are each dependent on the solution $\boldsymbol{R}$. White (1978) points out that Newton–Raphson is unacceptable here because of the cost of calculating the Jacobian and hence uses a predictor–corrector algorithm for the time-stepping. This corresponds to the form:

Predictor given by

$$\boldsymbol{R}_{n+1}^{*} - \boldsymbol{R}_n = \Delta t \boldsymbol{A}_n\{(1-\theta)\boldsymbol{R}_n + \theta \boldsymbol{R}_{n+1}^{*}\} + \Delta t \boldsymbol{V}_n \quad (8.2.26)$$

together with correctors given by

$$\boldsymbol{R}_{n+1}^{j+1} - \boldsymbol{R}_n = \Delta t \bar{\boldsymbol{A}}\{(1-\theta)\boldsymbol{R}_n + \theta \boldsymbol{R}_{n+1}^{j+1}\} + \Delta t \bar{\boldsymbol{V}}_n, \quad (8.2.27)$$

where

$$\bar{\boldsymbol{A}}_n = \boldsymbol{A}(\bar{\boldsymbol{R}}_n^j), \quad \bar{\boldsymbol{V}}_n = \boldsymbol{V}(\bar{\boldsymbol{R}}_n^j), \quad \bar{\boldsymbol{R}}_n^j = \gamma \boldsymbol{R}_{n+1}^j + (1-\gamma)\boldsymbol{R}_n. \quad (8.2.28)$$

It is possible to use this two-level scheme throughout the run or to use it to generate starting values for a three-level scheme. White (1978) discusses various possibilities such as Lees' algorithm (later referred to by Bettencourt *et al.* (1981) as 'notoriously oscillatory'). No guidance is given as to the best value of the parameter γ in eqn (8.2.28).

In the coupled problem of dynamic soil–structure interaction the motion of each is represented by a second order equation of the form

$$\boldsymbol{M}\ddot{\boldsymbol{x}} + \boldsymbol{C}\dot{\boldsymbol{x}} + \boldsymbol{K}\boldsymbol{x} = \boldsymbol{P},$$

where $\boldsymbol{x}$ is the vector of nodal displacements, $\boldsymbol{M}$, $\boldsymbol{C}$, and $\boldsymbol{K}$ are the mass, damping and stiffness matrices respectively and $\boldsymbol{P}$ contains the coupling terms. Smith (1975) reduces the problem to its first order form

$$\boldsymbol{A}\dot{\boldsymbol{u}} + \boldsymbol{B}\boldsymbol{u} = \boldsymbol{p}(t), \quad (8.2.29)$$

where

$$\boldsymbol{u}^{\mathrm{T}} = [\dot{\boldsymbol{x}}, \boldsymbol{x}], \quad \boldsymbol{A} = \begin{bmatrix} \boldsymbol{M} & \boldsymbol{0} \\ \boldsymbol{0} & \boldsymbol{I} \end{bmatrix}, \quad \boldsymbol{B} = \begin{bmatrix} \boldsymbol{C} & \boldsymbol{K} \\ -\boldsymbol{I} & \boldsymbol{0} \end{bmatrix}.$$

This first order form is then solved by the Nørsett method described in §5.3. Smith claims in the 1975 paper that the Nørsett method is more cost-effective than standard methods like Newmark and Wilson-θ. In a more recent comment (personal communication 1988) he says that he thought he had proved at that time that Nørsett was preferable to standard engineers' methods, but 'whether because of engineering conservatism or "not invented here" I never awakened much interest'. However, it seems worth mentioning here that the constitutive equations for the soil can be of such complexity that it is perhaps not surprising if engineers prefer very simple time-stepping schemes for the numerical integration.

Zienkiewicz and Bettess (1978) give a useful introduction to the formation of the equations of fluid–structure interaction. With finite element space discretization, the equation for the motion of the structure has the usual form

$$\bar{\boldsymbol{M}}\ddot{\boldsymbol{u}} + \bar{\boldsymbol{C}}\dot{\boldsymbol{u}} + \bar{\boldsymbol{K}}\boldsymbol{u} + \boldsymbol{f}_{\mathrm{I}} + \bar{\boldsymbol{f}}_0 = \mathbf{0}, \tag{8.2.30}$$

where $\boldsymbol{u}, \dot{\boldsymbol{u}}, \ddot{\boldsymbol{u}}$ are the vectors of nodal displacements, velocities and accelerations, $\bar{\boldsymbol{M}}, \bar{\boldsymbol{C}}, \bar{\boldsymbol{K}}$ are the mass, damping and stiffness matrices, respectively, $\boldsymbol{f}_{\mathrm{I}}$ is the vector of interface interaction, $\bar{\boldsymbol{f}}_0$ is the vector of external forces. Then in the Lagrangian form the motion of the fluid is represented by an equation of the same type, given by Zienkiewicz and Bettess as

$$\boldsymbol{M}\ddot{\boldsymbol{r}} + \boldsymbol{C}\dot{\boldsymbol{r}} + \boldsymbol{K}\boldsymbol{r} + \boldsymbol{f}_{\mathrm{I}} + \boldsymbol{f}_0 = \mathbf{0}, \tag{8.2.31}$$

where $\boldsymbol{r}$ is the vector of nodal displacements of the fluid which is assumed to be elastic with shear modulus equal to zero. Taking the shear modulus of the fluid as zero can give difficulties which Zienkiewicz and Bettess say can be overcome by introducing a small but finite shear modulus or introducing a condition of irrotational motion via a penalty function. Then a standard time-stepping scheme can be applied to the coupled systems of equations.

The Eulerian approach to the fluid flow part of the fluid–structure problem gives an equation in terms of the fluid pressure p. This is obtained via the Navier–Stokes equations and the equation of continuity neglecting the convective acceleration and viscosity effect. The result after space discretization by finite elements is an equation for the nodal pressures in the fluid of the form

$$\boldsymbol{E}\ddot{\boldsymbol{p}} + \boldsymbol{A}\dot{\boldsymbol{p}} + \boldsymbol{H}\boldsymbol{p} + \boldsymbol{h}_{\mathrm{I}} + \boldsymbol{h}_0 = 0, \tag{8.2.32}$$

where the matrix $\boldsymbol{E}$ includes a type of mass matrix plus a free surface integral, the matrix $\boldsymbol{A}$ comes from an exterior boundary integral modelling a radiation conditon, $\boldsymbol{H}$ is the stiffness matrix, and $\boldsymbol{h}_{\mathrm{I}}$ is the coupling term; $\boldsymbol{h}_0$ includes boundary conditions and external forces. The coupling term in eqn (8.2.30) is now given by the equation

$$f_{\mathrm{I}} = \int_{\Gamma_{\mathrm{I}}} \bar{\boldsymbol{N}}^{\mathrm{T}} \boldsymbol{N} \boldsymbol{p} \boldsymbol{n} \cdot \mathrm{d}\boldsymbol{\Gamma} = \boldsymbol{Q}\boldsymbol{p},$$

where $\bar{\boldsymbol{N}}$, $\boldsymbol{N}$ are the row matrices of basis functions for the structure and fluid respectively, $\boldsymbol{N}\boldsymbol{p} = \boldsymbol{p}$, the approximation for the pressure, Γ_{I} is the interface and $\boldsymbol{n}$ is the unit normal to this interface. The coupling term in eqn (8.2.32) is given by

$$\boldsymbol{h}_{\mathrm{I}} = \int_{\Gamma_{\mathrm{I}}} \rho \boldsymbol{N}^{\mathrm{T}} \bar{\boldsymbol{N}} \ddot{\boldsymbol{u}} \boldsymbol{n} \cdot \mathrm{d}\boldsymbol{\Gamma} = \rho \boldsymbol{Q}^{\mathrm{T}} \ddot{\boldsymbol{u}}.$$

Thus the complete system of fluid–structure equations can be written as

$$\begin{bmatrix} \bar{\boldsymbol{M}} & \boldsymbol{0} \\ \boldsymbol{Q}^{\mathrm{T}} & \boldsymbol{E} \end{bmatrix} \begin{bmatrix} \ddot{\boldsymbol{u}} \\ \ddot{\boldsymbol{p}} \end{bmatrix} + \begin{bmatrix} \bar{\boldsymbol{C}} & \boldsymbol{0} \\ \boldsymbol{0} & \boldsymbol{A} \end{bmatrix} \begin{bmatrix} \dot{\boldsymbol{u}} \\ \dot{\boldsymbol{p}} \end{bmatrix} + \begin{bmatrix} \bar{\boldsymbol{K}} & \boldsymbol{Q} \\ \boldsymbol{0} & \boldsymbol{H} \end{bmatrix} \begin{bmatrix} \boldsymbol{u} \\ \boldsymbol{p} \end{bmatrix} + \begin{bmatrix} \bar{f}_0 \\ \boldsymbol{h}_0 \end{bmatrix} = \boldsymbol{0}. \tag{8.2.33}$$

Zienkiewicz and Bettess (1978) comment that this coupled system of equations could be solved by the monolithic approach but the non-symmetric matrix multiplying the second derivative term can make difficulty. However, Zienkiewicz and Taylor (1985) show that with use of the SS22 time-stepping algorithm (§4.5) the monolithic system can be made such that the matrix to be inverted at each time step is symmetric. Their equations have precisely the same pattern as in the eqns (8.2.33) except that the matrix from the external boundary radiation condition is omitted and the coupling term matrix $\boldsymbol{Q}$ in the structure equation has changed its sign (due to a change in the convention used for the direction of the unit normal vector $\boldsymbol{n}$?). Thus in Zienkiewicz and Taylor (1985) the structure–fluid equations are given as

$$\begin{bmatrix} \boldsymbol{M} & \boldsymbol{0} \\ \boldsymbol{Q}^{\mathrm{T}} & \boldsymbol{S} \end{bmatrix} \begin{bmatrix} \ddot{\boldsymbol{u}} \\ \ddot{\boldsymbol{p}} \end{bmatrix} + \begin{bmatrix} \boldsymbol{C} & \boldsymbol{0} \\ \boldsymbol{0} & \boldsymbol{0} \end{bmatrix} \begin{bmatrix} \dot{\boldsymbol{u}} \\ \dot{\boldsymbol{p}} \end{bmatrix} + \begin{bmatrix} \boldsymbol{K} & -\boldsymbol{Q} \\ \boldsymbol{0} & \boldsymbol{H} \end{bmatrix} \begin{bmatrix} \boldsymbol{u} \\ \boldsymbol{p} \end{bmatrix} = \begin{bmatrix} f_1 \\ f_2 \end{bmatrix}. \tag{8.2.34}$$

Then $\boldsymbol{u}$ and $\boldsymbol{p}$ in eqns (8.2.34) are approximated by the SS22 expansion (§4.5) within the time step:

$$\boldsymbol{u} = \boldsymbol{u}_n + \tau \dot{\boldsymbol{u}}_n + \frac{\tau^2}{2} \boldsymbol{\alpha}_n^u, \qquad \boldsymbol{p} = \boldsymbol{p}_n + \tau \dot{\boldsymbol{p}}_n + \frac{\tau^2}{2} \boldsymbol{\alpha}_n^p. \tag{8.2.35}$$

The weighted residual form with θ_p values for the structure and $\bar{\theta}_p$ values for the fluid gives the system of equations for the vector $[\boldsymbol{\alpha}_n^u, \boldsymbol{\alpha}_n^p]^{\mathrm{T}}$ as

$$\begin{bmatrix} \boldsymbol{M} + \theta_1 \Delta t \boldsymbol{C} + \theta_2 \Delta t^2 \boldsymbol{K} & -\theta_2 \dfrac{\Delta t^2}{2} \boldsymbol{Q} \\ -\theta_2 \dfrac{\Delta t^2}{2} \boldsymbol{Q}^{\mathrm{T}} & -\theta_2 \dfrac{\Delta t^2}{2} \boldsymbol{S} - \bar{\theta}_2 \theta_2 \dfrac{\Delta t^4}{4} \boldsymbol{H} \end{bmatrix} \begin{bmatrix} \boldsymbol{\alpha}_n^u \\ \boldsymbol{\alpha}_n^p \end{bmatrix} = \boldsymbol{F}, \tag{8.2.36}$$

where the second equation has been multiplied by $-\theta_2 \Delta t^2/2$ in order to make the matrix to be inverted symmetric. The vector $\boldsymbol{F}$ on the right-hand side of eqn (8.2.36) contains all the known values from the beginning of the time step. In the Appendix to Zienkiewicz and Taylor (1985), Chan makes a substitution of the form

$$\begin{bmatrix} \boldsymbol{u}_{n+1} \\ \boldsymbol{p}_{n+1} \end{bmatrix} = r \begin{bmatrix} \boldsymbol{u}_n \\ \boldsymbol{p}_n \end{bmatrix},$$

and shows for the corresponding scalar equations that the SS22 algorithm given by eqns (8.2.35) and (8.2.36) is unconditionally stable when $\theta_2 \geq \theta_1 \geq 0.5$ and $\bar{\theta}_2 \geq \bar{\theta}_1 \geq 0.5$, as expected from the results in §4.5.

For the coupled problem of deformable soil and pore pressure (Lewis and Schrefler 1987), the displacement of the soil represented by the vector $\boldsymbol{u}$ is governed by a second order differential equation and the pore pressure $\boldsymbol{p}$ by a first order equation. The combined system can be written in the form

$$\begin{bmatrix} \boldsymbol{M} & \boldsymbol{0} \\ \boldsymbol{0} & \boldsymbol{0} \end{bmatrix} \begin{bmatrix} \ddot{\boldsymbol{u}} \\ \ddot{\boldsymbol{p}} \end{bmatrix} + \begin{bmatrix} \boldsymbol{C} & \boldsymbol{0} \\ \boldsymbol{Q}^{\mathrm{T}} & \boldsymbol{S} \end{bmatrix} \begin{bmatrix} \dot{\boldsymbol{u}} \\ \dot{\boldsymbol{p}} \end{bmatrix} + \begin{bmatrix} \boldsymbol{K} & -\boldsymbol{Q} \\ \boldsymbol{0} & \boldsymbol{H} \end{bmatrix} \begin{bmatrix} \boldsymbol{u} \\ \boldsymbol{p} \end{bmatrix} = \begin{bmatrix} \boldsymbol{f}_1 \\ \boldsymbol{f}_2 \end{bmatrix}, \tag{8.2.37}$$

where $\boldsymbol{M}$ and $\boldsymbol{S}$ are again mass matrix type, $\boldsymbol{K}$ and $\boldsymbol{H}$ are stiffness matrix type, $\boldsymbol{C}$ is a damping matrix and $\boldsymbol{Q}$ is the coupling matrix. Zienkiewicz and Taylor (1985) use an SS22 form for $\boldsymbol{u}$ and an SS11 form for $\boldsymbol{p}$,

$$\boldsymbol{u} = \boldsymbol{u}_n + \tau \dot{\boldsymbol{u}}_n + \frac{\tau^2}{2} \boldsymbol{\alpha}_n^u, \qquad \boldsymbol{p} = \boldsymbol{p}_n + \tau \boldsymbol{\alpha}_n^p.$$

The weighted residual equations for the $\boldsymbol{\alpha}_n$ values are given by

$$\begin{bmatrix} \boldsymbol{M} + \theta_1 \Delta t \boldsymbol{C} + \theta_2 \Delta t^2 \boldsymbol{K} & -\theta_1 \Delta t \boldsymbol{Q} \\ -\bar{\theta}_1 \Delta t \boldsymbol{Q}^{\mathrm{T}} & -\boldsymbol{S} - \bar{\theta}_1 \Delta t \boldsymbol{H} \end{bmatrix} \begin{bmatrix} \boldsymbol{\alpha}_n^u \\ \boldsymbol{\alpha}_n^p \end{bmatrix} = \boldsymbol{F}, \tag{8.2.38}$$

where the vector $\boldsymbol{F}$ contains the known values from the beginning of the time step. Zienkiewicz and Taylor propose to take $\bar{\theta}_1 = \theta_1$, but the matrix on the left of (8.2.38) can equally well be made symmetric if they are different. Chan in the appendix to Zienkiewicz and Taylor (1985) shows for the linear scalar case that there is unconditional stability with $\theta_2 \geq \theta_1 \geq 0.5$, and $\bar{\theta}_1 \geq 0.5$.

In their introduction to this 1985 paper describing a monolithic approach to the solution of fluid–structure and soil–pore pressure systems, Zienkiewicz and Taylor claim that 'Staggered procedures of a simple kind prove unstable and costly restructuring is needed to achieve satisfactory stability', with reference to the various schemes of Park (1977, 1980, 1983). However, Zienkiewicz *et al* (1988) come back with an unconditionally stable staggered scheme for the coupled soil–pore fluid problem. These staggered schemes are discussed in the next section.

8.3. Partitioning methods for coupled systems

The monolithic methods for solving coupled problems described in §8.2 seem very direct and simple to implement but they have disadvantages such as

1. For many practical problems the monolithic systems can be impossibly large.

2. It can be inappropriate to use the same basic formulation for two different media. For example, the displacement formulation for a structure and the fluid around it can mean that the condition of irrotational motion has to be introduced via a penalty function.

3. The two parts of the problem may have different time scales and hence require different time-stepping schemes. For instance, in the coupled problem of flow in a river or in the sea with effluent transport, it can be that the concentration of effluent changes slowly compared with the changes in the flow.

4. The two, or more, parts of the problem may have non-linearities which need to be handled in different ways.

Partitioning methods for the solution of coupled problems, as the name suggests, divide the problem into smaller units. The separation may be made at the differential equation stage, before discretization. The equations for two sets of physical function components such as the depth and the velocities in the shallow water equations (8.2.1) and (8.2.2) are time-stepped in turn and the results used to evaluate the coupling terms. This is a staggered solution procedure. With the problem divided in this way there is the possibility of using programs which already exist for the solution of the uncoupled equations with modifications to include the coupling effects. There are also the partitioning methods described in Chapter 5 which are applied after the space discretization and can be element-based or node-based. Park and Felippa (1983) give a useful general survey of partitioning methods for coupled systems from an engineering point of view.

We give a selection of examples using the staggered procedure for coupled systems of first order equations.

Connor and Wang (1973) have a finite element model of shallow water equations with vertically averaged flow which gives the equations in a slightly different form from (8.2.1), (8.2.2), using flux instead of velocity components. The equations can be written as

$$\boldsymbol{M}\dot{\boldsymbol{Q}} + \boldsymbol{P}(\boldsymbol{H}, \boldsymbol{Q}, t) = \boldsymbol{0} \tag{8.3.1}$$

and

$$\boldsymbol{M}_h\dot{\boldsymbol{H}} + \boldsymbol{P}_h(\boldsymbol{Q}) = \boldsymbol{0}, \tag{8.3.2}$$

where $\boldsymbol{H}$ is the vector of surface nodal elevations, $\boldsymbol{Q}$ is the vector of flux terms, such that $\boldsymbol{Q}^{\mathrm{T}} = [\boldsymbol{q}_x, \boldsymbol{q}_y]$, where $\boldsymbol{q}_x = \rho H\bar{\boldsymbol{u}}$, $\boldsymbol{q}_y = \rho H\bar{\boldsymbol{v}}$, ρ is the density, H is the depth and $\bar{\boldsymbol{u}}$, $\bar{\boldsymbol{v}}$ are the depth averaged velocity components. The matrices $\boldsymbol{M}$ and $\boldsymbol{M}_h$ are constant mass matrices and the non-linear terms are collected in the $\boldsymbol{P}(\boldsymbol{H}, \boldsymbol{Q}, t)$ and $\boldsymbol{P}_h(\boldsymbol{Q})$. Connor and Wang give a staggered iteration method using trapezium rule Crank–Nicolson type schemes for updating $\boldsymbol{H}$ and $\boldsymbol{Q}$, with relaxation factors to accelerate convergence which is useful for illustrating this kind of approach. The steps of the iteration are as follows:

$$\text{(1)}\quad \boldsymbol{M}_h(\boldsymbol{H}^* - \boldsymbol{H}_n) = \frac{\Delta t}{2}(\boldsymbol{P}^{j-1}_{h,n+1} + \boldsymbol{P}_{h,n}),\ \text{gives } \boldsymbol{H}^*; \tag{8.3.3}$$

$$\text{(2)}\quad \boldsymbol{H}^j_{n+1} = \theta_h\boldsymbol{H}^* + (1-\theta_h)\boldsymbol{H}^{j-1}_{n+1},\ \text{gives } \boldsymbol{H}^j_{n+1}; \tag{8.3.4}$$

(3) then $\boldsymbol{H}^j_{n+1}$ is used to give

$$\boldsymbol{P}^{j-1}_{n+1} = \boldsymbol{P}(\boldsymbol{H}^j_{n+1}, \boldsymbol{Q}^{j-1}_{n+1}, t_{n+1}); \tag{8.3.5}$$

(4) $$\boldsymbol{M}(\boldsymbol{Q}^* - \boldsymbol{Q}_n) = \frac{\Delta t}{2}(\boldsymbol{P}^{j-1}_{n+1} + \boldsymbol{P}_n), \text{ gives } \boldsymbol{Q}^*; \tag{8.3.6}$$

(5) and $$\boldsymbol{Q}^j_{n+1} = \theta_Q \boldsymbol{Q}^* + (1 - \theta_Q)\boldsymbol{Q}^{j-1}_{n+1}. \tag{8.3.7}$$

The last step, eqn (8.3.7), gives $\boldsymbol{Q}^j_{n+1}$ to substitute into $\boldsymbol{P}^j_{h,n+1}$, and the cycle is repeated. No information is given as to suitable values of the acceleration parameters θ_h and θ_Q. The convergence is defined by tolerances on the percentage changes in the Euclidean norms of both H and Q. Tested against problems for which analytical solutions are available, the authors say that the trapezium rule method requires an average of seven iterations per time step which could be accelerated by extrapolating to get the starting value for the iteration. However, because this means storage of a previous solution vector, the trapezium rule scheme is rejected in favour of a combination of a third order predictor–corrector and standard fourth order Runge–Kutta which is found to be more accurate and to converge faster. This appears to be employed on the monolithic system. The predictor–corrector as applied to the equation $\dot{y} = f(y, t)$ is given by

predictor: $$f^1_{n+1} = 3f_n - 3f_{n-1} + f_{n-2},$$

and

Corrector: $$\begin{aligned} y^* &= y_n + \Delta t(5f^{j-1}_{n+1} + 8f_n - f_{n-1})/12, \\ y^j_{n+1} &= \theta y^* + (1-\theta)y^{j-1}_{n+1}. \end{aligned} \tag{8.3.8}$$

The fourth order Runge–Kutta used by Connor and Wang is not the most popular Runge–Kutta method described in §5.4 but one given by Ralston (1965) as having the lowest bound on the error for this order of Runge–Kutta methods. The methods are applied to an analysis of the circulation in Massachusetts Bay. The authors' conclusion is that the performance of the Runge–Kutta method is satisfactory but the predictor–corrector method has to have some eddy viscosity introduced in order to control instability.

We can see why the predictor–corrector used by Connor and Wang is liable to give trouble. If the corrector is corrected to convergence it amounts to satisfying the equation

$$y_{n+1} = y_n + \Delta t(5f_{n+1} + 8f_n - f_{n-1})/12. \qquad (8.3.9)$$

Applied to the test equation $\dot{y} = -\lambda y$, with $\lambda > 0$, eqn (8.3.9) gives

$$y_{n+1} = y_n - \lambda\Delta t(5y_{n+1} + 8y_n - y_{n-1})/12. \qquad (8.3.10)$$

Substituting $y_{n+1} = ry_n = r^2 y_{n-1}$, we obtain the stability polynomial

$$(5q+1)r^2 + (8q-1)r - q = 0 \qquad (8.3.11)$$

where $q = \lambda\Delta t/12$. Making the Routh–Hurwitz substitution $r = (1+z)/(1-z)$, gives the equation

$$(1-2q)z^2 + (6q+1)z + 6q = 0. \qquad (8.3.12)$$

From Appendix 1, the Routh–Hurwitz conditions show that the algorithm (8.3.9) is stable if $\lambda\Delta t < 6$. But the roots of the quadratic (8.3.11) are real and the product is negative so we need to look at the possibility that the spurious negative root will dominate the solution and produce oscillation. From the formula for the roots of a quadratic, this happens when $8q - 1 > 0$, that is, when $\lambda\Delta t > \frac{3}{2}$, so it is not surprising that this method gives trouble.

The staggered iteration given by eqns (8.3.3) to (8.3.7) demonstrates the basic idea of staggered methods. One of the unknowns is extrapolated at the start, then the equations are solved in turn each providing the next value to insert into the coupling term of the other until convergence is attained. The rate of convergence of the iteration will depend on a good starting value, good acceleration technique and effective handling of the non-linear terms in the equations.

Pinder (1973) describes a finite element solution of the problem of groundwater contamination. He uses a two-dimensions in plan, vertically integrated form of the groundwater flow equations with transmissivity (Chapter 1). When spatially discretized by finite elements this gives the first order system:

$$\boldsymbol{R}\dot{\boldsymbol{h}} + \boldsymbol{P}\boldsymbol{h} + \boldsymbol{U} = \boldsymbol{0}, \qquad (8.3.13)$$

where $\boldsymbol{h}$ is the vector of values of the hydraulic head at the nodes, $\boldsymbol{R}$ is the mass matrix, $\boldsymbol{P}$ includes the stiffness matrix type terms and the leakage effect, and the vector $\boldsymbol{U}$ includes boundary and source terms.

The space-discretized form of the mass transport equation for the concentration, $\boldsymbol{c}$, of the contaminant, supposed here to be of the same density as the groundwater, is given by

$$\boldsymbol{M}\dot{\boldsymbol{c}} + \boldsymbol{N}\boldsymbol{c} + \boldsymbol{F} = \boldsymbol{0}, \tag{8.3.14}$$

where $\boldsymbol{M}$ is the mass matrix, the matrix $\boldsymbol{N}$ includes the diffusion and convection terms, and the vector $\boldsymbol{F}$ includes the source and boundary terms. The convection depends on the mass average flux q which is related to the hydraulic head h by the Darcy law relationship:

$$q_i = -\sum_j T_{ij} \frac{\partial h}{\partial x_j}, \tag{8.3.15}$$

where the T_{ij} are the components of the transmissivity. Equations (8.3.13) for the hydraulic head and (8.3.14) for the concentration of the contaminant are only coupled through eqn (8.3.15) for the mass flux which appears in the convection terms. Hence it is only necessary to solve (8.3.13) for h, then use eqn (8.3.15) for the mass flux and then (8.3.14) for the mass transport in each time step; there is no need to iterate. Pinder (1973) reports that although the Crank–Nicolson time-stepping algorithm gives a satisfactory solution to the groundwater flow equation (8.3.13), the backward difference algorithm gives much better results for the mass transport equation (8.3.14) and improves the accuracy of the solution of (8.3.13). Pinder uses the time-stepping scheme:

$$\begin{aligned} (\boldsymbol{R} + \Delta t\boldsymbol{P})\boldsymbol{h}_{n+1} &= \boldsymbol{R}\boldsymbol{h}_n - \Delta t\boldsymbol{U} \\ \text{and} \quad (\boldsymbol{M} + \Delta t\boldsymbol{N})\boldsymbol{c}_{n+1} &= \boldsymbol{M}\boldsymbol{c}_n - \Delta t\boldsymbol{F}. \end{aligned} \tag{8.3.16}$$

Pinder uses this method to investigate a chromium contamination problem on Long Island, New York.

We note that the values of the components of the mass flux q in eqn (8.3.15) can easily be obtained at the quadrature points in the elements using

$$\frac{\partial h}{\partial x_j} = \frac{\partial}{\partial x_j}\left(\sum_k h_k N_k\right), \tag{8.3.17}$$

where N_k are the basis functions, whose gradients have already been obtained for use in the entries in the stiffness matrix $\boldsymbol{P}$. The values of q_i at the quadrature points are then ready for use in calculating the entries in the matrix $\boldsymbol{N}$ for the second equation (8.3.16). Because of these convection terms the matrix to be inverted in this part of the time-stepping is not symmetric.

Taylor and Huyakorn (1976) describe a finite element analysis of three-dimensional groundwater flow with a contaminant such that the relationship between the density of the fluid and the concentration c of the contaminant is given by

$$\rho = \{\rho_1 c + \rho_0(c_1 - c)\}/c_1, \tag{8.3.18}$$

where ρ_1 is the density at the maximum concentration c_1 and ρ_0 is the density when c is zero. The three-dimensional groundwater flow equation is now represented by

$$\boldsymbol{H}\dot{\boldsymbol{h}} + \boldsymbol{A}\boldsymbol{h} = \boldsymbol{B}(\boldsymbol{c}), \tag{8.3.19}$$

where the hydraulic head $\boldsymbol{h} = \boldsymbol{x}_3 + \boldsymbol{p}/\rho_0 g$, $\boldsymbol{x}_3$ being height above datum (supposing the coordinate system to be (x_1, x_2, x_3)), $\boldsymbol{p}$ is the pressure and $\boldsymbol{g}$ is the gravitational acceleration vector. The matrices $\boldsymbol{H}$ and $\boldsymbol{A}$ are the mass and stiffness type matrices respectively and $\boldsymbol{B}(\boldsymbol{c})$ is the coupling term dependent on the concentration $\boldsymbol{c}$, present in the groundwater flow equation because of the density difference. The mass transport equation is represented by

$$\boldsymbol{M}\dot{\boldsymbol{c}} + \boldsymbol{P}(\boldsymbol{v}, \boldsymbol{c})\boldsymbol{c} = \boldsymbol{F}, \tag{8.3.20}$$

where $\boldsymbol{M}$ is the mass matrix, the matrix $\boldsymbol{P}(\boldsymbol{v}, \boldsymbol{c})$ includes the diffusion and convection terms, and the vector $\boldsymbol{F}$ includes the boundary conditions. The convection depends on the velocity v which has to be obtained from the flow equation by means of the formula

$$v_i = -\sum_j K_{ij}\left\{\frac{\partial h}{\partial x_j} + \frac{(\rho - \rho_0)}{\rho_0}\delta_{j3}\right\}, \tag{8.3.21}$$

where K_{ij} are the entries in the hydraulic conductivity tensor and δ_{j3} is the Kronecker delta. Bear (1979, Chapter 7) gives details of the theory for the formation of the equations for this kind of problem.

Taylor and Huyakorn propose to solve this coupled problem with a staggered procedure using backward time-stepping on

(8.3.19) and (8.3.20), which gives the equations

$$(\boldsymbol{H} + \Delta t\boldsymbol{A})\boldsymbol{h}_{n+1} = \boldsymbol{H}\boldsymbol{h}_n + \Delta \boldsymbol{b}_{n+1} \tag{8.3.22}$$

and

$$(\boldsymbol{M} + \Delta t\boldsymbol{P})\boldsymbol{c}_{n+1} = \boldsymbol{M}\boldsymbol{c}_n + \Delta t\boldsymbol{F}_{n+1}. \tag{8.3.23}$$

The order of the iteration in the time step from t_n to $t_n + \Delta t$ is given as

(1) compute $\boldsymbol{B}(\boldsymbol{c}_n)$;

(2) solve eqn (8.3.22) for $\boldsymbol{h}^1_{n+1}$;

(3) compute velocities from eqn (8.3.21) (these are easily obtained at quadrature points);

(4) calculate $\boldsymbol{P}(\boldsymbol{v}, \boldsymbol{c})$;

(5) solve eqn (8.3.23) for c^1_{n+1}, and repeat the iteration until the nodal values of c_{n+1} and $\boldsymbol{h}_{n+1}$ have converged to a prescribed tolerance.

Disappointingly, Taylor and Huyakorn only give numerical results for steady state problems.

Kawahara *et al.* (1982b) discuss the solution of the problem of the three-dimensional multiple level flow of water of variable density. Their method uses the two-step explicit scheme given in eqns (8.2.7) for the flow and standard Crank–Nicolson for the first order concentration equation (8.2.10). The time step for the Crank–Nicolson is much larger than for the flow because the concentration is changing much more slowly. Kawahara *et al.* (1982b) give results from the application of their method to the prediction of water quality resulting from the development of a coastal industrial estate. A steady state solution is computed as the limit of a transient solution starting from zero initial values of velocity and water elevation, with the density assumed known on the boundary.

8.4. Coupled first order and second order systems

In this section we include examples of staggered solution procedures used on problems where a first order system is coupled to a second order system.

We start with the problem of pore fluid–soil interaction which is important in the study of the safety of dams and the foundations of offshore structures. The equation for the soil has the form of the familiar dynamic vibration equation for the soil displacement $\boldsymbol{u}$, with the inclusion of the coupling term representing the influence of the pore pressure $\boldsymbol{p}$, as follows:

$$\boldsymbol{M\ddot{u}} + \boldsymbol{C\dot{u}} + \boldsymbol{Ku} - \boldsymbol{Qp} = \boldsymbol{f}_u. \tag{8.4.1}$$

The form of the equation for the pore pressure $\boldsymbol{p}$ depends on the assumptions made. Zienkiewicz *et al.* (1988) give a list of references to the background of the formulation. Park (1983) starts with an equation for $\boldsymbol{p}$ of the form

$$\boldsymbol{\hat{M}\ddot{u}} + \boldsymbol{Q}^{\mathrm{T}}\boldsymbol{\dot{u}} + \boldsymbol{S\dot{p}} + \boldsymbol{Hp} = \boldsymbol{f}_p \tag{8.4.2}$$

(which was suggested to him by Zienkiewicz). Park calls $\boldsymbol{\hat{M}}$ the permeable mass matrix, and $\boldsymbol{S}$ and $\boldsymbol{H}$ the pressure capacitance and resistance matrices. Zienkiewicz *et al.* do not include the matrix $\boldsymbol{\hat{M}}$, and they refer to $\boldsymbol{S}$ and $\boldsymbol{H}$ as the compressibility and permeability matrices respectively of the pore fluid. The matrix $\boldsymbol{S}$ is a mass matrix type including the fluid compressibility and the matrix $\boldsymbol{H}$ is a stiffness matrix type including the permeability.

Park (1983) omits the damping term $\boldsymbol{C\dot{u}}$ from eqn (8.4.1) and starts with the monolithic system

$$\begin{bmatrix} \boldsymbol{M} & \boldsymbol{0} \\ \boldsymbol{\hat{M}} & \boldsymbol{0} \end{bmatrix}\begin{bmatrix} \boldsymbol{\ddot{u}} \\ \boldsymbol{\ddot{p}} \end{bmatrix} + \begin{bmatrix} \boldsymbol{0} & \boldsymbol{0} \\ \boldsymbol{Q}^{\mathrm{T}} & \boldsymbol{S} \end{bmatrix}\begin{bmatrix} \boldsymbol{\dot{u}} \\ \boldsymbol{\dot{p}} \end{bmatrix} + \begin{bmatrix} \boldsymbol{K} & -\boldsymbol{Q} \\ \boldsymbol{0} & \boldsymbol{H} \end{bmatrix}\begin{bmatrix} \boldsymbol{u} \\ \boldsymbol{p} \end{bmatrix} = \begin{bmatrix} \boldsymbol{f}_s \\ \boldsymbol{f}_p \end{bmatrix}. \tag{8.4.3}$$

Park proposes to time-step eqns (8.4.3) with the trapezium rule (equivalent to Newmark with $2\beta = \gamma = 0.5$ or SS22 with $\theta_1 = \theta_2 = 0.5$), and uses what he calls a convenient staggered solution procedure which involved partitioning the matrix

$$\begin{bmatrix} \boldsymbol{K} & -\boldsymbol{Q} \\ \boldsymbol{0} & \boldsymbol{H} \end{bmatrix}$$

into an implicit part:

$$\begin{bmatrix} \boldsymbol{K} & \boldsymbol{0} \\ \boldsymbol{0} & \boldsymbol{H} \end{bmatrix}$$

and an explicit part:

$$\begin{bmatrix} \boldsymbol{0} & -\boldsymbol{Q} \\ \boldsymbol{0} & \boldsymbol{0} \end{bmatrix}.$$

Park performs a stability analysis on the equivalent scalar (linear) system and shows that the method cannot be stable unless $\hat{\boldsymbol{M}} \equiv 0$ because the term with $\hat{\boldsymbol{M}}$ as coefficient has the wrong sign. The term $\hat{\boldsymbol{M}}\ddot{\boldsymbol{u}}$ then disappears from the subsequent analysis with a reference to Paul (1982) having shown that in certain circumstances it can be neglected. A suspicion that the difficulty here may be due to the fact that the $\hat{\boldsymbol{M}}$ term has the wrong sign in eqn (8.4.2) can be answered by substituting $u = \exp(\lambda t)\bar{u}$, $p = \exp(\lambda t)\bar{p}$ into the scalar homogeneous form of eqns (8.4.3) to give

$$\begin{aligned} (m\lambda^2 + k)\bar{u} - q\bar{p} &= 0, \\ (\hat{m}\lambda^2 + q\lambda)\bar{u} + (s\lambda + h)\bar{p} &= 0, \end{aligned} \tag{8.4.4}$$

using lower case letters to represent the scalars corresponding to the capitals for the matrices. Eliminating $\bar{u}$ and $\bar{p}$ then gives the stability polynomial of the differential equation system:

$$ms\lambda^3 + (mh + \hat{m}q)\lambda^2 + (ks + q^2)\lambda + kh = 0. \tag{8.4.5}$$

For stability we want the roots of eqn (8.4.5) to have negative real parts and the Routh–Hurwitz conditions from Appendix 1 show that this is so provided

$$ms > 0, \qquad mh + \hat{m}q > 0, \qquad ks + q^2 > 0, \qquad kh > 0,$$

and

$$mhq^2 + \hat{m}qks + \hat{m}q^3 > 0. \tag{8.4.6}$$

The conditions (8.4.6) are satisfied if m, s, k, h, are each positive and $\hat{m}$ and q have the same sign, so that there is no necessity for $\hat{m}$ to be zero from this point of view. Even with the assumption that the $\hat{\boldsymbol{M}}$ term is absent Park's staggered procedure is only conditionally stable so he produces a matrix augmentation process which is unconditionally stable. Zienkiewicz *et al.* (1988) say that Park's matrix augmentation process is unacceptably costly and they introduce a much simpler staggered method which is also unconditionally stable. The equations used by Zienkiewicz *et al.* (1988) are the same as those in Zienkiewicz and Taylor (1985) which gives the monolithic approach to their solution.

The obvious staggered procedure to use on the equations for pore fluid–soil interaction is to take a predicted value of the pressure, $\boldsymbol{p}_{n+1}^{p}$, and substitute it in the dynamic equation for the soil to give

$$M\ddot{u}_{n+1} + C\dot{u}_{n+1} + Ku_{n+1} = (f_u)_{n+1} + Qp^p_{n+1}. \qquad (8.4.7)$$

The values of u_n, $\dot{u}_n$, and $\ddot{u}_n$ are known from the last time step and hence eqn (8.4.7) can be time-stepped using Newmark to obtain u_{n+1}, $\dot{u}_{n+1}$, $\ddot{u}_{n+1}$. The value obtained for $\dot{u}_{n+1}$ is then substituted into the equation for the pore pressure

$$S\dot{p}_{n+1} + Hp_{n+1} = (f_p)_{n+1} - Q^{\mathrm{T}}\dot{u}_{n+1}, \qquad (8.4.8)$$

which can be time-stepped by the θ-method to obtain the new value for p_{n+1}. With $2\beta \geq \gamma \geq 0.5$, and $\theta \geq 0.5$, this procedure, if iterated to convergence, would be unconditionally stable. but with only one pass it is only conditionally stable. Zienkiewicz *et al.* state that this is true also with higher order time-stepping schemes and with the order reversed, that is, start by predicting $\dot{u}_{n+1}$. Chan (1988) gives a detailed analysis. The engineers want a method which will be unconditionally stable with only one pass through the equations and require no iterations for linear problems. Sufficient accuracy is to be achieved by a control on the time step dependent on an estimate of the error. The stability with one pass is achieved by adjusting the soil equation to the form

$$\begin{aligned} M\ddot{u}_{n+1} + C\dot{u}_{n+1} + (K_d + QS^{-1}Q^{\mathrm{T}})u_{n+1} \\ = (f_u)_{n+1} + Qp^p_{n+1} + QS^{-1}Q^{\mathrm{T}}u^p_{n+1}, \qquad (8.4.9) \end{aligned}$$

where K_{d} is identified as the stiffness matrix for the drained soil and $K_{\mathrm{u}} = K_{\mathrm{d}} + QS^{-1}Q^{\mathrm{T}}$ as the undrained stiffness matrix. Equation (8.4.9) now requires predicted values for both the displacement and the pressure. With the Newmark-trapezium rule formulae $(2\beta = \gamma = 0.5)$, for both the displacement u and the pore pressure p, and the predictor values taken as

$$u^p_{n+1} = u_n + \Delta t\dot{u}_n + \Delta t^2\ddot{u}_n/4, \quad \text{and } p^p_{n+1} = p_n + \Delta t\dot{p}_n + \Delta t^2\ddot{p}_n/4,$$

neglecting the damping term in the soil equation (8.4.9), Zienkiewicz *et al.* obtain the corresponding Routh–Hurwitz stability polynomial. This shows that the one-pass method is now unconditionally stable.

Leaving in the damping term the Routh–Hurwitz polynomial

for the corresponding scalar system is

$$8msz^3 + \Delta tz^2\left(4mh + 4cs + \frac{q^2}{s}h\Delta t^2\right) + 2\Delta t^2 z(ch + sk + q^2) + kh\Delta t^3 = 0. \quad (8.4.10)$$

The Routh–Hurwitz conditions (Appendix 1) for this polynomial to have roots with negative real parts are that the coefficients should all be positive and

$$\left(4mh + 4cs + \frac{q^2}{s}h\Delta t^2\right)(ch + sk + q^2) - 4mskh \geq 0.$$

The presence of the damping parameter c makes no difference to the satisfaction of these conditions as expected with Newmark.

Zienkiewicz *et al.* (1988) also show that unconditional stability is achieved with one pass using SS22 for the displacement and SS11 for the pore pressure and $\boldsymbol{u}_n$ and $\boldsymbol{p}_n$ as the predictors, with the parameters each equal to 0.5. Chan (1988) shows that both the Newmark and the SS*pj* one pass schemes used on the adjusted eqn (8.4.9) are unconditionally stable with the parameters which would make each individually stable, that is, $2\beta \geq \gamma \geq 0.5$ for Newmark and $\theta_2 \geq \theta_1 \geq 0.5$ for SS22, $\theta_1 \geq 0.5$ for SS11. Zienkiewicz *et al.* (1988) also claim that using Newmark for the displacement and the θ-method for the pore pressure here cannot be unconditionally stable with the same path of computation.

Zienkiewicz *et al.* (1988) note that the identification of the augmented stiffness matrix in eqn (8.4.9) as the 'undrained' stiffness matrix $\boldsymbol{K}_\mathrm{u}$, is reminiscent of some intuitive methods used in soil mechanics where the pore fluid effect is computed first as undrained then as drained. The analogy would appear to be closer if a second pass were made through the system with the displacement equation in its original form (8.4.1) with $\boldsymbol{K} = \boldsymbol{K}_\mathrm{d}$, the stiffness matrix for the drained behaviour.

The soil mechanicians' idea of solving for the undrained and drained behaviour alternatively is reminiscent of the method of Adey and Brebbia (1973) for the finite element solution of the convection–diffusion of effluent. Adey and Brebbia treat the convection and diffusion alternately. They have the equation for the diffusion–convection of effluent in the form:

$$\boldsymbol{M}\dot{\boldsymbol{c}} + \boldsymbol{A}\boldsymbol{c} + \boldsymbol{K}\boldsymbol{c} = \boldsymbol{f}, \quad (8.4.11)$$

where $\boldsymbol{c}$ is the concentration vector, $\boldsymbol{M}$ is the mass matrix, the matrix $\boldsymbol{K}$ includes the diffusion and the matrix $\boldsymbol{A}$ the convection effect. The matrix $\boldsymbol{A}$ is not symmetric. The velocities for the convection terms in the matrix $\boldsymbol{A}$ are obtained from site observations. Equation (8.4.11) is first time stepped with $\boldsymbol{A}=\boldsymbol{0}$ by the Crank–Nicolson trapezium rule:

$$\frac{1}{\Delta t}\boldsymbol{M}(\boldsymbol{c}_{n+1}-\boldsymbol{c}_n)=\tfrac{1}{2}\boldsymbol{K}(\boldsymbol{c}_{n+1}+\boldsymbol{c}_n)+\tfrac{1}{2}(\boldsymbol{f}_{n+1}+\boldsymbol{f}_n).$$

This gives the effect of diffusion alone. Then it is assumed that there is a fixed mass of fluid in each element and this moves during the time step according to the given velocity. The concentration is thus known at a new set of points and the corresponding values at the original nodes are calculated by interpolation ready for the next step. Adey and Brebbia give results from the use of this method for a study of the movement of effluent in the Solent. It was also used for making a very effective demonstration film showing the effluent moving to and fro with the tides in the Solent.

Another example of a staggered solution procedure for coupled systems of first and second order equations is given in Park *et al.* (1977) on fluid–structure interaction. The structure is represented in the displacement formulation by the equation

$$\boldsymbol{M}\ddot{\boldsymbol{u}}+\boldsymbol{K}\boldsymbol{u}+\boldsymbol{T}\boldsymbol{A}\dot{\boldsymbol{q}}=\boldsymbol{f}_1, \tag{8.4.12}$$

with the damping term omitted. The term $\boldsymbol{q}$ is a pressure integral vector and $\boldsymbol{TA}$ is the coupling matrix, $\boldsymbol{f}_1$ represents external effects. The motion of the fluid is represented by a first order equation in $\boldsymbol{q}$ using a 'doubly asymptotic approximation' (DAA) from Geers (1971), which produces the equation

$$\boldsymbol{M}_{\mathrm{f}}\dot{\boldsymbol{q}}+\rho c\boldsymbol{A}\boldsymbol{q}-\rho c\boldsymbol{M}_{\mathrm{f}}\boldsymbol{T}^{\mathrm{T}}\dot{\boldsymbol{u}}=\boldsymbol{f}_2, \tag{8.4.13}$$

where $\boldsymbol{M}_{\mathrm{f}}$ is the fluid mass matrix, ρ and c are the density and speed of sound in the fluid. The discretization of the fluid is taken to be by the boundary integral technique (Zienkiewicz 1977), so that the number of unknowns in the fluid system of equations is comparatively small. An obvious approach to the staggered solution of eqns (8.4.12) and (8.4.13) is, with the values at the end of the nth time step known:

(1) extrapolate the pressure term, $\dot{\boldsymbol{q}}$, in eqn (8.4.12);

(2) solve eqn (8.4.12) for $\boldsymbol{u}_{n+1}$;

(3) hence obtain $\dot{\boldsymbol{u}}_{n+1}$ to substitute into eqn (8.4.13);

(4) solve eqn (8.4.13) for $\boldsymbol{q}_{n+1}$;

and iterate if required. Park *et al.* show by using the equivalent scalar forms of eqns (8.4.12) and (8.4.13), that the one-pass procedure is conditionally stable, that the limit on the time step decreases with the number of passes but as this tends to infinity the convergence of the iteration itself limits the time step. Park *et al.* conclude that with this procedure there is no point in iterating; it would be better to reduce the time-step size. The same kind of result comes from extrapolating the velocity term in eqn (8.4.13) and proceding from there. As a result of extensive experiments with the equivalent scalar equations, Park *et al.* conclude that the best procedure is to differentiate eqn (8.4.13) to produce the equation

$$\boldsymbol{M}_{\mathrm{f}}\ddot{\boldsymbol{q}} + \rho c\boldsymbol{A}\dot{\boldsymbol{q}} - \rho c\boldsymbol{M}_{\mathrm{f}}\boldsymbol{T}^{\mathrm{T}}\ddot{\boldsymbol{u}} = \dot{\boldsymbol{f}}_2, \tag{8.4.14}$$

so that there are now two coupled second order systems. The acceleration term is eliminated by using eqn (8.4.12) to give an equation of the form

$$\boldsymbol{M}_{\mathrm{f}}\ddot{\boldsymbol{q}} + \rho c\boldsymbol{A}\dot{\boldsymbol{q}} + \rho c\boldsymbol{M}_{\mathrm{f}}\boldsymbol{T}^{\mathrm{T}}\boldsymbol{M}^{-1}(\boldsymbol{K}\boldsymbol{u} + \boldsymbol{T}\boldsymbol{A}\dot{\boldsymbol{q}}) = \boldsymbol{f}_3. \tag{8.4.15}$$

With the structure displacement $\boldsymbol{u}_{n+1}$ extrapolated, eqn (8.4.15) is solved for $\boldsymbol{q}_{n+1}$, and so on. We can see that the success of this manoeuvre depends on neglecting the structural damping, otherwise there would be a $\boldsymbol{C}\dot{\boldsymbol{u}}$ term in eqn (8.4.15). Park *et al.* find that unconditional stability is achieved with a single pass using as the displacement predictor either

$$\boldsymbol{u}^{p}_{n+1} = \boldsymbol{u}_n \quad \text{or} \quad \boldsymbol{u}^{p}_{n+1} = 2\boldsymbol{u}_n - \boldsymbol{u}_{n-1}, \tag{8.4.16}$$

together with a trapezium rule algorithm for obtaining $\boldsymbol{q}_{n+1}$ and $\boldsymbol{u}_{n+1}$. Park *et al.* attribute this success to the effective introduction of an extra damping term in eqn (8.4.15). Intuition suggests that extra damping helps to stabilize an algorithm but of course this term is there due to a rearrangement of the equations; it is not an artificially introduced damping term which could seriously affect the accuracy.

Hitchings and Gossain (1985) also have a first order system representing the fluid motion in a flow induced vibration problem. Here the Navier–Stokes equations are solved with the pressure eliminated by a penalty function formulation using selective reduced integration (Hughes *et al.* 1979). The result is that the fluid motion is represented by a system of equations of the form

$$\boldsymbol{M}_{\mathrm{f}}\dot{\boldsymbol{u}} + \boldsymbol{K}_f\boldsymbol{u} + \boldsymbol{N}(\boldsymbol{u})\boldsymbol{u} = \boldsymbol{f}_1, \tag{8.4.17}$$

where $\boldsymbol{M}_{\mathrm{f}}$ is the fluid mass matrix, $\boldsymbol{K}_f$ is a stiffness matrix type representing the Stokes flow and $\boldsymbol{N}(\boldsymbol{u})\boldsymbol{u}$ represents the non-linear $\boldsymbol{u} \cdot \nabla u$ terms. At the beginning of a new time step the fluid equation (8.4.17) is solved first by a simple predictor–corrector algorithm suggested by Hughes *et al.* (1979) which puts the non-linear terms over on the right-hand side:

$$\frac{1}{\Delta t}\boldsymbol{M}_f(\boldsymbol{u}_{n+1}^{i+1} - \boldsymbol{u}_n) + \boldsymbol{K}_f\boldsymbol{u}_{n+1}^{i+1} = (\boldsymbol{f}_1)_{n+1} - \boldsymbol{N}(\boldsymbol{u}_{n+1}^i)\boldsymbol{u}_{n+1}^i. \tag{8.4.18}$$

The iteration can be started with the predictor value

$$\boldsymbol{u}_{n+1}^p = \{\boldsymbol{u}_n - (1-\gamma)\boldsymbol{u}_{n-1}\}/\gamma, \tag{8.4.19}$$

where γ is a positive parameter. (The values $\gamma = 1$ and 0.5 give the predictors recommended by Park *et al.* (1977), eqn (8.4.16).) The interface velocities required in eqn (8.4.18) are given by the structural solution from the previous time step. The fluid flow equations are solved for some chosen number of iterations. The relative change in the solution in a time step is monitored. The time step is chosen so that it approximately satisfies the Courant condition

$$\Delta t \le 1 \bigg/ \left(\frac{|\boldsymbol{u}|}{\Delta x} + \frac{|\boldsymbol{v}|}{\Delta y}\right), \tag{8.4.20}$$

where $\boldsymbol{u}$ and $\boldsymbol{v}$ are the average velocities and Δx and Δy the dimensions in an element in the x and y directions respectively. The structure equations include the damping matrix and are solved using Newmark with $2\beta = \gamma = 0.5$ and the forces from the fluid pressures calculated from the fluid velocities. The same time step is used for both fluid and structure. This is taken to be the minimum of that given by the Courant condition (8.4.20) and one-tenth of the smallest period of the structure (one version of the engineer's 'rule of thumb'). The starting conditions are taken to be the velocities from the Stokes flow solution given by $\boldsymbol{Ku} = \boldsymbol{f}$ and a zero–zero start for the structure, that is, with displacement, velocity and acceleration each zero. Hitchings and Gossain use their method for the solution of the problem of vortex shedding from a vibrating circular cylinder and give results for Reynolds numbers up to 1000.

8.5. Coupled systems of second order–second order equations—fluid–structure

In this section, we illustrate the solution of coupled second order–second order problems with some examples of the interaction of a structure with an exterior fluid. The motion of the fluid is sometimes represented by a first order equation as in Park *et al.* (1977), and sometimes by a second order equation as in this section, depending on the assumptions made. Here we use the Eulerian pressure formulation as in Zienkiewicz and Bettess (1978) to represent the fluid motion by an equation of the form

$$\boldsymbol{Q}^{\mathrm{T}}\ddot{\boldsymbol{u}} + \boldsymbol{M}_{\mathrm{f}}\ddot{\boldsymbol{p}} + \boldsymbol{A}\dot{\boldsymbol{p}} + \boldsymbol{H}\boldsymbol{p} = \boldsymbol{f}_{\mathrm{f}}, \tag{8.5.1}$$

where $\boldsymbol{p}$ is the fluid pressure, $\boldsymbol{u}$ is the displacement vector for the structure, $\boldsymbol{M}_{\mathrm{f}}$ is the mass matrix, the matrix $\boldsymbol{A}$ comes from the external radiation boundary condition, $\boldsymbol{H}$ is the stiffness matrix, $\boldsymbol{Q}$ is the coupling matrix, and $\boldsymbol{f}_{\mathrm{f}}$ includes the given boundary conditions. The structure equation has the usual form

$$\boldsymbol{M}_{\mathrm{s}}\ddot{\boldsymbol{u}} + \boldsymbol{C}\dot{\boldsymbol{u}} + \boldsymbol{K}\boldsymbol{u} - \boldsymbol{Q}\boldsymbol{p} = \boldsymbol{f}_{\mathrm{s}}, \tag{8.5.2}$$

where $\boldsymbol{M}_{\mathrm{s}}$, $\boldsymbol{C}$, $\boldsymbol{K}$, are the mass, damping, and stiffness matrices respectively and the vector $\boldsymbol{f}_{\mathrm{s}}$ includes the given boundary conditions and external forcing function.

Babu and Reddy (1981) use the Wilson-θ method to solve the coupled equations (8.5.1) and (8.5.2). We recall that the single-step equivalent of Wilson-θ is SS32 with $\theta_1 = \theta$, $\theta_2 = \theta^2$, $\theta_3 = \theta^3$, as described in §4.7. The Babu and Reddy method is equivalent to using eqns (4.5.3) to substitute for the displacement

$\boldsymbol{u}$ and its derivatives from

$$\boldsymbol{u}(\tau)=\boldsymbol{u}_n+\tau\overset{(1)}{\boldsymbol{u}_n}+\frac{\tau^2}{2}\overset{(2)}{\boldsymbol{u}_n}+\frac{\tau^3}{3!}\boldsymbol{\alpha}_n \tag{8.5.3}$$

and the pressure $\boldsymbol{p}$ and its derivatives from

$$\boldsymbol{p}(\tau)=\boldsymbol{p}_n+\tau\overset{(1)}{\boldsymbol{p}_n}+\frac{\tau^2}{2}\overset{(2)}{\boldsymbol{p}_n}+\frac{\tau^3}{3!}\boldsymbol{\beta}_n \tag{8.5.4}$$

into eqns (8.5.1) and (8.5.2). The results can be substituted into the weighted residual equation as in §4.5 and the resultant equations are of the form

$$\boldsymbol{B}_\alpha\boldsymbol{\alpha}_n-\boldsymbol{Q}\boldsymbol{\beta}_n=\boldsymbol{F}_1 \tag{8.5.5}$$

and

$$\boldsymbol{Q}^{\mathrm{T}}\boldsymbol{\alpha}_n+\boldsymbol{B}_\beta\boldsymbol{\beta}_n=\boldsymbol{F}_2, \tag{8.5.6}$$

where $\boldsymbol{F}_1$, $\boldsymbol{F}_2$ are known values from the beginning of the time step. The matrices $\boldsymbol{B}_\alpha$, $\boldsymbol{B}_\beta$ are given by

$$\boldsymbol{B}_\alpha=\frac{1}{\theta_3\Delta t^2}(6\theta_1\boldsymbol{M}_{\mathrm{s}}+3\theta_2\Delta t\boldsymbol{C}+\theta_3\Delta t^2\boldsymbol{K}) \tag{8.5.7}$$

and

$$\boldsymbol{B}_\beta=\frac{1}{6\theta_1}(6\theta_1\boldsymbol{M}_{\mathrm{f}}+3\theta_2\Delta t\boldsymbol{A}+\theta_3\Delta t^2\boldsymbol{H}), \tag{8.5.8}$$

supposing the same θ values are used for the displacement and the pressure. Then the method used by Babu and Reddy corresponds to eliminating $\boldsymbol{\beta}_n$ between eqns (8.5.5) and (8.5.6) to give the equation

$$(\boldsymbol{B}_\alpha+\boldsymbol{Q}\boldsymbol{B}_\beta^{-1}\boldsymbol{Q}^{\mathrm{T}})\boldsymbol{\alpha}_n=\boldsymbol{F}_3, \tag{8.5.9}$$

where the vector $\boldsymbol{F}_3$ is known and the matrix on the left-hand side is symmetric. Then $\boldsymbol{\alpha}_n$ is substiututed into eqn (8.5.6) and $\boldsymbol{\beta}_n$ obtained, again with a symmetric matrix to invert. Babu and Reddy report finding Wilson-θ, $\theta\geq 1.37$, unconditionally stable and use this method for the response of a floating liquid petroleum gas storage vessel to earthquake acceleration transmitted through the sea.

Paul *et al.* (1981) use the same form of eqns (8.5.1) and (8.5.2) to represent the interaction of a reservoir and dam. They solve the problem by using Newmark in a predictor–corrector form.

Either the structure acceleration is predicted using $\ddot{\boldsymbol{u}}^p_{n+1} = \ddot{\boldsymbol{u}}_n$ in order to solve eqn (8.5.2) for $\boldsymbol{p}_{n+1}$, then this is substituted into (8.5.1) to obtain the improved value of $\boldsymbol{u}_{n+1}$, or the other way round using $\boldsymbol{p}^p_{n+1} = \boldsymbol{p}_n + \Delta t\dot{\boldsymbol{p}}_n + (\Delta t^2/2)\ddot{\boldsymbol{p}}_n$. The solution can be iterated to convergence. Paul *et al.* make the interesting observation that the two solution paths give identical results but the convergence is faster when using the predicted value of the pressure to solve for the displacement first.

Chan (1988) uses the structure–fluid equations in a form which omits the first derivative terms:

$$\boldsymbol{M}_s\ddot{\boldsymbol{u}} + \boldsymbol{K}\boldsymbol{u} - \boldsymbol{Q}\boldsymbol{p} = \boldsymbol{f}_s \tag{8.5.10}$$

and

$$\boldsymbol{Q}^{\mathrm{T}}\ddot{\boldsymbol{u}} + \boldsymbol{M}_f\ddot{\boldsymbol{p}} + \boldsymbol{H}\boldsymbol{p} = \boldsymbol{f}_f. \tag{8.5.11}$$

Chan shows that a one-pass procedure using Newmark, $2\beta = \gamma = 0.5$, and a predictor value for the pressure in eqn (8.5.10) given by

$$\boldsymbol{p}^p_{n+1} = \boldsymbol{p}_n + \Delta t\dot{\boldsymbol{p}}_n + \tfrac{1}{4}\Delta t^2\ddot{\boldsymbol{p}}_n \tag{8.5.12}$$

is only conditionally stable, but it can be made unconditionally stable by a modification similar to that with the soil–pore pressure problem in eqn (8.4.9). Equation (8.5.10) at $t = t_{n+1}$ is rewritten as

$$\boldsymbol{M}_s\ddot{\boldsymbol{u}}_{n+1} + (\boldsymbol{K} + \boldsymbol{Q}\boldsymbol{M}_f^{-1}\boldsymbol{Q}^{\mathrm{T}})\boldsymbol{u}_{n+1} = (\boldsymbol{f}_s)_{n+1} + \boldsymbol{Q}\boldsymbol{p}^p_{n+1} + \boldsymbol{Q}\boldsymbol{M}_f^{-1}\boldsymbol{Q}^{\mathrm{T}}\boldsymbol{u}^p_{n+1}. \tag{8.5.13}$$

The predictor values $\boldsymbol{p}^p_{n+1}$ for the pressure and $\boldsymbol{u}^p_{n+1}$ for the structure displacement can again be given by eqn (8.5.12) and the corresponding equation

$$\boldsymbol{u}^p_{n+1} = \boldsymbol{u}_n + \Delta t\dot{\boldsymbol{u}}_n + \tfrac{1}{4}\Delta t^2\ddot{\boldsymbol{u}}_n. \tag{8.5.14}$$

Then the Newmark approximations for the other displacement and pressure terms give an unconditionally stable one-pass procedure provided the usual stability conditions hold, that is, $2\beta \geq \gamma \geq 0.5$. There seems to be no physical interpretation of the augmentation device used in eqn (8.5.13) as there is with the soil–pore pressure problem discussed in §8.4. With the fluid–structure problem, the matrix $\boldsymbol{Q}$ only comes from the nodes on the interface and hence is conveniently sparse.

References

Adey, R. A. and Brebbia, C. A. (1973). Finite element solution for effluent dispersion. In *Numerical methods in fluid dynamics* (eds. C. A. Brebbia, and J. J. Connor). Pentech Press.

Babu, P. V. T. and Reddy, D. V. (1981). A numerical integration scheme for solving coupled equations of fluid-structure interaction systems. In *Numerical methods for coupled problems* (eds. E. Hinton, P. Bettess and R W. Lewis). Pineridge Press.

Bear, J. (1979). *Hydraulics of groundwater.* McGraw-Hill.

Bettencourt, J. M., Zienkiewicz, O. C., and Cantin, G. (1981). Consistent use of finite elements in time and the performance of various recurrence schemes for the heat diffusion equation. *International Journal of Numerical Methods in Engineering,* **17,** 931–938.

Chan, A. H. C. (1988). A unified finite element solution to state and dynamic problems of geomechanics. *Ph.D. Thesis.* University College of Swansea.

Connor, J. J. and Wang, J. (1973). Finite element modelling of hydrodynamic circulation. In *Numerical methods in fluid dynamics* (eds. C. A. Brebbia and J. J. Connor). Pentech Press, London.

Geers, T. L. (1971). Residual potential and approximate methods for three-dimensional fluid–structure interaction problems. *Journal of the Acoustical Society of America,* **49,** 1505–1510.

Hitchings, D. and Gossain, S. R. (1985). The analysis of flow induced vibrations using combined fluid and structural finite element models. *Proceedings of NUMETA 1985 Conference, Swansea* pp. 191–201.

Hughes, T. J. R., Liu, W. K., and Brooks, A. (1979). Finite element analysis of incompressible viscous flows by the penalty function formulation. *Journal of Computational Physics,* **30,** 1–60.

Hwang, C. T., Morgenstern, N. R., and Murray, D. W. (1971). On solution of plane strain consolidation problems by finite element method. *Canadian Geotechnical Journal* **8,** 109–118.

Kawahara, M. and Hirano, H. (1983). A finite element method for high Reynolds' number viscous fluid flow using two-step explicit scheme. *International Journal of Numerical Methods in Engineering,* **3,** 137–163.

Kawahara, M., Takeuchi, N., and Yoshida, T. (1978). Two-step explicit finite element method for tsunami wave propagation analysis. *International Journal of Numerical Methods in Engineering,* **12,** 331–351.

Kawahara, M., Hirano, H., Tsubota, K., and Inagaki, K. (1982a). Selective lumping finite element method for shallow water flow. *International Journal of Numerical Methods in Engineering,* **2,** 89–112.

Kawahara, M., Kobayashi, M., and Nakata, K. (1982b). A three dimensional multiple level finite element method considering variable water density. In *Finite Elements in Fluids,* Vol. 4, (eds. R. H. Gallagher, D. H. Norrie, J. T. Oden, and O. C. Zienkiewicz). Wiley.

Lewis, R. W. and Schrefler, B. A. (1987). *The finite element method in the deformation and consolidation of porous media.* Wiley.

Park, K. C. (1977). Stabilization of staggered solution procedures for fluid-structure interaction analysis. In *Computational methods for fluid interaction analysis* (eds. T. Belytschko and T. L. Geers) ASME, AMD Vol. 26, pp. 94–124.

Park, K. C. (1982). Partitioned transient analysis procedures for coupled field problems; stability analysis. *Journal of Applied Mechanics,* **47,** 370–376.

Park, K. C. (1983). Stabilization of partitioned solution procedure for pore fluid–soil interaction analysis. *International Journal of Numerical Methods in Engineering,* **19,** 1669–1673.

Park, K. C. and Felippa, C. A. (1983). *Computational methods for transient analysis* (eds. T. Belytschko and T. J. R. Hughes). North-Holland.

Park, K. C., Felippa, C. A., and De Runtz, J. A. (1977). Stabilization of staggered solution procedures for fluid–structure interaction analysis. In *Computer methods for fluid–structure interaction problems* (eds. T. Belytschko and T. L. Geers). ASME, ADM, pp. 95–124.

Paul, D. K. (1982). Efficient dynamic solutions for single and coupled multiple field problems. *Ph.D. Thesis 64/82.* Department of Civil Engineering, Swansea.

Paul, D. K., Zienkiewicz, O. C., and Hinton, E. (1981). *Numerical methods for coupled problems* (eds. E. Hinton P. Bettess, and R. W. Lewis). Pineridge Press.

Pinder, G. F. (1973). Simulation of groundwater contamination using a Galerkin finite element technique. In *Numerical methods in fluid dynamics* (eds. C. A. Brebbia and J. J. Connor). Pentech Press, London.

Ralston, A. (1965). *A first course in numerical analysis.* McGraw-Hill.

Sandhu, R. S. (1968). Fluid flow in saturated porous elastic media. *Ph.D. Thesis.* University of California, Berkeley.

Smith, I. M. (1975). Some time-dependent soil-structure interaction problems. In *Finite elements in geomechanics* (ed. G. Gudehus), Ch. 8. Wiley.

Taylor, C. and Davis, J. M. (1974a). A finite element model of tides in estuaries. In *Finite element methods in flow problems* (eds. J. T. Oden, *et al.*). UAH Press, Huntsville, Alabama,

Taylor, C. and Davis, J. M. (1974b). A numerical model of dispersion in estuaries. In *Finite element methods in flow problems* (eds. J. T. Oden *et al.*). UAH Press, Huntsville Alabama.

Taylor, C. and Huyakorn, P. S. (1976). Finite element analysis of three-dimensional groundwater flow with convective dispersion. In *Finite elements in fluids,* Vol. 3 (eds. R. H. Gallagher, O. C. Zienkiewicz, J. T. Oden, M. Morandi Cecchi, and C. Taylor). Wiley.

Taylor, C., Al-Mashidani, G., and Davis, J. M. (1972). A finite element approach to watershed run-off. *Report C/R/185/72.* Department of Civil Engineering, University of Wales, Swansea.

Verner, E. A., Lewis, R. W., and Zienkiewicz, O. C. (1974). Finite elements for two-phase flow in porous media. In *Finite element methods in flow problems* (eds. Oden *et al.*). UAH Press, Huntsville, Alabama.

White, I. R. (1978). The finite element analysis of multiphase flow in porous media and its application to reservoir engineering. *Ph.D. Thesis C/Ph/47/78*. University of Wales.

White, I. R., Lewis, R. W., and Wood, W. L. (1981). The numerical simulation of multiphase flow through a porous medium and its application to reservoir engineering. *Applied Mathematical Modelling,* **5,** 165–172.

Zienkiewicz, O. C. (1977). The *finite element method* (3rd edn). McGraw-Hill.

Zienkiewicz, O. C. and Bettess, P. (1978). Fluid–structure dynamic interaction and wave forces. An introduction to numerical treatment. *International Journal of Numerical Methods in Engineering,* **13,** 1–16.

Zienkiewicz, O. C. and Taylor, R. L. (1985). Coupled Problems—a simple time-stepping procedure. *International Journal of Communications in Applied Numerical Methods,* **1,** 233–239.

Zienkiewicz, O. C., Paul, D. K., and Chan, A. C. H. (1988). Unconditionally stable staggered solution procedure for soil-pore fluid interaction problems. *International Journal of Numerical Methods in Engineering,* **26,** 1039–1055.

Appendix 1: The Routh–Hurwitz Conditions

The Routh–Hurwitz conditions (Routh 1877, 1905, and Hurwitz 1895) were originally produced in connection with the stability of the solution of differential equations. For example, with the differential equation

$$a_0\frac{\mathrm{d}^n x}{\mathrm{d}t^n}+a_1\frac{\mathrm{d}^{n-1}x}{\mathrm{d}t^{n-1}}+\cdots+a_{n-1}\frac{\mathrm{d}x}{\mathrm{d}t}+a_n x=0, \tag{1.1}$$

if we put $x=\exp(\lambda t)$, we have

$$a_0\lambda^n+a_1\lambda^{n-1}+\cdots+a_{n-1}\lambda+a_n=0. \tag{1.2}$$

Equation (1.1) then has a decaying solution if the roots of eqn (1.2) have negative real parts. We can also put

$$\frac{\mathrm{d}x}{\mathrm{d}t}=x_1,\qquad \frac{\mathrm{d}x_1}{\mathrm{d}t}=\frac{\mathrm{d}^2x}{\mathrm{d}t^2}=x_2,\ldots,\qquad \frac{\mathrm{d}x_{n-2}}{\mathrm{d}t}=\frac{\mathrm{d}^{n-1}x}{\mathrm{d}t^{n-1}}=x_{n-1},$$

and

$$\frac{\mathrm{d}x_{n-1}}{\mathrm{d}t}=\frac{\mathrm{d}^n x}{\mathrm{d}t^n}=-\frac{a_1}{a_0}x_{n-1}-\frac{a_2}{a_0}x_{n-2}-\cdots-\frac{a_n}{a_0}x,$$

and form the equivalent system of equations:

$$\frac{\mathrm{d}\boldsymbol{x}}{\mathrm{d}t}=\boldsymbol{A}\boldsymbol{x} \tag{1.3}$$

where $\boldsymbol{x}^{\mathrm{T}}=[x, x_1,\ldots, x_{n-1}]$. The stability of the solution of eqn 1.3 now depends on the eigenvalues of the matrix $\boldsymbol{A}$ whose characteristic polynomial must be that in eqn (1.2).

Gantmacher (1959) gives a detailed discussion of the theory of the Routh–Hurwitz conditions which initially are the conditions for the roots of eqn (1.2) to have negative real parts and they are all strict inequalities. Perhaps the easiest way to remember these conditions is that the principal minors of the determinant

$$\begin{vmatrix} a_1 & a_3 & a_5 & \cdots \\ a_0 & a_2 & a_4 & \cdots \\ 0 & a_1 & a_3 & \cdots \\ 0 & a_0 & a_2 & \cdots \end{vmatrix} \tag{1.4}$$

should be positive. But these conditions are usually quoted as $a_0 > 0$ and the other principal minors ≥ 0, as the conditions for the polynomial (1.2) to have roots with non-positve real parts. From the point of view of the theory required in this book where the coefficients of the terms in the polynomial will be functions of certain parameters, we can regard the 'some roots on the imaginary axis' as the limiting case of the 'all roots with negative real parts" situation. If there is any doubt about the implication of equalities they can be examined separately.

References

Gantmacher, F. R. (1959). *The theory of matrices,* Vol. 2. Chelsea Publishing Company.

Hurwitz, A. (1895). Über die Bedingungen, unter welchen eine Gleichung nur Wurzeln mit negativen reellen Teilen besitzt. *Mathematische Annalen,* **46,** 273–84.

Routh, E. J. (1877). A *treatise on the stability of a given state of motion.* Macmillan, London.

Routh, E. J. (1905). *The advanced part of a treatise on the dynamics of a rigid body* (6th edn). Macmillan, London: reprinted (1959) Dover, New York.

Solutions to Exercises

Chapter 1

1.1.

$$\boldsymbol{M}_{\mathrm{D}}^{(e)} = \frac{h_j}{6}\begin{bmatrix} 2 & 1 \\ 1 & 2 \end{bmatrix}, \qquad \boldsymbol{M}_{\mathrm{D}}^{(e)^{-1}} = \frac{1}{h_j}\begin{bmatrix} 4 & -2 \\ -2 & 4 \end{bmatrix},$$

and

$$\boldsymbol{M}_{\mathrm{D}}^{(e)^{-1}}\boldsymbol{K}^{(e)} = \frac{6}{h_j^2}\begin{bmatrix} 1 & -1 \\ -1 & 1 \end{bmatrix},$$

which has eigenvalues 0 and $12h_j^{-2}$. Since the eigenvalues of $\boldsymbol{K}^{(e)}$ are 0 and $2h_j^{-1}$ and the minimum eigenvalue of $\boldsymbol{M}_{\mathrm{D}}^{(e)}$ is $h_j/6$ this corresponds to equality in (1.4.7a).

1.2. Inverting $\boldsymbol{M}_{\mathrm{D}}^{(e)}$ we have

$$\boldsymbol{M}_{\mathrm{D}}^{(e)^{-1}} = \frac{3}{\Delta}\begin{bmatrix} 3 & -1 & -1 \\ -1 & 3 & -1 \\ -1 & -1 & 3 \end{bmatrix}$$

and $\boldsymbol{M}_{\mathrm{D}}^{(e)^{-1}}\boldsymbol{K}^{(e)} = \dfrac{12}{\Delta}\boldsymbol{K}^{(e)}$. From the Gerschgorin theorem, the maximum eigenvalue of $\dfrac{12}{\Delta}\boldsymbol{K}^{(e)}$ is $\leq 24\max\{(\cot B + \cot C)/bc \sin A)\} = 24\max\{1/b^2 \sin^2 C\} = 24h_{\min}^{-2}$ where h is the element height.

1.3. For the bilinear square of side h the distributed mass matrix is given by

$$\boldsymbol{M}_{\mathrm{D}}^{(e)} = \frac{h^2}{36}\begin{bmatrix} 4 & 2 & 1 & 2 \\ 2 & 4 & 2 & 1 \\ 1 & 2 & 4 & 2 \\ 2 & 1 & 2 & 4 \end{bmatrix}.$$

The eigenvalues of $\boldsymbol{M}_{\mathrm{D}}^{(e)}$ are $h^2/36$, $h^2/12$, $h^2/12$, $h^2/4$. The four eigenvalues of $\boldsymbol{M}_{\mathrm{L}}^{(e)}$ are equal to $h^2/4$. Hence (1.4.7a) gives an estimate of the difference in the bound on λ in eqn (1.4.1) as 9:1 and on ω in eqn (1.4.2) as 3:1.

1.4. With the distributed mass matrix and $h = 1/3$,

$$-\omega^2 \boldsymbol{M}_D + \boldsymbol{K} = -\frac{\omega^2}{18}\begin{bmatrix} 4 & 1 \\ 1 & 4 \end{bmatrix} + 3\begin{bmatrix} 2 & -1 \\ -1 & 2 \end{bmatrix},$$

hence $\omega_1 = \sqrt{(54/5)} = 3.286 > \pi$, and $\omega_2 = \sqrt{54} = 7.348 > 2\pi$.
With the lumped mass matrix.

$$-\omega^2 \boldsymbol{M}_L + \boldsymbol{K} = -\frac{\omega^2}{3}\begin{bmatrix} 1 & 0 \\ 0 & 1 \end{bmatrix} + 3\begin{bmatrix} 2 & -1 \\ -1 & 2 \end{bmatrix},$$

hence $\omega_1 = 3 < \pi$ and $\omega_2 = \sqrt{27} = 5.196 < 2\pi$.

Chapter 2

2.1. The characteristic polynomial of $\boldsymbol{A}$ is given by

$$\begin{vmatrix} 1-\lambda & 1 & 0.5 \\ A_{21} & A_{22}-\lambda & 0.5A_{22} \\ 2A_{21} & A_{32} & 0.5A_{32}-\lambda \end{vmatrix} = 0$$

$C_3 - \frac{1}{2}C_2$ gives

$$\begin{vmatrix} 1-\lambda & 1 & 0 \\ A_{21} & A_{22}-\lambda & 0.5\lambda \\ 2A_{21} & A_{32} & -\lambda \end{vmatrix} = 0$$

$R_2 + \frac{1}{2}R_3$ gives

$$\begin{vmatrix} 1-\lambda & 1 & 0 \\ 2A_{21} & A_{22}+0.5A_{32}-\lambda & 0 \\ 2A_{21} & A_{32} & -\lambda \end{vmatrix} = 0.$$

Hence $\lambda = 0$ or $\lambda^2 - \lambda(1 + A_{22} + 0.5A_{32}) + A_{22} + 0.5A_{32} - 2A_{21} = 0$.
Substituting for A_{21}, A_{22}, A_{32} gives

$$(m + \tfrac{1}{2}\Delta t\mu)\lambda^2 + \lambda(-2m + \Delta t^2 k) + m - \tfrac{1}{2}\Delta t\mu = 0.$$

Hence the corresponding two-step scheme in the displacements is the central difference scheme

$$(m + \tfrac{1}{2}\Delta t\mu)x_{n+1} + (-2m + \Delta t^2 k)x_n + (m - \tfrac{1}{2}\Delta t\mu)x_{n-1} = 0.$$

2.2. The first order form of $m\ddot{x} + \mu\dot{x} + kx = 0$ is

$$\begin{bmatrix} \dot{x} \\ \dot{v} \end{bmatrix} = \begin{bmatrix} 0 & 1 \\ -\dfrac{k}{m} & -\dfrac{\mu}{m} \end{bmatrix}\begin{bmatrix} x \\ v \end{bmatrix}.$$

The explicit ($\theta = 0$) scheme on this gives

$$\begin{bmatrix} x_{n+1} \\ v_{n+1} \end{bmatrix} = \begin{bmatrix} 1 & \Delta t \\ -\dfrac{\Delta t k}{m} & 1 - \dfrac{\Delta t \mu}{m} \end{bmatrix} \begin{bmatrix} x_n \\ v_n \end{bmatrix}.$$

The characteristic polynomial of the amplification matrix is $mr^2 + r(-2m + \Delta t\mu) + m - \Delta t\mu + \Delta t^2 k = 0$. Hence the corresponding two-step scheme is

$$m(x_{n+1} - 2x_n + x_{n-1}) + \Delta t\mu(x_n - x_{n-1}) + \Delta t^2 k x_{n-1} = 0.$$

By the Routh–Hurwitz conditions the scheme is stable if

(1) $4m - 2\Delta t\mu + \Delta t^2 k > 0$, (2) $\Delta t\mu - \Delta t^2 k \geq 0$, (3) $\Delta t^2 k \geq 0$.

(1) is equivalent to $k(\Delta t - \mu/k)^2 + 4m - \mu^2/k > 0$, which is satisfied if the damping is less than critical. (3) is satisfied and (2) gives the limit on the time step as $\Delta t \leq \mu/k$. With $\mu = 2\nu m\omega$ and $k = m\omega^2$ this is the same as in eqn (2.6.10).

2.3. The stiffness matrix for a bilinear square element is

$$\boldsymbol{K}_{\mathrm{B}}^{(e)} = \tfrac{1}{6}\begin{bmatrix} 4 & -1 & -2 & -1 \\ -1 & 4 & -1 & -2 \\ -2 & -1 & 4 & -1 \\ -1 & -2 & -1 & 4 \end{bmatrix}.$$

The maximum eigenvalue is 1, hence the maximum eigenvalue of $\boldsymbol{M}_{\mathrm{BL}}^{(e)^{-1}}\boldsymbol{K}_{\mathrm{B}}^{(e)}$ is $4h^{-2}$. The stiffness matrix for a 45°, 45°, 90° triangle is

$$\boldsymbol{K}_{\mathrm{T}}^{(e)} = \tfrac{1}{2}\begin{bmatrix} 1 & -1 & 0 \\ -1 & 2 & -1 \\ 0 & -1 & 1 \end{bmatrix}$$

eigenvalues, 0, $\tfrac{1}{2}$, $\tfrac{3}{2}$. Hence the maximum eigenvalue of $\boldsymbol{M}_{\mathrm{TL}}^{(e)^{-1}}\boldsymbol{K}_{\mathrm{T}}^{(e)}$ is $9h^{-2}$, larger than for the bilinear square.

2.4. The characteristic polynomial has conjugate complex eigenvalues $p \pm iq$, and

$$\boldsymbol{S}^{-1} = \frac{1}{2iq}\begin{bmatrix} p + iq & 1 \\ p - iq & 1 \end{bmatrix}.$$

From eqn (2.7.7), $q^2 = 4(1 - \nu^2)\omega^2\Delta t^2$, and the real part of the eigenvalues is $1 + \mathrm{O}(\Delta t)$. Hence the result.

Chapter 3

3.1. From eqn (3.2.19) the roots are complex if $\mu^2 + \frac{2}{3}k^2\Delta t^2 < 4mk$. If $\mu = 0$, this is $\omega^2\Delta t^2 < 6$, which is the same as the inequality part of the stability condition, (3.4.21).

3.2. No.

3.3. If $\mu^2 < 4mk$ and $\mu \neq 0$ the algorithm is only stable if $k\Delta t < \mu$, and this cannot be satisfied if $\mu = 0$. If $\mu^2 > 4mk$ the roots are real and the algorithm is stable if

$$\Delta t < \{\mu - \surd(\mu^2 - 4mk)\}/k.$$

Chapter 4

4.1.

$$\mathbf{A} = D^{-1}\begin{bmatrix} m + \frac{1}{2}(\theta_2 - 1)\Delta t^2 k & m + \frac{1}{2}(\theta_2 - \theta_1)\Delta t^2 k \\ -\Delta t^2 k & m + \frac{1}{2}(\theta_2 - 2\theta_1)\Delta t^2 k \end{bmatrix}$$

where $D = m + \frac{1}{2}\theta_2\Delta t^2 k$. Comparing with the θ-method amplification matrix in eqn (3.3.8), the two are the same if $\theta_1 = \theta_2 = \theta = 0.5$. The characteristic polynomial of $\mathbf{A}$ is

$$(m + \tfrac{1}{2}\theta_2\Delta t^2 k)r^2 - r\{2m + \tfrac{1}{2}(2\theta_2 - 2\theta_1 - 1)\Delta t^2 k\} \\ + m + \tfrac{1}{2}(\theta_2 - 2\theta_1 + 1)\Delta t^2 k = 0.$$

(N.B. the factor D cancels.) The two methods (a) and (b) give the same results with $\theta_1 = \theta_2 = \theta = 0.5$ provided they are given the same starting conditions.

4.2. If the roots of the stability polynomial in eqn (4.1) are r_1, r_2, then $r_1 - r_2 = 2D^{-1}\surd(b^2 - ac)$, where

$$\begin{aligned} b^2 - ac &= \{m + \tfrac{1}{4}(2\theta_2 - 2\theta_1 - 1)\Delta t^2 k\}^2 \\ &\quad - (m + \tfrac{1}{2}\theta_2\Delta t^2 k)\{m + \tfrac{1}{2}(\theta_2 - 2\theta_1 + 1)\Delta t^2 k\} \\ &= -m\Delta t^2 k + \mathrm{O}(\Delta t^4). \end{aligned}$$

Hence $r_1 - r_2$ is $\mathrm{O}(\Delta t)$.

4.3.

$$L_{\Delta t}(x_n) = \frac{1}{\Delta t^2}(x_{n+1} - 2x_n + x_{n-1}) + \omega^2(\theta_2 x_n + \theta_1 x_{n-1}) = 0$$

and

$$L_{\Delta t}(x) = \ddot{x} + \frac{\Delta t^2}{12}\overset{(4)}{x} + \omega^2\{(\theta_1 + \theta_2)x - \theta_1\Delta t\dot{x} + \tfrac{1}{2}\theta_1\Delta t^2\ddot{x} + \cdots\}.$$

Hence for consistency we must have $\theta_1 + \theta_2 = 1$ and the error is $O(\Delta t)$ unless $\theta_1 = 0$, when it is $O(\Delta t^2)$.

4.4. From eqns (4.11.20) θ_1, θ_2 must satisfy $\theta_1 - 1 + \frac{1}{4}(\theta_2 - 2\theta_1 + 1) = 0$, i.e. $\theta_2 = 3 - 2\theta_1$, and

$$(\theta_1 + \tfrac{1}{4}\theta_2)\exp(-0.5) = -\{1 - 2\theta_1 + \tfrac{1}{4}(2\theta_1 - 2\theta_2 + 1)\}$$

which give $\theta_1 = 1.0414$, $\theta_2 = 0.9172$ to 4 d.p. This does not satisfy the stability condition $\theta_2 > \theta_1$. From eqn (4.11.9), the stability limit is given by $\tilde{\lambda}\Delta t < (2\theta_1 - 1)/(\theta_1 - \theta_2) = 8.7$ approximately.

4.5. Substituting into eqn (2.2.4) we have

$$L_{\Delta t}(x_n) = \frac{1}{\Delta t^2}(x_{n+1} - 2x_n + x_{n-1})$$
$$+ \frac{\nu\omega}{\Delta t}(x_{n+1} - x_{n-1}) + \frac{\omega^2}{12}(x_{n+1} + 10x_n + x_{n-1}) = 0.$$

Hence

$$L_{\Delta t}(x) = \frac{\Delta t^2}{12}(\overset{(4)}{x} + 4\nu\omega\overset{(3)}{x} + \omega^2\ddot{x}) + O(\Delta t^4).$$

Thus if $\nu \neq 0$, from eqn (2.9.21), the cumulative error is $O(\Delta t^2)$. If $\nu = 0$ the error is $O(\Delta t^4)$.

4.6. The SS32 equations give $\alpha_0 = 3\theta_2\omega^4\Delta t a/(6\theta_1 + \theta_3\omega^2\Delta t^2)$. Hence

$$x_1 = a - \frac{\Delta t^2}{2}\omega^2 a + \tfrac{1}{2}a\theta_2\omega^4\Delta t^4/(6\theta_1 + \theta_3\omega^2\Delta t^2) < -a,$$

if $(\theta_3 - \theta_2)\Omega^4 + 2\Omega^2(3\theta_1 - 2\theta_3) - 24\theta_1 > 0$, i.e. since $\theta_3 > \theta_2$ for (i) and (ii), we require $\Omega^2 >$ the larger root of $\cdots = 0$.

(i) With Wilson-$\theta = 1.4$, $\theta_1 = 1.4$, $\theta_2 = 1.96$, $\theta_3 = 2.744$, and $\Omega > 2.9$, approximately, i.e. $\omega\Delta t/2\pi > 0.46$;

(ii) With Houbolt, $\theta_1 = 2$, $\theta_2 = 11/3$, $\theta_3 = 6$, and $\Omega > 2.8$, approximately, i.e. $\omega\Delta t/2\pi > 0.44$ for overshoot on the displacement on the first time step.

Chapter 5

5.1. Using $\dfrac{d}{dt} = \dfrac{1}{\Delta t}\dfrac{d}{d\zeta}$, and substituting as required, we have $(-6 + \Omega^2)x_{n+1} + 4\Delta t v_{n+1} + 6x_n + 2\Delta t v_n = 0$, and

$$-12x_{n+1} + (6 + \Omega^2)\Delta t v_{n+1} + 12x_n + 6\Delta t v_n = 0,$$

where $\Omega = \omega\Delta t$. Putting $x_{n+1} = rx_n$, $v_{n+1} = rv_n$, the condition for a non-trivial solution is

$$r^2(\Omega^4 + 12) + 12r(\Omega^2 - 2) + 12 = 0. \qquad (1)$$

The Routh–Hurwitz conditions are satisfied and the method is stable for all Ω, but the condition for complex roots is

$$36(\Omega^2-2)^2<12(\Omega^4+12),$$

i.e. $\Omega^2<6$. Hence the method cannot be C-stable, Expanding the solution of eqn (1) in powers of Ω we have for the approximation of $\exp(\mathrm{i}\Omega)$,

$$\begin{aligned} r &= \{12-6\Omega^2+12\Omega\mathrm{i}\surd(1-\Omega^2/6)\}/(12+\Omega^4) \\ &= 1+\mathrm{i}\Omega-\frac{1}{2}\Omega^2-\frac{1}{12}\mathrm{i}\Omega^3+\cdots \end{aligned}$$

Hence the error is $O(\Omega^3)$. For negligible Ω^4 eqn (1) is the same as the stability polynomial for the central difference algorithm.

5.2. (ω assumed real)

$(0, 1)$, $\left|\dfrac{1}{1+\mathrm{i}\omega}\right|\leq 1$ if $1+\omega^2\geq 1$, true for all ω.

$(0, 2)$, $\left|\dfrac{1}{1+\mathrm{i}\omega-\frac{1}{2}\omega^2}\right|\leq 1$ if ... $\omega^4\geq 0$, hence stable.

$(1, 1)$, $\left|\dfrac{1-\mathrm{i}\omega/2}{1+\mathrm{i}\omega/2}\right|=1$, hence stable.

$(1, 2)$, $\left|\dfrac{1-\mathrm{i}\omega/3}{1+2\mathrm{i}\omega/3-\omega^2/6}\right|\leq 1$ if ... $\omega^4\geq 0$, hence stable.

$(2, 1)$, $\left|\dfrac{1-2\mathrm{i}\omega/3-\omega^2/6}{1+\mathrm{i}\omega/3}\right|\leq 1$ if $\omega^4\leq 0$, hence unstable for all ω.

5.3. From eqn (5.3.39) we have

$$\left|\frac{1+(2\alpha-1)\mathrm{i}\omega}{(1+\mathrm{i}\alpha\omega)^2}\right|\leq 1 \quad \text{if} \ldots 2\alpha^2-4\alpha+1\leq\omega^2\alpha^4,$$

but $2\alpha^2-4\alpha+1=0$ from eqn (5.3.30) with $k=1$. Hence stable.

5.4. Substituting in eqns 5.4.12 we have

$$z_{n+1}=(1-\lambda\Delta t+\tfrac{1}{2}\lambda^2\Delta t^2-\tfrac{1}{6}\lambda^3\Delta t^3+\tfrac{1}{24}\lambda^4\Delta t^4)z_n=\{\mathrm{e}^{-\lambda\Delta t}+O(\Delta t^5)\}z_n,$$

and the cumulative error is $O(\Delta t^4)$.

5.5. Here $f(y)=-\lambda y$, $g_1=-\lambda\Delta t y_n$, $g_2=-\lambda\Delta t(1-c\lambda\Delta t)y_n$. Put $b_1+b_2=\beta$ and $b_2c=\alpha$, then we have

$$\left(1-\frac{\alpha}{\beta}\lambda\Delta t\right)y_{n+1}=\left\{1-\frac{(1+\alpha)}{\beta}\lambda\Delta t\right\}y_n.$$

For stability we require $(1+2\alpha)/(\beta\lambda\Delta t) \leq 2$, hence unconditional if $\alpha \leq -0.5$. Second order if it corresponds to Crank–Nicolson, i.e. $\alpha/\beta = -0.5$, $(1+\alpha)/\beta = 0.5$, hence $\alpha = -0.5$, $\beta = 1$.

5.6. If $\alpha = k\Delta t^2/16$, the stability polynomial is

$$(m+\alpha)^2 r^2 + 2r(6m\alpha - m^2 - \alpha^2) + (m+\alpha)^2 = 0,$$

and the Routh–Hurwitz stability conditions are satisfied for all Δt.

Author index

Numbers in bold type refer to the references at the end of a chapter.

Subject index